Endemic Mycoses: Recent Advances in Epidemiology, Diagnosis and Treatment

Endemic Mycoses: Recent Advances in Epidemiology, Diagnosis and Treatment

Editor

Alessandro C. Pasqualotto

MDPI

Basel • Beijing • Wuhan • Barcelona • Belgrade • Novi Sad • Cluj • Manchester

Editor
Alessandro C. Pasqualotto
Federal University of Health
Sciences of Porto Alegre (UFCSPA)
Porto Alegre
Brazil

Editorial Office
MDPI AG
Grosspeteranlage 5
4052 Basel, Switzerland

This is a reprint of articles from the Special Issue published online in the open access journal *Journal of Fungi* (ISSN 2309-608X) (available at: https://www.mdpi.com/journal/jof/special_issues/Endemic_Mycoses).

For citation purposes, cite each article independently as indicated on the article page online and as indicated below:

Lastname, A.A.; Lastname, B.B. Article Title. *Journal Name* **Year**, *Volume Number*, Page Range.

ISBN 978-3-7258-1443-5 (Hbk)
ISBN 978-3-7258-1444-2 (PDF)
doi.org/10.3390/books978-3-7258-1444-2

Contents

Preface

Endemic fungal infections present diagnostic challenges as they can mimic various other illnesses such as tuberculosis, pneumonia, and cancer, especially in regions where they are prevalent. Despite being familiar diseases in endemic areas, these fungal infections are now emerging in new geographic locations due to increased travel. It is crucial for everyone to recognize their presence.

Although the microscopic appearance of these fungal agents often provides clues, diagnosing them requires significant expertise, and test accuracy varies. The gold standard for diagnosis is culturing samples from potentially affected areas, but this process can take a long time. Severe cases typically require treatment with amphotericin B or its lipid formulations, while itraconazole is effective for oral treatment. Clinical trials are limited, and access to modern medications remains a challenge, especially in developing countries.

To consolidate recent research on the epidemiology, diagnosis, and treatment of endemic mycoses, we present a dedicated reprint. We hope this resource proves valuable and insightful to you.

Alessandro C. Pasqualotto
Editor

Journal of *Fungi*

Editorial

Endemic Mycoses: Novel Findings for the Clinician

Alessandro C. Pasqualotto [1,2,3]

1 Santa Casa de Misericordia de Porto Alegre, Porto Alegre 90035-075, Brazil; pasqualotto@santacasa.org.br; Tel.: +55-51-999951614
2 Universidade Federal de Ciencias da Saude de Porto Alegre, Porto Alegre 90035-075, Brazil
3 Molecular Biology Laboratory, Hospital Dom Vicente Scherer, Heliponto, Av Independencia 155, Centro, Porto Alegre 90035-075, Brazil

Endemic mycoses are difficult-to-diagnose conditions that may mimic several other diseases, particularly tuberculosis, community-acquired pneumonia, and cancer. Even though these are well-known diseases for physicians living in endemic areas, fungal diseases have now reached novel geographic areas, in a world where extensive travelling occurs. We should therefore all be aware of their existence.

The microscopic appearance of the agents of endemic mycoses is frequently suggestive, but their diagnoses require considerable expertise, and test sensitivities are variable. Culture from potentially involved sites remains the diagnostic gold-standard; however, plates need to be incubated for up to 6–8 weeks. Treatments of severe clinical forms usually rely on amphotericin B and its lipid formulations, while itraconazole is the most active oral antifungal agent. Controlled trials are rare in the field, and access to modern medicines is usually a challenge in developing countries.

With the purpose of summarizing recent findings involving the epidemiology, diagnosis and treatment of these conditions, the *Journal of Fungi* is publishing a Special Issue dedicated to endemic mycoses. A summary of the novel findings associated with some of these conditions is presented herein.

Citation: Pasqualotto, A.C. Endemic Mycoses: Novel Findings for the Clinician. *J. Fungi* **2022**, *8*, 1184. https://doi.org/10.3390/jof8111184

Received: 2 November 2022
Accepted: 7 November 2022
Published: 10 November 2022

1. Blastomycosis

Blastomycosis is endemic in the Mississippi and Ohio River basins, in the United States. The range of disease may be expanding, with cases now commonly occurring in the state of New York, and other areas previously considered outside the traditional region of endemicity. Regarding the diagnosis of blastomycosis, an antigen detection assay is available in North America and has acceptable sensitivity (85–93%) and specificity (79–99%) [1]. Testing of the urine is preferable over other sample types.

2. Coccidioidomycosis

The majority of cases of coccidioidomycosis are reported in Arizona and California; the number of coccidioidomycosis cases is continuing to increase yearly in the United States. Other arid areas in Latin America are also involved. Culturing *Coccidioides* spp. from clinical samples presents biosafety concerns. Most patients are diagnosed with serology, but weeks are also required for antibodies to be produced and detected. Coccidioidal antigen testing is also available and is best performed in urine samples [2]. Antigenuria should be tested only when the diagnosis is in question.

3. Cryptococcosis Due to *Cryptococcus gattii*

In contrast to the opportunistic agent *C. neoformans*, which has a global distribution and is frequently isolated from bird droppings, *C. gattii* is reported in tropical and subtropical areas of the globe, and is associated with trees. Moreover, *C. gattii* may cause infection in immunocompetent individuals. Infections caused by *C. gattii* are more likely to result in mass-like lesions, either pulmonary or cerebral. *C. gattii* shows higher minimum inhibitory concentrations to several antifungals than *C. neoformans*.

4. Emergomycosis

Since the introduction of molecular-based identification methods of dimorphic fungi in some South African laboratories, emergomycosis has been recognized as the most frequently diagnosed endemic mycosis in South Africa [3]. To date, nearly all cases of emergomycosis have involved immunocompromised patients. *Emergomyces* spp. can cross-react with the *Histoplasma* galactomannan antigen test; therefore, urine samples can be sent for *Histoplasma* antigen testing.

5. Histoplasmosis

Once considered a disease restricted to the Mississippi and Ohio rivers in the United States, histoplasmosis is now a global disease, with hundreds of cases being reported in China, India, Oceania, Africa, and Europe. The highest burden of disease occurs in Latin America. As for the other agents of endemic mycoses, the growth of *Histoplasma* species in culture is slow and the detection of *Histoplasma* antigens may enable the rapid diagnosis of histoplasmosis, particularly in patients with acute or disseminated disease. Antigen testing has recently been evaluated using a lateral flow technique. Urine antigen testing is preferred. Histoplasmosis is the only endemic mycosis for which a randomized controlled trial has been conducted, revealing the superiority of liposomal amphotericin B (L-AmB) over deoxycholate amphotericin B in disseminated disease [4]. An ongoing study is evaluating the effectiveness and safety of single-dose regimens of L-AmB in AIDS patients with histoplasmosis (clinical trials identification NCT04059770), in a similar fashion to what has been performed with leishmaniasis and cryptococcosis.

6. Lobomycosis

Lobomycosis is a neglected fungal disease that mostly occurs in the Amazon rainforest in Brazil. Infected patients present with keloidal nodules after fungal inoculation following traumatic exposure. Even though this is not a notifiable disease, an increase in the number of cases has been observed in recent years. Diagnosis—which still relies on histopathology—is challenging, because lesions are quite often mistaken for leishmaniosis, atypical mycobacteria including leprosy, sporothrychosis, and other dermatological fungal diseases. No drug is effective for the treatment of lobomycosis, and surgical resection has usually been followed by disease recurrence. Preliminary data suggest that posaconazole might be effective in lobomycosis; however, clinical trials are lacking.

7. Paracoccidioidomycosis

Outside endemic areas in Latin America, paracoccidioidomycosis is a disease carried by travelers who have lived in those areas for extended times. An expansion of endemicity has been associated with changes in agricultural practices, as well as the emergence of the newly recognized species *P. lutzii* [5]. The majority of patients with chronic disease are diagnosed using serology, or the microscopy of infected tissues or fluids. Antigen detection is not yet applicable to clinical practice, due to the lack of standardized or commercially available tests.

8. Sporothrychosis

Until recently, it was believed that only members of the pathogenic clade of *Sporothrix schenckii* were able to cause disease, but additional species such as *S. brasiliensis*, *S. globosa*, *S. mexicana*, and *S. pallida* have gained importance. The clinical relevance of identifying *Sporothrix* at the species level remains unclear. Serologic testing is promising, but currently limited by availability [6]. Antigen testing has the potential to be used in sporothrychosis.

9. Talaromycosis

Another disease which has had its name changed in mycology: penicilliosis is now talaromycosis. Initially a disease solely associated with AIDS patients, talaromycosis is now increasingly recognized in association with other immunocompromising conditions [7].

The region of endemicity is expanding, from Vietnam and southern China to northern and eastern China (Beijing and Shanghai) and in northeastern India. Mp1p antigen testing is now strongly recommended in the diagnosis of talaromycosis (superior sensitivity compared with culture: specificity > 95%) in plasma and urine specimens.

10. Conclusions

Most significant changes involving endemic mycoses have affected diagnostic tools. With such slow-growing fungi, this is fertile terrain for point-of-care tests to grow. Molecular assays are also in place for diagnosis, but standardization is required. With advances in fungal DNA sequencing, novel fungal species have been described, with some being of clinical or epidemiological relevance. For most of these fungi, the value of antifungal susceptibility testing is unknown, and serology remains largely unstandardized. Most pharmacological interventions in endemic mycoses are based on uncontrolled studies. Modern trials involving antifungal drugs are required in endemic mycosis. Additionally, above all, the development of drugs and diagnostic tests is only justifiable if these are made available to the countries in which these conditions are endemic, to reduce the burden of these diseases.

Funding: This research received no external funding.

Institutional Review Board Statement: Not applicable.

Informed Consent Statement: Not applicable.

Data Availability Statement: Not applicable.

Conflicts of Interest: Pasqualotto has received research grants from Gilead, Pfizer, and MSD.

References

1. Frost, H.M.; Novicki, T.J. Blastomyces Antigen Detection for Diagnosis and Management of Blastomycosis. *J. Clin. Microbiol.* **2015**, *53*, 3660–3662. [CrossRef]
2. Durkin, M.; Connolly, P.; Kuberski, T.; Myers, R.; Kubak, B.M.; Bruckner, D.; Pegues, D.; Wheat, L.J. Diagnosis of coccidioidomycosis with use of the Coccidioides antigen enzyme immunoassay. *Clin. Infect. Dis.* **2008**, *47*, e69–e73. [CrossRef] [PubMed]
3. Maphanga, T.G.; Britz, E.; Zulu, T.G.; Mpembe, R.S.; Naicker, S.D.; Schwartz, I.S.; Govender, N.P. In vitro antifungal susceptibility of yeast and mold phases of isolates of dimorphic fungal pathogen *Emergomyces africanus* (formerly *Emmonsia* sp.) from HIV-infected South African patients. *J. Clin. Microbiol.* **2017**, *55*, 1812–1820. [PubMed]
4. Johnson, P.C.; Wheat, L.J.; Cloud, G.A.; Goldman, M.; Lancaster, D.; Bamberger, D.M.; Powderly, W.; Hafner, R.; Kauffman, C.A.; Dismukes, W.E. Safety and efficacy of liposomal amphotericin B compared with conventional amphotericin B for induction therapy of histoplasmosis in patients with AIDS. *Ann. Intern Med.* **2002**, *137*, 105–109. [CrossRef] [PubMed]
5. Martinez, R. New Trends in Paracoccidioidomycosis Epidemiology. *J. Fungi* **2017**, *3*, 1. [CrossRef] [PubMed]
6. Bernardes-Engemann, A.R.; Costa, R.C.O.; Miguens, B.P.; Penha, C.V.L.; Neves, E.; Pereira, B.A.S.; Dias, C.M.P.; Mattos, M.; Gutierrez, M.C.; Schubach, A.; et al. Development of an enzyme-linked immunosorbent assay for the serodiagnosis of several clinical forms of sporotrichosis. *Med. Mycol.* **2005**, *43*, 487–493. [CrossRef] [PubMed]
7. Chan, J.F.; Lau, S.K.; Yuen, K.Y.; Woo, P.C. Talaromyces (Penicillium) marneffei infection in non-HIV-infected patients. *Emerg. Microbes Infect.* **2016**, *5*, e19. [CrossRef] [PubMed]

Journal of
Fungi

MDPI

Review

Molecular Diagnosis of Endemic Mycoses

Clara Valero [1], María Teresa Martín-Gómez [2,*] and María José Buitrago [3,4,*]

1 Manchester Fungal Infection Group, Division of Evolution, Infection, and Genomics, Faculty of Biology, Medicine and Health, University of Manchester, Manchester M13 9WU, UK
2 Microbiology Department, Hospital Universitari Vall d'Hebron, 08035 Barcelona, Spain
3 Mycology Reference Laboratory, Centro Nacional de Microbiología, Instituto de Salud Carlos III, Carretera Majadahonda-Pozuelo Km 2, Majadahonda, 28220 Madrid, Spain
4 CIBERINFEC, ISCIII-CIBER de Enfermedades Infecciosas, Instituto de Salud Carlos III, Majadahonda, 28220 Madrid, Spain
* Correspondence: mariateresa.martin@vallhebron.cat (M.T.M.-G.); buitrago@isciii.es (M.J.B.)

Abstract: Diagnosis of endemic mycoses is still challenging. The moderated availability of reliable diagnostic methods, the lack of clinical suspicion out of endemic areas and the limitations of conventional techniques result in a late diagnosis that, in turn, delays the implementation of the correct antifungal therapy. In recent years, molecular methods have emerged as promising tools for the rapid diagnosis of endemic mycoses. However, the absence of a consensus among laboratories and the reduced availability of commercial tests compromises the diagnostic effectiveness of these methods. In this review, we summarize the advantages and limitations of molecular methods for the diagnosis of endemic mycoses.

Keywords: endemic mycoses; molecular diagnosis; PCR; NGS

Citation: Valero, C.; Martín-Gómez, M.T.; Buitrago, M.J. Molecular Diagnosis of Endemic Mycoses. *J. Fungi* **2023**, *9*, 59. https://doi.org/10.3390/jof9010059

Academic Editor: Alessandro C. Pasqualotto

Received: 29 November 2022
Revised: 22 December 2022
Accepted: 24 December 2022
Published: 30 December 2022

1. Introduction

The common term "endemic fungi" usually refers to fungal species within the Onygenales order sharing, among others, four distinctive features: (i) thermal dimorphism, (ii) geographical distribution restricted to specific regions of the world, (iii) ability to cause a disease in otherwise healthy humans, although illness tends to be more severe in immunocompromised individuals, and (iv) high mortality rates if the illness fails to be timely diagnosed and treated [1]. Recently, the WHO has released the fungal priority pathogen list to strengthen the global response to fungal infections. Several endemic fungi are listed within the high and medium priority groups [2].

1.1. Epidemiology of Endemic Mycoses

Endemic mycoses (EM) are caused by species of the genera *Histoplasma*, *Blastomyces*, *Coccidioides*, *Paracoccidioides*, *Talaromyces*, *Sporothrix*, *Lacazia*, and the recently described *Emergomyces*. The distribution area of endemic cases encompasses countries across the five continents. Coccidiomycosis, paracoccidiomycosis and lobomycosis are restricted to the American continent, whereas sporothrichosis and histoplasmosis have a cosmopolite distribution with high presence in the Americas and Africa. Blastomycosis extends mainly across Africa, the western basins of United States of America (USA), and the south-western Canadian border. Talaromycosis cases are typically found in south-eastern Asia, while emergomycosis is frequently diagnosed in South Africa, but cases have also been reported in North America, Europe, Asia and India [3,4]. Certain species within endemic genera can be found only in specific areas of the world, usually associated to particular environmental conditions of heat, moisture, pH or nutrients, among others [5].

The true epidemiology of endemic fungal infections is unknown. Many primary infections are asymptomatic or present with mild self-resolving symptoms not requiring the search of medical care, and frequently the etiological agent of the infection fails to be identified due to lack of awareness and limited access to the appropriate diagnostic tools. This is particularly concerning outside hyperendemic territories, where the vast majority of EM cases are imported and associated to immigration and travels from endemic areas [6]. Despite EM global burden is increasing, clinical infections are not subjected to mandatory notification to Public Health systems with exceptions restricted to specific areas [7,8].

The incidence of histoplasmosis has been estimated to range between 0.1–100 cases-/100,000 inhabitants, with lowest rates observed in areas with temperate climates, and the highest incidence in humid tropical territories [9]. Serologic studies indicate that up to 40% of the population living in highly endemic areas may have been exposed to the fungus, with seropositivity reaching up to 87% in specific populations [7]. The real number of coccidioidomycosis cases has been estimated to exceed 350,000 per year in the USA, with an increasing trend observed over the last years [10–12]. In Brazil, paracoccidioidomycosis is estimated to affect 3–4 new patients/100,000 inhabitants/year, with an incidence that may reach up to 40 patients/100,000 inhabitants depending on the location. In this country, paracoccidioidomycosis represents the main cause of hospitalization and death among the overall systemic mycoses [3]. Talaromycosis is one of the most neglected and underrecognized EM as its prevalence is largely unknown. This disease is strongly associated to poverty and uncontrolled advanced HIV disease, especially in areas where the access to healthcare is limited. Some reports describe that the burden of this disease could exceed 17,000 cases/year, being lethal in as much as 1 in 3 cases [13]. With approximately 40,000 new cases every year sporotrichosis is considered the most prevalent EM in South America, and the most frequent EM in regions of Southern Brazil [14]. It is also endemic in Mexico and Northern China, and has been responsible of large outbreaks in North-America, Australia, and South Africa [15]. Emergomycosis has been described as an HIV-associated infection in South Africa, where it ranked the second most frequent EM only after sporotrichosis in a recent review of contemporary cases spanning 10 years [16]. Scattered reports locate *Emergomyces* spp. also in Europe, North America, and Asia. The true incidence of emergomycosis is unknown, but after the introduction of molecular techniques, many cases initially classified as histoplasmosis on the basis of histopathology have been demonstrated to be emergomycosis, indicating that its prevalence may be more frequent than previously thought [17]. Lobomycosis prevalence is unknown but an increase in new cases has been observed in recent years [18].

EM incidence has been reported to be on the rise in recent years [3,7,19,20]. This has been mainly attributed to environmental changes, travels, and expansion of at-risk populations, along with increased awareness and wider access to improved diagnostic techniques. As more research and educational actions are undertaken, new areas devoted to combatting these diseases will be uncovered [7,13,21,22]. Thus, a wide availability of sensitive, specific, rapid, and versatile diagnostic techniques will become an immediate necessity.

1.2. Diagnosis of Endemic Mycoses

To date, the laboratory diagnosis of EM is an unsolved issue [23]. The diagnostic yield of currently available microbiological techniques has been extensively reviewed by an international team of experts in a joint ECMM-ISHAM initiative, resulting in evidence-based recommendations of use recently published [4]. Conventional techniques, such as histopathology and culture, are not difficult to implement, but exhibit a number of limitations that should be taken into consideration. Firstly, these techniques require a high level of expertise and special caution is needed when handling specimens and cultures, as some species are classified as BSL-3 microorganisms; depending on the specimen and phase of the illness [24]. Moreover, cultures may delay the diagnosis up to four-to-six weeks, as these fungal species are slow-growing, and the confirmation of the dimorphism may be required for the final identification. In addition, their diagnostic yield is hampered by lack of sensitivity, particularly in non-invasive chronic forms.

Culture independent commercial assays, which rely on the detection of antigens or antibodies in clinical samples, are only available for the most prevalent EM. Antibody-based diagnosis is determined by the immune status of the host, as immunosuppresed patients fail to produce high antibody titers and seropositivity remains long time after the infection [25]. Moreover, these tests exhibit cross-reactivity among EM-causing species and with other human fungal pathogens. Antigen tests have been proved to be useful for rapid diagnosis in populations generally affected by severe immunosuppresion and disseminated forms of disease [26], but little information is available on their applications in other contexts. Specificity of antigen tests is reduced by cross-reactivity issues with other fungi [27]. During the diagnostic process, the possibility of cross-reactivity of antigens and antibodies shall be considered in areas where endemic genera co-exist. Point-of-care methods have been developed for the detection of *Coccidioides* spp. and *H. capsulatum*, although studies to date are limited, results seem to be promising [28,29].

Molecular techniques have been key for taxonomic placement and to uncover cryptic species withing the EM-causing species [30], but their application to clinical diagnosis is far from being of routine use. Most specific PCR techniques have been developed by reference laboratories without a consensus about the technology used (conventional PCR, quantitative PCR, LAMP etc.) or the genomic regions targeted by the assays. Despite several techniques for the detection of EM have been reported to be useful for diagnosis, only a *Coccidioides*-specific PCR is commercially available ((accessed on 28 December 2022)).

Despite the efforts on validating molecular assays in diverse types of patients and samples, the limited presence of molecular techniques in international diagnostic guidelines is due to the need of further standardization, and the lack of solid multicentered studies involving large populations. Recently, an European initiative has established a working group devoted to perform intercomparison multicenter diagnostic studies with the objective of improving EM diagnosis and acquiring a better knowledge about the epidemiology of these neglected fungal infections (https://www.ecmm.info/working-groups/working-group-on-the-diagnosis-and-the-epidemiology-of-endemic-mycoses (accessed on 28 December 2022)). Notwithstanding these limitations, molecular techniques currently seem to represent a good immediate alternative for a fast and specific diagnosis of such infections, as well as a feasible tool to go deeper into the knowledge of their epidemiology [31].

This review is intended to summarize the techniques, targets, applications of molecular techniques to the diagnosis of endemic mycoses, covering the full spectrum of techniques, from the most traditional PCR protocols to the most advanced sequencing methods (Figure 1).

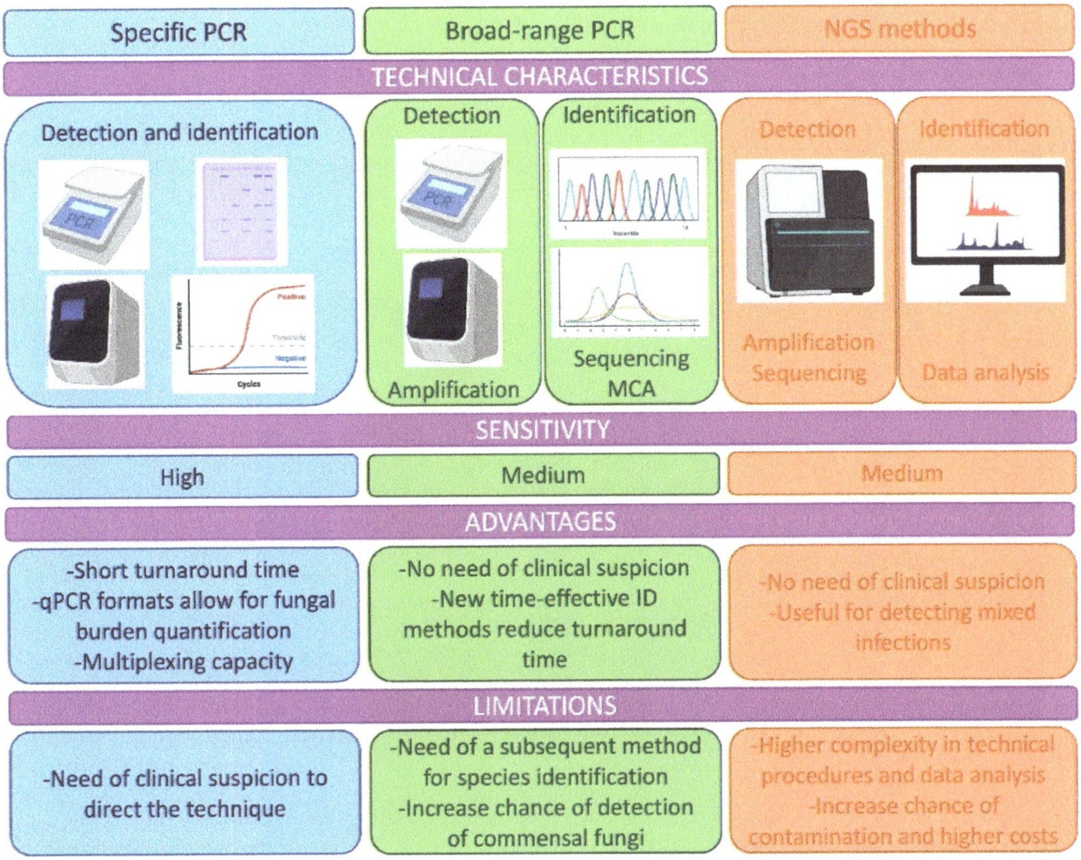

Figure 1. Description of the technical characteristics, advantages and limitations of the molecular methods used for the diagnosis of endemic mycoses including specific PCR methods, methods based on broad range PCR and new methods as those based on Next Generation Sequencing (NGS). MCA: melting curve analysis.

2. Specific PCR Assays

Specific PCR assays have been developed last years in reference laboratories mainly focused on the detection of *H. capsulatum* and *Coccidioides* spp. For the remaining EM species, there are considerably fewer studies. In general, commercial tests and inter-comparison studies are lacking. The global SARS-CoV-2 pandemic has allowed the implementation of conventional and real-time PCR (qPCR) technology in several laboratories worldwide, including endemic regions, which offers an excellent opportunity to expand the application of molecular techniques for the detection of these neglected pathogens in a near future.

2.1. Histoplasmosis

Methods based on PCR (conventional or real time) for the detection of *H. capsulatum* target different genomic regions: (i) ribosomal DNA (rDNA) multicopy regions as 18S [32], ITS1 and ITS2 regions [33–36] and the ribosomal small subunit RNA [37], or (ii) unicopy targets as genes coding the 100-kDa-like protein or the M antigen [38–42] and, more recently, *PPK* and *CFP4 genes* [43].

DNA from clinical and reference isolates or/and clinical samples has been used for validation of these specific PCR assays. The type of clinical samples varies including respiratory secretions, biopsies, bone marrow, blood, or sera. In general, better sensitivity values were reported using clinical specimens sampled at the site of the infection, such as respiratory and biopsy samples. However, less invasive samples, such as sera and blood, were often preferred in disseminated infections [33,37,41]. Methods for the nucleic acid extraction from clinical samples also differed depending on the assay with some including sample pretreatment and others using total nucleic acids, the latter introducing a reverse transcription step before the amplification of the target to improve sensitivity [37]. Although the number of clinical samples in some publications was very limited, sensitivity values reported in these studies ranged from 70–100% [35,38]. A recent meta-analysis focused on HIV+ patients with progressive disseminated histoplasmosis reported an overall sensitivity and specificity of (95% CI) of 95.4% (88.8–101.9) and 98.7% (95.7–101.7), respectively, in different type of samples including respiratory, biopsies, blood and bone marrow [44].

LAMP methods described for the diagnosis of histoplamosis are scarce. These assays have been designed to target the ITS region [45] or the 100-kDa-like protein [46] showing variable sensitivity results.

Regarding the establishment of a consensus about histoplasmosis PCR diagnostic methods, to date, only one multicenter study involving laboratories from four Latin American countries and Spain has been published [47]. In this work, seven different PCR protocols were compared using the same DNA panel for testing the assays. Although the overall sensitivity and specificity was 86 and 100%, respectively, PCR real-time based protocols were demonstrated to be the most sensitive and reproducible approaches compared to conventional PCR assays. Methods targeting unicopy genes showed the poorest sensitivity.

2.2. *Coccidiomycosis*

Molecular techniques for the detection of *Coccidioides* spp. have been developed to be used on both clinical [48,49] and environmental settings, such as endemic regions from USA, where a steady rise in coccidioidomycosis infections has been reported [50,51]. These assays were designed to target the ITS region of the ribosomal DNA and genes encoding both Antigen 2 and Proline rich Antigen, with sensitivity ranging from 74 to 100%. Clinical samples used to test these assays were mainly respiratory, fresh and paraffin embedded biopsies and cerebrospinal fluid. When comparing different clinical samples, the best performance was obtained when using respiratory samples, fresh tissues reached 93% sensitivity, and paraffin-embedded tissues sensitivity was reported to be around 73% [48]. In 2018, the FDA authorized a commercial assay for the rapid detection of coccidioidomycosis, the Genestat MDX *Coccidioides* (https://www.aacc.org/cln/articles/2018/march/fda-clears-first-molecular-test-for-valley-fever (accessed on 28 December 2022)). In a multicenter study, this method reached a 100% sensitivity, with a specificity that ranged between 93.8% and 100% depending on the sample tested [52].

Of interest, *Coccidioides* spp. is the only fungal genus included in the international lists of potential bioterrorism agents [53], making essential to be able to face this contingency with the aid of a rapid detection method. In this line, molecular techniques represent an excellent option to be included in preparedness and response protocols due to their short turnaround response time and remarkable sensitivity and specificity scores. However, further standardization and consensus are needed.

2.3. *Paracoccidioidomycosis*

Several "in house" molecular techniques have been described for the detection of *Paracoccidioides* spp., especially in laboratories from Brazil and other non-endemic regions (Table 1). Most of these assays were based on conventional PCR methodologies [54–57]. On the other hand, two methods based on qPCR for their use in paracoccidioidomycosis diagnosis have been published [31,58]. Targets selected for amplification were the multicopy ITS rDNA region and the genes encoding the proteins Gp43 or Pb27. The clinical samples tested in theses assays were mainly respiratory, biopsies, blood and sera. Of note, sera samples were not recommended in two of these studies [31,56] as authors never detected DNA in these kinds of samples. The overall sensitivity ranges reported were 91–100%, showing a great potential of these techniques for clinical use.

Only one LAMP method has been described to date targeting the gene encoding the Gp43 protein; however, the sensitivity reported on sputum samples was moderate (61%) [59].

Paracoccidioides spp. are considered fastidious microorganisms as recovering these pathogens from culture is hard and time-consuming, commercial antigen tests are still not available and serological methods have strong limitations. Considering all the above, the inclusion of these molecular diagnostic techniques in the routine of clinical microbiology laboratories is substantially justified. This is even more imperative in non-endemic regions where the delay in diagnosis has fatal consequences in paracoccidioidomycosis patients [60–62].

2.4. Blastomycosis

Fewer assays have been described for the diagnosis of blastomycosis. The *BAD1* gene, an important conserved adhesion-promoting protein and virulence factor of *Blastomyces* spp. has been chosen as target in several assays developed for the detection of the fungus in soil [63] or in clinical samples [64]. Other targets as *DRK1* gene have also been used [65]. Although there is little evidence of the usefulness of theses assays in a clinical setting, the results obtained were very promising with high specificity and sensitivity values reported.

2.5. Talaromycosis

A recent meta-analysis has reviewed the methods based on PCR developed for the rapid diagnosis of talaromycosis [66]. Most of them have been published by authors from endemic regions (China, Vietnam, Thailand) which used conventional nested PCR [67] or real-time PCR [68,69] targeting the ribosomal DNA or other gene encoding regions [70]. Samples tested included plasma, blood, serum and bone marrow reporting an overall sensitivity and specificity of 84% and 99%, respectively. A LAMP assay has been published recently showing a suitable sensitivity and detecting all the biopsy samples tested [71].

2.6. Conclusions

Although data are very heterogeneous among works, specific PCR assays are rapid sensitive and specifics. Some studies used a limited number of samples for the validation of the assays, and studies focus on blastomycosis and talaromycosis are scarce. Reaching consensus about targets and kind of samples should be a priority (Table 1).

Table 1. Details of the studies where specific PCR assays were used to diagnose endemic mycoses.

PCR Technology	Target	Sample	Sensitivity (Cases)/Specificity	Specificity	Ref
		Histoplasmosis			
Conventional (nested)	18S rDNA	Blood, spleen, lung (mice)	83.1%	ND	[32]
Conventional (nested)	100-kDa-like protein gene	Biopsy	70%	100%	[72]
Conventional	M antigen gene	ND	100%	100%	[39]
Conventional (semi-nested)	M antigen gene	Biopsy, blood, mucose, BM	ND (30)	ND	[38]
Real-time	ITS rDNA	BAL, lung biopsy, BM	100% (3)	100%	[35]
Conventional (nested)	100-kDa-like protein gene	Blood, serum, BAL, BAS, biopsy, CSF, others	100% (40)	100%	[41]

Table 1. *Cont.*

PCR Technology	Target	Sample	Sensitivity (Cases)/Specificity	Specificity	Ref
Real-time	ITS rDNA	Blood, serum, BM, sputum, BAS, BAL, biopsy, CSF, others	89% Proven H (54) 60% Probable H (13)	100%	[31]
Real-time	ITS rDNA	BAL, biopsy, BM, CSF	95.4% (348)	96%	[36]
Real-time (multiplex)	ITS rDNA	BAL, biopsy, serum, BM	92.5% (72)	100%	[34]
Real-time	*mtSSU* gene	Blood, serum, BAL, BAS, biopsy, CSF, others	97.7% (44)	ND	[37]
Conventional Real-time	*PPK, CFP4*	FFPE tissue	100% (2)	ND	[43]
			Paracoccidioidomycosis		
Conventional (nested)	Gp43	Biopsy (mice)	91% (23)	ND	[57]
LAMP	Gp43	Sputum	60% (18)	ND	[59]
Conventional (semi-nested)	ITS rDNA	Biopsy (mice)	100% (4)	100%	[54]
Real-time	ITS rDNA	Serum, blood, sputum	100% (6)	ND	[73]
Conventional	ITS rDNA	Serum, biopsy	ND	ND	[56]
Conventional (semi-nested)	ITS rDNA	Sputum	100% (14)	ND	[74]
Conventional (nested)	GP43 gene	BAL, biopsy, sputum	100% (25)	100%	[55]
Real-time	Pb27 gene	Blood, serum, biopsy and others	94% (78)	100%	[58]
			Coccidioidomycosis		
Conventional (nested)/real-time	Antigen2/Proline-Rich Antigen,	FFPE- biopsy	100% (3)	ND	[75]
Real-time	ITS rDNA	Respiratory, biopsy, FFPE-biopsy	89% (480)	98%	[48]
Real-time	ITS rDNA	Mice samples	98% (44)	100%	[49]
Real-time	GeneSTAT *Coccidioides* assay	BAL/BW	100% (332)	93.85–100%	[52]
			Blastomycosis		
Conventional (nested)	*WI-1 (BAD 1)*	PE-biopsy (dogs)	ND (73)	ND	[76]
Real-time	*DRK-1*	Respiratory, biopsy and others	86% (14)	99.4%	[65]
Real-time	*BAD-1*	FFPE-biopsy	83% (12)	100%	[64]
Real-time (duplex)	*BAD-1*	FFPE-biopsy, respiratory and others	ND (33)	ND	[77]
			Talaromycosis		
Real-time	5.8S rDNA	Blood	60% (20)	100%	[78]
Conventional (nested)	18S rDNA	Serum	68.6% (35)	100%	[67]
LAMP	ITS rDNA	Biopsy	100% (12)	100%	[71]
Conventional (nested)/ real-time	ITS rDNA	Blood, serum	82% (22)/91% (22)	75%/63%	[68]
Real-time	ITS rDNA	Serum	86.11% (36)	ND	[69]

ND: no data; FFPE-biopsy: formalin-fixed paraffin-embedded biopsy; BAL: brochoalveolar lavage; BAS: brochoaspirate; BW: bronchial wash; CSF: cerebrospinal fluid; BM: bone marrow.

3. Broad-Range PCRs

Broad-range or panfungal PCR assays are especially useful for EM diagnosis, generally used when there is not a clear suspicion of the fungal agent causing the disease, which is one of the hallmarks of EM, or when the infection is not frequent in the setting of the diagnostic laboratory, as it is in non-endemic areas. This approach relies on the use of fungal (or fungal group)-specific primers to amplify fungal DNA directly from clinical samples followed by an identification method, mainly Sanger sequencing, to confirm the causative agent [79]. With the aim of improving sensitivity, classic multi-copy targets as the ribosomal operon [37] are often selected for panfungal amplification, while fresh tissue samples are preferred over formalin-fixed, paraffin-embedded samples [80].

Sample contamination, detection of commensal fungi, PCR bias due to primer mismatches and, the lack of adequate reference databases for fungi identification are the main limitations of panfungal PCR assays. However, the limitation of delay in response time associated to species determination has been addressed by replacing Sanger sequencing identification with other time-saving post-PCR methods such as melting curve analysis, DNA microarray, electrospray-ionization mass spectrometry analysis and T2 magnetic resonance [81].

In conclusion, although proper studies directed to EM diagnosis by using broad-range PCRs are still missing, there are plenty reports in the literature showing the ability of this technique to provide a definite diagnosis when paired with other reference methods. This technique has the advantage of being cost-effective and can be an alternative to specific PCR considering their limitations (Table 2).

Table 2. Details of the studies where broad-range PCR was used to diagnose endemic mycoses.

Target	Sample	Post-PCR ID Method	Notes	Ref
Histoplasmosis				
rDNA (18S)	BM	Sanger sequencing	Confirmed by histopathology and culture	[82]
rDNA (ITS1)	BM	Sanger sequencing	Confirmed by culture	[83]
rDNA (ITS, 28S)	Lung tissue	Sanger sequencing	Confirmed by histopathology	[84]
rDNA (28S)	Mucosal biopsy	Sanger sequencing	Confirmed by specific PCR	[85]
rDNA (28S)	FFPE tissue	Sanger sequencing	Confirmed by histopathology and specific qPCR	[86]
Coccidioidomycosis				
rDNA (ITS)	Biopsy	Sanger sequencing	Confirmed by histopathology, qPCR format	[87]
Blastomycosis				
rDNA (ITS2 and D2)	FFPE tissue	Sanger sequencing	Confirmed by histopathology	[88]
Emergomycosis				
rDNA (28S, ITS2)	FFPE tissue	Sanger sequencing	Confirmed by histopathology	[89]
Lobomycosis				
rDNA (ITS1-4)	Biopsy	Sanger sequencing	Confirmed by histopathology	[90]
Multiple EM identified				
rDNA (ITS2)	Biopsies	MCA and sanger sequencing	Histoplasmosis, coccidioidomycosis, paracoccidioidomycosis. Confirmed by histopathology	[91]
rDNA (28S, ITS2, D1-D2)	FFPE and fresh tissue	Sanger sequencing	Histoplasmosis, talaromycosis, blastomycosis. Some cases confirmed by histopathology	[92]
rDNA (ITS2, D2)	FFPE tissue	Sanger sequencing	Histoplasmosis, coccidioidomycosis. Confirmed by histopathology, qPCR format	[93]
rDNA (ITS1-2)	FFPE and fresh tissue	Sanger sequencing	Histoplasmosis, paracoccidioidomycosis. Confirmed by culture or histopathology	[94]

BM: bone marrow; FFPE: formalin-fixed paraffin-embedded; MCA: melting curve analysis.

4. Next Generation Sequencing (NGS)

NGS has revolutionized the diagnosis of fungal and other microbial infections and it is already considered the future replacement for the current broad-range PCR methods. The most used NGS approach for diagnosis nowadays is targeted amplicon sequencing or metabarcoding. By using fungal-specific primers, thousands of copies of different DNA templates are amplified and sequenced simultaneously, reducing turnaround time and costs [95]. However, shotgun metagenomic sequencing can be also used to target most parts of the genomes of the microorganisms present in the sample. This approach is more expensive and computationally demanding, but allows for further characterization of the infecting agent as other features, such as identifying the subtype or the antimicrobial resistance profile, could be retrieved from the sequenced data [96]. In general, NGS methods face the same limitations as broad-range PCR assays but with the additional requirement of expertise in data analysis and increasing complexity in the technical procedures [97].

NGS technologies were originally standardized as an exploratory tool to study the fungal community profile (mycobiome) of human specimens. As an example, McTaggart LR and colleagues developed an NGS-based method for the analysis of the lung mycobiome during *Blastomyces dermatitidis/gilchristii* infection [98]. The successful detection of the causative agent as well as other fungal pathogens indicated the potential of this method for the diagnosis of EM. However, proper standardization and retrospective studies including a substantial number of clinical isolates are still missing, it not being currently possible to recommend or suggest a method or to consider NGS as a suitable tool for EM diagnosis. Most studies reported in the literature describe brief case reports or anecdotical presence of EM samples in bigger specimen sets. Nevertheless, NGS methods have already been employed successfully in the differential diagnosis of infections with similar clinical symptoms and the identification of the biological source of an outbreak (Table 3). Recently, the assessment of the clinical performance of NGS for the rapid diagnosis of talaromycosis in HIV patients has been evaluated [99]. The sensitivity of the new method was significantly higher than culture and serum galactomannan determination (98.3% vs. 66.7% and 83.3%, respectively) underlining the potential use of NGS for EM diagnosis. In conclusion, although the NGS-based method seems to be promising, more studies need to be able of consider it as a tool for the diagnosis of EM (Table 3).

Table 3. Details of the studies where NGS was used to diagnose endemic mycoses.

Target	Samples	Aim	Notes	Ref
		Talaromycosis		
Total DNA	BAL, CSF and BM	Diagnosis of a patient with a 3-months record of undiagnosed disease	Confirmed by histopathology and positive culture in skin lesion	[100]
Total DNA	CSF	Diagnosis of a patient with meningoencephalitis		[101]
Not mentioned	BAL	Diagnosis of a patient with chronic pneumonia	Confirmed by culture in BAL	[102]
Total DNA	Peripheral blood	Diagnosis of HIV febrile patient	Confirmed by panfungal PCR on lymph node biopsy	[103]
Not mentioned	BAL	Diagnosis of a patient with chronic pneumonia	Confirmed by culture in BAL	[104]
Total DNA	Skin tissue and eye aqueous humor	Diagnosis of a patient with eye tumor	Confirmed by PCR in the aqueous humor	[105]
Not mentioned	BAL and blood	Diagnosis of a patient with chronic pneumonia	Confirmed by culture in sputum	[106]

Table 3. *Cont.*

Target	Samples	Aim	Notes	Ref
Total DNA	FFPE tissue	Differential diagnosis of a patient with peritonitis		[107]
Not mentioned	BAL	Diagnosis of a patient with chronic pneumonia	Confirmed by culture in BAL	[108]
Total DNA	BAL, blood, and BM	Assessment of clinical performance of NGS for talaromycosis diagnosis	Sensitivity and specificity values were 98.3 and 98.6%, respectively. The clinical final diagnosis was used as the reference standard.	[99]
		Histoplasmosis		
Total RNA	CSF	Differential diagnosis of meningitis	Statistical framework supported by environmental and non-infected control samples	[109]
Total DNA	Miscellaneous	Identification of the causative agent causing an outbreak		[110]
Not mentioned	Not mentioned	Diagnosis of a patient with chronic progressive lung lesions		[111]
DNA (ITS region)	FFPE tissue	Diagnosis of a patient with a skin lesion	Confirmed by histopathology	[112]
Not mentioned	BM	Diagnosis of non-HIV febrile patient	Confirmed by direct visualization	[113]
		Blastomycosis		
Cell-free DNA	Plasma	Diagnosis of a patient with chronic pneumonia		[114]
Not mentioned	BAL and biopsy	Diagnosis of a patient with chronic pneumonia	Confirmed by histopathology of BAL	[115]
		Multiple EM identified		
Not mentioned	Peripheral blood and BM	Differential diagnosis in immunocompromised patients	Histoplasmosis (confirmed by histopathology), talaromycosis	[116]
DNA (ITS region)	FFPE tissue	Retrospective evaluation of the NGS clinical utility	Confirmed by histopathology	[117]

BAL: brochoalveolar lavage; CSF: cerebrospinal fluid; FFPE: formalin-fixed paraffin-embedded; BM: bone marrow.

5. Conclusions and Perspectives

Diagnosis of EM is still difficult in endemic regions and even more complicated out of these regions, where the lack of suspicion and expertise are the major shortcomings. Molecular techniques have shown their great potential for the rapid diagnosis of EM in several studies performed in reference laboratories in the last years. The recent COVID-19 pandemic has not only increased the awareness on how critical a rapid diagnosis is but paved the way to the generalized implementation of the molecular diagnosis of infectious diseases. As summarized in this review, several molecular techniques developed in recent years show a great potential for the rapid diagnosis of EM. In non-endemic countries, where the availability of some other useful techniques, as antigen detection, is limited, qPCR-based molecular assays have been developed to this purpose, extending their usefulness to difficult-to-diagnose forms of infection [34,37]. The introduction of multiplex formats also allows for performing a differential diagnosis with other pathogens causing similar clinical patterns reducing costs [118]. In endemic areas, especially in resource-limited settings, cost-effective molecular methods such as LAMP could be a promising alternative. However, in general terms, there is still great variability in published methods to date and commercial kits are practically non-existent. An effort to standardize and achieve a consensus should be performed among the different laboratories. Technical issues such as the selection of genomic targets or nucleic acid extraction methods, coupled with the implementation of inter-comparison studies should be prioritized to include these techniques in the future guidelines for patient management. Panfungal assays stand for an interesting alternative to specific assays as these techniques are easy to implement and more cost-effective; however, limitations of these tests should be considered when performing a final diagnosis. Recently, NGS has emerged as an alternative to overcome some of these limitations soon. As a conclusion, the implementation of molecular techniques in clinical settings will revolutionize the rapid diagnosis of EM, especially in countries where laboratories use diagnostic PCR routinely.

Author Contributions: C.V. review the panfugal and NGS methods, elaborated the figures and Tables, organized the bibliography and participated in the redactions. M.J.B. and M.T.M.-G. designed the structure of the paper and reviewed the final version. M.T.M.-G. elaborated the introduction and participated in the redaction. M.J.B. elaborated the specific PCR section and the Table associate. All authors have read and agreed to the published version of the manuscript.

Funding: This research was funded by research project PI21CIII/00007 from Spanish Fondo de Investigaciones Sanitarias of the Instituto de Salud Carlos III.

Institutional Review Board Statement: Not applicable.

Informed Consent Statement: Not applicable.

Data Availability Statement: Not applicable.

Conflicts of Interest: M.J.B. is a founding partner and holds shares of Micologia Molecular S.L. She has received grant support from the Instituto de Salud Carlos III and has been paid for talks on behalf of United Medical LTDA.

References

1. Gnat, S.; Lagowski, D.; Nowakiewicz, A.; Dylag, M. A global view on fungal infections in humans and animals: Infections caused by dimorphic fungi and dermatophytoses. *J. Appl. Microbiol.* **2021**, *131*, 2688–2704. [CrossRef] [PubMed]
2. WHO. *Fungal Priority Pathogens List to Guide Research, Development and Public Health Action*; World Health Organization: Geneva, Switzerland, 2022.
3. Shikanai-Yasuda, M.A.; Mendes, R.P.; Colombo, A.L.; Queiroz-Telles, F.; Kono, A.S.G.; Paniago, A.M.M.; Nathan, A.; Valle, A.; Bagagli, E.; Benard, G.; et al. Brazilian guidelines for the clinical management of paracoccidioidomycosis. *Rev. Soc. Bras. Med. Trop.* **2017**, *50*, 715–740. [CrossRef] [PubMed]
4. Thompson, G.R., 3rd; Le, T.; Chindamporn, A.; Kauffman, C.A.; Alastruey-Izquierdo, A.; Ampel, N.M.; Andes, D.R.; Armstrong-James, D.; Ayanlowo, O.; Baddley, J.W.; et al. Global guideline for the diagnosis and management of the endemic mycoses: An initiative of the European Confederation of Medical Mycology in cooperation with the International Society for Human and Animal Mycology. *Lancet Infect. Dis.* **2021**, *21*, e364–e374. [CrossRef] [PubMed]
5. Gorris, M.E.; Cat, L.A.; Zender, C.S.; Treseder, K.K.; Randerson, J.T. Coccidioidomycosis Dynamics in Relation to Climate in the Southwestern United States. *Geohealth* **2018**, *2*, 6–24. [CrossRef] [PubMed]
6. Salzer, H.J.F.; Stoney, R.J.; Angelo, K.M.; Rolling, T.; Grobusch, M.P.; Libman, M.; Lopez-Velez, R.; Duvignaud, A.; Asgeirsson, H.; Crespillo-Andujar, C.; et al. Epidemiological aspects of travel-related systemic endemic mycoses: A GeoSentinel analysis, 1997-2017. *J. Travel Med.* **2018**, *25*, tay055. [CrossRef] [PubMed]
7. Ashraf, N.; Kubat, R.C.; Poplin, V.; Adenis, A.A.; Denning, D.W.; Wright, L.; McCotter, O.; Schwartz, I.S.; Jackson, B.R.; Chiller, T.; et al. Re-drawing the Maps for Endemic Mycoses. *Mycopathologia* **2020**, *185*, 843–865. [CrossRef]
8. Vallabhaneni, S.; Mody, R.K.; Walker, T.; Chiller, T. The Global Burden of Fungal Diseases. *Infect. Dis. Clin. N. Am.* **2016**, *30*, 1–11. [CrossRef]
9. Rodrigues, A.M.; Beale, M.A.; Hagen, F.; Fisher, M.C.; Terra, P.P.D.; de Hoog, S.; Brilhante, R.S.N.; de Aguiar Cordeiro, R.; de Souza Collares Maia Castelo-Branco, D.; Rocha, M.F.G.; et al. The global epidemiology of emerging *Histoplasma* species in recent years. *Stud. Mycol.* **2020**, *97*, 100095. [CrossRef]
10. Benedict, K.; McCotter, O.Z.; Brady, S.; Komatsu, K.; Sondermeyer Cooksey, G.L.; Nguyen, A.; Jain, S.; Vugia, D.J.; Jackson, B.R. Surveillance for Coccidioidomycosis—United States, 2011-2017. *MMWR Surveill Summ.* **2019**, *68*, 1–15. [CrossRef]

11. Sondermeyer Cooksey, G.L.; Nguyen, A.; Vugia, D.; Jain, S. Regional Analysis of Coccidioidomycosis Incidence—California, 2000-2018. *MMWR Morb. Mortal. Wkly. Rep.* **2020**, *69*, 1817–1821. [CrossRef]
12. Van Dyke, M.C.C.; Thompson, G.R.; Galgiani, J.N.; Barker, B.M. The Rise of Coccidioides: Forces Against the Dust Devil Unleashed. *Front. Immunol.* **2019**, *10*, 2188. [CrossRef] [PubMed]
13. Narayanasamy, S.; Dat, V.Q.; Thanh, N.T.; Ly, V.T.; Chan, J.F.; Yuen, K.Y.; Ning, C.; Liang, H.; Li, L.; Chowdhary, A.; et al. A global call for talaromycosis to be recognised as a neglected tropical disease. *Lancet Glob. Health* **2021**, *9*, e1618–e1622. [CrossRef] [PubMed]
14. Bongomin, F.; Gago, S.; Oladele, R.O.; Denning, D.W. Global and Multi-National Prevalence of Fungal Diseases-Estimate Precision. *J. Fungi* **2017**, *3*, 57. [CrossRef]
15. Chakrabarti, A.; Bonifaz, A.; Gutierrez-Galhardo, M.C.; Mochizuki, T.; Li, S. Global epidemiology of sporotrichosis. *Med. Mycol.* **2015**, *53*, 3–14. [CrossRef] [PubMed]
16. Mapengo, R.E.; Maphanga, T.G.; Grayson, W.; Govender, N.P. Endemic mycoses in South Africa, 2010-2020: A decade-long description of laboratory-diagnosed cases and prospects for the future. *PLoS Negl. Trop. Dis.* **2022**, *16*, e0010737. [CrossRef]
17. Schwartz, I.S.; Kenyon, C.; Feng, P.; Govender, N.P.; Dukik, K.; Sigler, L.; Jiang, Y.; Stielow, J.B.; Munoz, J.F.; Cuomo, C.A.; et al. 50 Years of *Emmonsia* Disease in Humans: The Dramatic Emergence of a Cluster of Novel Fungal Pathogens. *PLoS Pathog.* **2015**, *11*, e1005198. [CrossRef]
18. Goncalves, F.G.; Rosa, P.S.; Belone, A.F.F.; Carneiro, L.B.; de Barros, V.L.Q.; Bispo, R.F.; Sbardelott, Y.; Neves, S.; Vittor, A.Y.; Woods, W.J.; et al. Lobomycosis Epidemiology and Management: The Quest for a Cure for the Most Neglected of Neglected Tropical Diseases. *J. Fungi* **2022**, *8*, 494. [CrossRef]
19. Baker, J.; Setianingrum, F.; Wahyuningsih, R.; Denning, D.W. Mapping histoplasmosis in South East Asia—Implications for diagnosis in AIDS. *Emerg. Microbes Infect.* **2019**, *8*, 1139–1145. [CrossRef]
20. McCotter, O.Z.; Benedict, K.; Engelthaler, D.M.; Komatsu, K.; Lucas, K.D.; Mohle-Boetani, J.C.; Oltean, H.; Vugia, D.; Chiller, T.M.; Sondermeyer Cooksey, G.L.; et al. Update on the Epidemiology of coccidioidomycosis in the United States. *Med. Mycol.* **2019**, *57*, S30–S40. [CrossRef]
21. Amona, F.M.; Denning, D.W.; Moukassa, D.; Develoux, M.; Hennequin, C. Histoplasmosis in the Republic of Congo dominated by African histoplasmosis, *Histoplasma capsulatum* var. *duboisii*. *PLoS Negl. Trop. Dis.* **2021**, *15*, e0009318. [CrossRef]
22. Rakislova, N.; Hurtado, J.C.; Palhares, A.E.M.; Ferreira, L.; Freire, M.; Lacerda, M.; Monteiro, W.; Navarro, M.; Casas, I.; Teixeira, M.M.; et al. High prevalence and mortality due to *Histoplasma capsulatum* in the Brazilian Amazon: An autopsy study. *PLoS Negl. Trop. Dis.* **2021**, *15*, e0009286. [CrossRef]
23. Caceres, D.H.; Echeverri Tirado, L.C.; Bonifaz, A.; Adenis, A.; Gomez, B.L.; Flores, C.L.B.; Canteros, C.E.; Santos, D.W.; Arathoon, E.; Soto, E.R.; et al. Current situation of endemic mycosis in the Americas and the Caribbean: Proceedings of the first international meeting on endemic mycoses of the Americas (IMEMA). *Mycoses* **2022**, *65*, 1179–1187. [CrossRef]
24. Wheat, L.J. Approach to the diagnosis of the endemic mycoses. *Clin. Chest Med.* **2009**, *30*, 379–389, viii. [CrossRef] [PubMed]
25. Richardson, M.; Page, I. Role of Serological Tests in the Diagnosis of Mold Infections. *Curr. Fungal Infect. Rep.* **2018**, *12*, 127–136. [CrossRef] [PubMed]
26. Kassis, C.; Durkin, M.; Holbrook, E.; Myers, R.; Wheat, L. Advances in Diagnosis of Progressive Pulmonary and Disseminated Coccidioidomycosis. *Clin. Infect. Dis.* **2021**, *72*, 968–975. [CrossRef] [PubMed]
27. Bongomin, F.; Govender, N.P.; Chakrabarti, A.; Robert-Gangneux, F.; Boulware, D.R.; Zafar, A.; Oladele, R.O.; Richardson, M.D.; Gangneux, J.P.; Alastruey-Izquierdo, A.; et al. Essential in vitro diagnostics for advanced HIV and serious fungal diseases: International experts' consensus recommendations. *Eur. J. Clin. Microbiol. Infect. Dis.* **2019**, *38*, 1581–1584. [CrossRef]
28. Cáceres, D.H.; Gómez, B.L.; Tobon, A.M.; Minderman, M.; Bridges, N.; Chiller, T.; Lindsley, M.D. Validation and Concordance Analysis of a New Lateral Flow Assay for Detection of *Histoplasma* Antigen in Urine. *J. Fungi* **2021**, *7*, 799. [CrossRef]
29. Donovan, F.M.; Ramadan, F.A.; Khan, S.A.; Bhaskara, A.; Lainhart, W.D.; Narang, A.T.; Mosier, J.M.; Ellingson, K.D.; Bedrick, E.J.; Saubolle, M.A.; et al. Comparison of a Novel Rapid Lateral Flow Assay to Enzyme Immunoassay Results for Early Diagnosis of Coccidioidomycosis. *Clin. Infect. Dis.* **2021**, *73*, e2746–e2753. [CrossRef]
30. Van Dyke, M.C.C.; Teixeira, M.M.; Barker, B.M. Fantastic yeasts and where to find them: The hidden diversity of dimorphic fungal pathogens. *Curr. Opin. Microbiol.* **2019**, *52*, 55–63. [CrossRef]
31. Buitrago, M.J.; Bernal-Martinez, L.; Castelli, M.V.; Rodriguez-Tudela, J.L.; Cuenca-Estrella, M. Histoplasmosis and paracoccidioidomycosis in a non-endemic area: A review of cases and diagnosis. *J. Travel Med.* **2011**, *18*, 26–33. [CrossRef]
32. Bialek, R.; Fischer, J.; Feucht, A.; Najvar, L.K.; Dietz, K.; Knobloch, J.; Graybill, J.R. Diagnosis and monitoring of murine histoplasmosis by a nested PCR assay. *J. Clin. Microbiol.* **2001**, *39*, 1506–1509. [CrossRef] [PubMed]

33. Buitrago, M.J.; Berenguer, J.; Mellado, E.; Rodriguez-Tudela, J.L.; Cuenca-Estrella, M. Detection of imported histoplasmosis in serum of HIV-infected patients using a real-time PCR-based assay. *Eur. J. Clin. Microbiol. Infect. Dis.* **2006**, *25*, 665–668. [CrossRef] [PubMed]
34. Gago, S.; Esteban, C.; Valero, C.; Zaragoza, O.; Puig de la Bellacasa, J.; Buitrago, M.J. A multiplex real-time PCR assay for identification of *Pneumocystis jirovecii, Histoplasma capsulatum,* and *Cryptococcus neoformans/Cryptococcus gattii* in samples from AIDS patients with opportunistic pneumonia. *J. Clin. Microbiol.* **2014**, *52*, 1168–1176. [CrossRef]
35. Martagon-Villamil, J.; Shrestha, N.; Sholtis, M.; Isada, C.M.; Hall, G.S.; Bryne, T.; Lodge, B.A.; Reller, L.B.; Procop, G.W. Identification of *Histoplasma capsulatum* from culture extracts by real-time PCR. *J. Clin. Microbiol.* **2003**, *41*, 1295–1298. [CrossRef] [PubMed]
36. Simon, S.; Veron, V.; Boukhari, R.; Blanchet, D.; Aznar, C. Detection of *Histoplasma capsulatum* DNA in human samples by real-time polymerase chain reaction. *Diagn. Microbiol. Infect. Dis.* **2010**, *66*, 268–273. [CrossRef] [PubMed]
37. Alanio, A.; Gits-Muselli, M.; Lanternier, F.; Sturny-Leclère, A.; Benazra, M.; Hamane, S.; Rodrigues, A.M.; García-Hermoso, D.; Lortholary, O.; Dromer, F.; et al. Evaluation of a New *Histoplasma* spp. Quantitative RT-PCR Assay. *J. Mol. Diagn.* **2021**, *23*, 698–709. [CrossRef] [PubMed]
38. Bracca, A.; Tosello, M.E.; Girardini, J.E.; Amigot, S.L.; Gomez, C.; Serra, E. Molecular detection of *Histoplasma capsulatum* var. *capsulatum* in human clinical samples. *J. Clin. Microbiol.* **2003**, *41*, 1753–1755. [CrossRef]
39. Guedes, H.L.; Guimaraes, A.J.; Muniz Mde, M.; Pizzini, C.V.; Hamilton, A.J.; Peralta, J.M.; Deepe, G.S., Jr.; Zancopé-Oliveira, R.M. PCR assay for identification of *Histoplasma capsulatum* based on the nucleotide sequence of the M antigen. *J. Clin. Microbiol.* **2003**, *41*, 535–539. [CrossRef]
40. López, L.F.; Munoz, C.O.; Cáceres, D.H.; Tobon, A.M.; Loparev, V.; Clay, O.; Chiller, T.; Litvintseva, A.; Gade, L.; Gonzalez, A.; et al. Standardization and validation of real time PCR assays for the diagnosis of histoplasmosis using three molecular targets in an animal model. *PLoS ONE* **2017**, *12*, e0190311. [CrossRef]
41. Maubon, D.; Simon, S.; Aznar, C. Histoplasmosis diagnosis using a polymerase chain reaction method. Application on human samples in French Guiana, South America. *Diagn. Microbiol. Infect. Dis.* **2007**, *58*, 441–444. [CrossRef]
42. Rickerts, V.; Bialek, R.; Tintelnot, K.; Jacobi, V.; Just-Nubling, G. Rapid PCR-based diagnosis of disseminated histoplasmosis in an AIDS patient. *Eur. J. Clin. Microbiol. Infect. Dis.* **2002**, *21*, 821–823. [CrossRef]
43. Gallo, J.E.; Torres, I.; Gomez, O.M.; Rishishwar, L.; Vannberg, F.; Jordan, I.K.; McEwen, J.G.; Clay, O.K. New *Histoplasma* Diagnostic Assays Designed via Whole Genome Comparisons. *J. Fungi* **2021**, *7*, 544. [CrossRef] [PubMed]
44. Caceres, D.H.; Knuth, M.; Derado, G.; Lindsley, M.D. Diagnosis of Progressive Disseminated Histoplasmosis in Advanced HIV: A Meta-Analysis of Assay Analytical Performance. *J. Fungi* **2019**, *5*, 76. [CrossRef] [PubMed]
45. Zatti, M.D.S.; Arantes, T.D.; Fernandes, J.A.L.; Bay, M.B.; Milan, E.P.; Naliato, G.F.S.; Theodoro, R.C. Loop-mediated Isothermal Amplification and nested PCR of the Internal Transcribed Spacer (ITS) for *Histoplasma capsulatum* detection. *PLoS Negl. Trop. Dis.* **2019**, *13*, e0007692. [CrossRef] [PubMed]
46. Scheel, C.M.; Zhou, Y.; Theodoro, R.C.; Abrams, B.; Balajee, S.A.; Litvintseva, A.P. Development of a loop-mediated isothermal amplification method for detection of *Histoplasma capsulatum* DNA in clinical samples. *J. Clin. Microbiol.* **2014**, *52*, 483–488. [CrossRef]
47. Buitrago, M.J.; Canteros, C.E.; Frias De Leon, G.; Gonzalez, A.; Marques-Evangelista De Oliveira, M.; Munoz, C.O.; Ramirez, J.A.; Toranzo, A.I.; Zancope-Oliveira, R.; Cuenca-Estrella, M. Comparison of PCR protocols for detecting *Histoplasma capsulatum* DNA through a multicenter study. *Rev. Iberoam Micol.* **2013**, *30*, 256–260. [CrossRef]
48. Binnicker, M.J.; Buckwalter, S.P.; Eisberner, J.J.; Stewart, R.A.; McCullough, A.E.; Wohlfiel, S.L.; Wengenack, N.L. Detection of *Coccidioides* species in clinical specimens by real-time PCR. *J. Clin. Microbiol.* **2007**, *45*, 173–178. [CrossRef]
49. Gago, S.; Buitrago, M.J.; Clemons, K.V.; Cuenca-Estrella, M.; Mirels, L.F.; Stevens, D.A. Development and validation of a quantitative real-time PCR assay for the early diagnosis of coccidioidomycosis. *Diagn. Microbiol. Infect. Dis.* **2014**, *79*, 214–221. [CrossRef]
50. Bowers, J.R.; Parise, K.L.; Kelley, E.J.; Lemmer, D.; Schupp, J.M.; Driebe, E.M.; Engelthaler, D.M.; Keim, P.; Barker, B.M. Direct detection of *Coccidioides* from Arizona soils using CocciENV, a highly sensitive and specific real-time PCR assay. *Med. Mycol.* **2019**, *57*, 246–255. [CrossRef]
51. Lauer, A.; Baal, J.D.; Baal, J.C.; Verma, M.; Chen, J.M. Detection of *Coccidioides immitis* in Kern County, California, by multiplex PCR. *Mycologia* **2012**, *104*, 62–69. [CrossRef]
52. Saubolle, M.A.; Wojack, B.R.; Wertheimer, A.M.; Fuayagem, A.Z.; Young, S.; Koeneman, B.A. Multicenter Clinical Validation of a Cartridge-Based Real-Time PCR System for Detection of *Coccidioides* spp. in Lower Respiratory Specimens. *J. Clin. Microbiol.* **2018**, *56*, e01277-17. [CrossRef] [PubMed]
53. Warnock, D.W. *Coccidioides* species as potential agents of bioterrorism. *Future Microbiol.* **2007**, *2*, 277–283. [CrossRef] [PubMed]
54. Koishi, T.; Yasuoka, K.; Zeng, X.C.; Fujikawa, S. Molecular dynamics simulations of urea-water binary droplets on flat and pillared hydrophobic surfaces. *Faraday Discuss* **2010**, *146*, 185–193; discussion 195–215, 395–403. [CrossRef] [PubMed]

55. Gaviria, M.; Rivera, V.; Munoz-Cadavid, C.; Cano, L.E.; Naranjo, T.W. Validation and clinical application of a nested PCR for paracoccidioidomycosis diagnosis in clinical samples from Colombian patients. *Braz. J. Infect. Dis.* **2015**, *19*, 376–383. [CrossRef]
56. Dias, L.; de Carvalho, L.F.; Romano, C.C. Application of PCR in serum samples for diagnosis of paracoccidioidomycosis in the southern Bahia-Brazil. *PLoS Negl. Trop. Dis.* **2012**, *6*, e1909. [CrossRef]
57. Bialek, R.; Ibricevic, A.; Aepinus, C.; Najvar, L.K.; Fothergill, A.W.; Knobloch, J.; Graybill, J.R. Detection of *Paracoccidioides brasiliensis* in tissue samples by a nested PCR assay. *J. Clin. Microbiol.* **2000**, *38*, 2940–2942. [CrossRef]
58. Rocha-Silva, F.; Maria de Figueiredo, S.; Rutren La Santrer, E.F.; Machado, A.S.; Fernandes, B.; Assuncao, C.B.; Goes, A.M.; Caligiorne, R.B. Paracoccidioidomycosis: Detection of *Paracoccidioides brasiliensis* genome in biological samples by quantitative chain reaction polymerase (qPCR). *Microb. Pathog.* **2018**, *121*, 359–362. [CrossRef]
59. Tatibana, B.T.; Sano, A.; Uno, J.; Kamei, K.; Igarashi, T.; Mikami, Y.; Miyaji, M.; Nishimura, K.; Itano, E.N. Detection of *Paracoccidioides brasiliensis* gp43 gene in sputa by loop-mediated isothermal amplification method. *J. Clin. Lab. Anal.* **2009**, *23*, 139–143. [CrossRef]
60. Onda, H.; Komine, M.; Murata, S.; Ohtsuki, M. Letter: Imported paracoccidioidomycosis in Japan. *Dermatol. Online J.* **2011**, *17*, 11. [CrossRef]
61. Ginarte, M.; Pereiro, M., Jr.; Toribio, J. Imported paracoccidioidomycosis in Spain. *Mycoses* **2003**, *46*, 407–411. [CrossRef]
62. Botas-Velasco, M.; Jover-Diaz, F.; Ortiz de la Tabla-Duccase, V.; Martinez-Garcia, C. [Imported paracoccidioidomycosis in Spain]. *Enferm. Infecc. Microbiol. Clin.* **2010**, *28*, 259–260. [CrossRef] [PubMed]
63. Burgess, J.W.; Schwan, W.R.; Volk, T.J. PCR-based detection of DNA from the human pathogen *Blastomyces dermatitidis* from natural soil samples. *Med. Mycol.* **2006**, *44*, 741–748. [CrossRef] [PubMed]
64. Sidamonidze, K.; Peck, M.K.; Perez, M.; Baumgardner, D.; Smith, G.; Chaturvedi, V.; Chaturvedi, S. Real-time PCR assay for identification of *Blastomyces dermatitidis* in culture and in tissue. *J. Clin. Microbiol.* **2012**, *50*, 1783–1786. [CrossRef] [PubMed]
65. Babady, N.E.; Buckwalter, S.P.; Hall, L.; Le Febre, K.M.; Binnicker, M.J.; Wengenack, N.L. Detection of *Blastomyces dermatitidis* and *Histoplasma capsulatum* from culture isolates and clinical specimens by use of real-time PCR. *J. Clin. Microbiol.* **2011**, *49*, 3204–3208. [CrossRef] [PubMed]
66. Ning, C.; Lai, J.; Wei, W.; Zhou, B.; Huang, J.; Jiang, J.; Liang, B.; Liao, Y.; Zang, N.; Cao, C.; et al. Accuracy of rapid diagnosis of *Talaromyces marneffei*: A systematic review and meta-analysis. *PLoS ONE* **2018**, *13*, e0195569. [CrossRef]
67. Pongpom, M.; Sirisanthana, T.; Vanittanakom, N. Application of nested PCR to detect *Penicillium marneffei* in serum samples. *Med. Mycol.* **2009**, *47*, 549–553. [CrossRef]
68. Lu, S.; Li, X.; Calderone, R.; Zhang, J.; Ma, J.; Cai, W.; Xi, L. Whole blood Nested PCR and Real-time PCR amplification of *Talaromyces marneffei* specific DNA for diagnosis. *Med. Mycol.* **2016**, *54*, 162–168. [CrossRef]
69. Li, X.; Zheng, Y.; Wu, F.; Mo, D.; Liang, G.; Yan, R.; Khader, J.A.; Wu, N.; Cao, C. Evaluation of quantitative real-time PCR and Platelia galactomannan assays for the diagnosis of disseminated *Talaromyces marneffei* infection. *Med. Mycol.* **2020**, *58*, 181–186. [CrossRef]
70. Hien, H.T.A.; Thanh, T.T.; Thu, N.T.M.; Nguyen, A.; Thanh, N.T.; Lan, N.P.H.; Simmons, C.; Shikuma, C.; Chau, N.V.V.; Thwaites, G.; et al. Development and evaluation of a real-time polymerase chain reaction assay for the rapid detection of *Talaromyces marneffei* MP1 gene in human plasma. *Mycoses* **2016**, *59*, 773–780. [CrossRef]
71. Sun, J.; Li, X.; Zeng, H.; Xie, Z.; Lu, C.; Xi, L.; de Hoog, G.S. Development and evaluation of loop-mediated isothermal amplification (LAMP) for the rapid diagnosis of *Penicillium marneffei* in archived tissue samples. *FEMS Immunol. Med. Microbiol.* **2010**, *58*, 381–388. [CrossRef]
72. Bialek, R.; Feucht, A.; Aepinus, C.; Just-Nubling, G.; Robertson, V.J.; Knobloch, J.; Hohle, R. Evaluation of two nested PCR assays for detection of *Histoplasma capsulatum* DNA in human tissue. *J. Clin. Microbiol.* **2002**, *40*, 1644–1647. [CrossRef] [PubMed]
73. Buitrago, M.J.; Merino, P.; Puente, S.; Gomez-Lopez, A.; Arribi, A.; Zancope-Oliveira, R.M.; Gutierrez, M.C.; Rodriguez-Tudela, J.L.; Cuenca-Estrella, M. Utility of real-time PCR for the detection of *Paracoccidioides brasiliensis* DNA in the diagnosis of imported paracoccidioidomycosis. *Med. Mycol.* **2009**, *47*, 879–882. [CrossRef] [PubMed]
74. Pitz Ade, F.; Koishi, A.C.; Tavares, E.R.; Andrade, F.G.; Loth, E.A.; Gandra, R.F.; Venancio, E.J. An optimized one-tube, semi-nested PCR assay for *Paracoccidioides brasiliensis* detection. *Rev. Soc. Bras. Med. Trop.* **2013**, *46*, 783–785. [CrossRef] [PubMed]
75. Bialek, R.; Kern, J.; Herrmann, T.; Tijerina, R.; Cecenas, L.; Reischl, U.; Gonzalez, G.M. PCR assays for identification of *Coccidioides posadasii* based on the nucleotide sequence of the antigen 2/proline-rich antigen. *J. Clin. Microbiol.* **2004**, *42*, 778–783. [CrossRef]
76. Bialek, R.; Cirera, A.C.; Herrmann, T.; Aepinus, C.; Shearn-Bochsler, V.I.; Legendre, A.M. Nested PCR assays for detection of *Blastomyces dermatitidis* DNA in paraffin-embedded canine tissue. *J. Clin. Microbiol.* **2003**, *41*, 205–208. [CrossRef]
77. Kaplan, M.; Zhu, Y.; Kus, J.V.; McTaggart, L.; Chaturvedi, V.; Chaturvedi, S. Development of a Duplex Real-Time PCR Assay for the Differentiation of *Blastomyces dermatitidis* and *Blastomyces gilchristii* and a Retrospective Analysis of Culture and Primary Specimens from Blastomycosis Cases from New York (2005 to 2019). *J. Clin. Microbiol.* **2021**, *59*, e02078-20. [CrossRef]

78. Pornprasert, S.; Praparattanapan, J.; Khamwan, C.; Pawichai, S.; Pimsarn, P.; Samleerat, T.; Leechanachai, P.; Supparatpinyo, K. Development of TaqMan real-time polymerase chain reaction for the detection and identification of *Penicillium marneffei*. *Mycoses* **2009**, *52*, 487–492. [CrossRef]
79. Kidd, S.E.; Chen, S.C.; Meyer, W.; Halliday, C.L. A New Age in Molecular Diagnostics for Invasive Fungal Disease: Are We Ready? *Front. Microbiol.* **2019**, *10*, 2903. [CrossRef]
80. White, P.L.; Alanio, A.; Brown, L.; Cruciani, M.; Hagen, F.; Gorton, R.; Lackner, M.; Millon, L.; Morton, C.O.; Rautemaa-Richardson, R.; et al. An overview of using fungal DNA for the diagnosis of invasive mycoses. *Expert. Rev. Mol. Diagn.* **2022**, *22*, 169–184. [CrossRef]
81. Buitrago, M.J.; Valero, C. *Diagnosis of Fungal Infections*; Elsevier: Amsterdam, The Netherlands, 2018. [CrossRef]
82. Imhof, A.; Schaer, C.; Schoedon, G.; Schaer, D.J.; Walter, R.B.; Schaffner, A.; Schneemann, M. Rapid detection of pathogenic fungi from clinical specimens using LightCycler real-time fluorescence PCR. *Eur. J. Clin. Microbiol. Infect. Dis.* **2003**, *22*, 558–560. [CrossRef]
83. Trubiano, J.A.; Dennison, A.M.; Morrissey, C.O.; Chua, K.Y.; Halliday, C.L.; Chen, S.C.; Spelman, D. Clinical utility of panfungal polymerase chain reaction for the diagnosis of invasive fungal disease: A single center experience. *Med. Mycol.* **2016**, *54*, 138–146. [CrossRef] [PubMed]
84. Ala-Houhala, M.; Koukila-Kahkola, P.; Antikainen, J.; Valve, J.; Kirveskari, J.; Anttila, V.J. Clinical use of fungal PCR from deep tissue samples in the diagnosis of invasive fungal diseases: A retrospective observational study. *Clin. Microbiol. Infect.* **2018**, *24*, 301–305. [CrossRef] [PubMed]
85. Lindner, A.K.; Rickerts, V.; Kurth, F.; Wilmes, D.; Richter, J. Chronic oral ulceration and lip swelling after a long term stay in Guatemala: A diagnostic challenge. *Travel. Med. Infect. Dis.* **2018**, *23*, 103–104. [CrossRef] [PubMed]
86. Wilmes, D.; McCormick-Smith, I.; Lempp, C.; Mayer, U.; Schulze, A.B.; Theegarten, D.; Hartmann, S.; Rickerts, V. Detection of *Histoplasma* DNA from Tissue Blocks by a Specific and a Broad-Range Real-Time PCR: Tools to Elucidate the Epidemiology of Histoplasmosis. *J. Fungi* **2020**, *6*, 319. [CrossRef]
87. Buitrago, M.J.; Bernal-Martinez, L.; Castelli, M.V.; Rodriguez-Tudela, J.L.; Cuenca-Estrella, M. Performance of panfungal–and specific-PCR-based procedures for etiological diagnosis of invasive fungal diseases on tissue biopsy specimens with proven infection: A 7-year retrospective analysis from a reference laboratory. *J. Clin. Microbiol.* **2014**, *52*, 1737–1740. [CrossRef]
88. Morjaria, S.; Otto, C.; Moreira, A.; Chung, R.; Hatzoglou, V.; Pillai, M.; Banaei, N.; Tang, Y.W.; Figueroa, C.J. Ribosomal RNA gene sequencing for early diagnosis of *Blastomyces dermatitidis* infection. *Int. J. Infect. Dis.* **2015**, *37*, 122–124. [CrossRef]
89. Rooms, I.; Mugisha, P.; Gambichler, T.; Hadaschik, E.; Esser, S.; Rath, P.M.; Haase, G.; Wilmes, D.; McCormick-Smith, I.; Rickerts, V. Disseminated Emergomycosis in a Person with HIV Infection, Uganda. *Emerg. Infect. Dis.* **2019**, *25*, 1750–1751. [CrossRef]
90. Beltrame, A.; Danesi, P.; Farina, C.; Orza, P.; Perandin, F.; Zanardello, C.; Rodari, P.; Staffolani, S.; Bisoffi, Z. Case Report: Molecular Confirmation of Lobomycosis in an Italian Traveler Acquired in the Amazon Region of Venezuela. *Am. J. Trop. Med. Hyg.* **2017**, *97*, 1757–1760. [CrossRef]
91. Valero, C.; de la Cruz-Villar, L.; Zaragoza, O.; Buitrago, M.J. New Panfungal Real-Time PCR Assay for Diagnosis of Invasive Fungal Infections. *J. Clin. Microbiol.* **2016**, *54*, 2910–2918. [CrossRef]
92. Gade, L.; Hurst, S.; Balajee, S.A.; Lockhart, S.R.; Litvintseva, A.P. Detection of mucormycetes and other pathogenic fungi in formalin fixed paraffin embedded and fresh tissues using the extended region of 28S rDNA. *Med. Mycol.* **2017**, *55*, 385–395. [CrossRef]
93. Gomez, C.A.; Budvytiene, I.; Zemek, A.J.; Banaei, N. Performance of Targeted Fungal Sequencing for Culture-Independent Diagnosis of Invasive Fungal Disease. *Clin. Infect. Dis.* **2017**, *65*, 2035–2041. [CrossRef] [PubMed]
94. Sabino, R.; Simoes, H.; Verissimo, C. Detection of deep fungal infections: A polyphasic approach. *J. Med. Microbiol.* **2019**, *68*, 81–86. [CrossRef] [PubMed]
95. Lefterova, M.I.; Suarez, C.J.; Banaei, N.; Pinsky, B.A. Next-Generation Sequencing for Infectious Disease Diagnosis and Management: A Report of the Association for Molecular Pathology. *J. Mol. Diagn.* **2015**, *17*, 623–634. [CrossRef] [PubMed]
96. Forbes, J.D.; Knox, N.C.; Ronholm, J.; Pagotto, F.; Reimer, A. Metagenomics: The Next Culture-Independent Game Changer. *Front. Microbiol.* **2017**, *8*, 1069. [CrossRef] [PubMed]
97. Boers, S.A.; Jansen, R.; Hays, J.P. Understanding and overcoming the pitfalls and biases of next-generation sequencing (NGS) methods for use in the routine clinical microbiological diagnostic laboratory. *Eur. J. Clin. Microbiol. Infect. Dis.* **2019**, *38*, 1059–1070. [CrossRef]
98. McTaggart, L.R.; Copeland, J.K.; Surendra, A.; Wang, P.W.; Husain, S.; Coburn, B.; Guttman, D.S.; Kus, J.V. Mycobiome Sequencing and Analysis Applied to Fungal Community Profiling of the Lower Respiratory Tract During Fungal Pathogenesis. *Front. Microbiol.* **2019**, *10*, 512. [CrossRef]
99. Mao, Y.; Shen, H.; Yang, C.; Jia, Q.; Li, J.; Chen, Y.; Hu, J.; Huang, W. Clinical performance of metagenomic next-generation sequencing for the rapid diagnosis of talaromycosis in HIV-infected patients. *Front. Cell Infect Microbiol.* **2022**, *12*, 962441. [CrossRef]

100. Zhu, Y.M.; Ai, J.W.; Xu, B.; Cui, P.; Cheng, Q.; Wu, H.; Qian, Y.Y.; Zhang, H.C.; Zhou, X.; Xing, L.; et al. Rapid and precise diagnosis of disseminated *T. marneffei* infection assisted by high-throughput sequencing of multifarious specimens in a HIV-negative patient: A case report. *BMC Infect. Dis.* **2018**, *18*, 379. [CrossRef]
101. Wang, D.M.; Ma, H.L.; Tan, M.Q.; Wu, Y.M.; Wang, S.N. Next-generation sequencing confirmed the diagnosis of isolated central nervous system infection caused by *Talaromyces marneffei* in an immunocompetent patient. *Chin. Med. J.* **2020**, *133*, 374–376. [CrossRef]
102. Zhang, W.; Ye, J.; Qiu, C.; Wang, L.; Jin, W.; Jiang, C.; Xu, L.; Xu, J.; Li, Y.; Wang, L.; et al. Rapid and precise diagnosis of *T. marneffei* pulmonary infection in a HIV-negative patient with autosomal-dominant STAT3 mutation: A case report. *Ther. Adv. Respir. Dis.* **2020**, *14*, 1753466620929225. [CrossRef]
103. Du, R.; Feng, Y.; Liu, L.N.; Liu, Y.B.; Ye, H.; Lu, X.J.; Wang, X.H.; Zong, Z.Y. [Pathogen Diagnosis of a Febrile HIV Case by the Metagenomic Next-generation Sequencing]. *Sichuan Da Xue Xue Bao Yi Xue Ban* **2020**, *51*, 257–260. [CrossRef]
104. Zhang, J.; Zhang, D.; Du, J.; Zhou, Y.; Cai, Y.; Sun, R.; Zhou, J.; Tian, J.; Wu, H.; Lu, M.; et al. Rapid diagnosis of *Talaromyces marneffei* infection assisted by metagenomic next-generation sequencing in a HIV-negative patient. *IDCases* **2021**, *23*, e01055. [CrossRef] [PubMed]
105. Shi, T.; Wu, L.; Cai, J.; Chen, H. An Iris Tumor Secondary to *Talaromyces marneffei* Infection in a Patient with AIDS and Syphilis. *Ocul. Immunol. Inflamm.* **2022**, *30*, 1129–1132. [CrossRef] [PubMed]
106. Shi, J.; Yang, N.; Qian, G. Case Report: Metagenomic Next-Generation Sequencing in Diagnosis of Talaromycosis of an Immunocompetent Patient. *Front. Med.* **2021**, *8*, 656194. [CrossRef] [PubMed]
107. Zhou, Y.; Liu, Y.; Wen, Y. Gastrointestinal manifestations of *Talaromyces marneffei* infection in an HIV-infected patient rapidly verified by metagenomic next-generation sequencing: A case report. *BMC Infect. Dis.* **2021**, *21*, 376. [CrossRef]
108. Shen, Q.; Sheng, L.; Zhou, J. HIV-negative case of *Talaromyces marneffei* pulmonary infection with a TSC2 mutation. *J. Int. Med. Res.* **2021**, *49*, 3000605211016761. [CrossRef]
109. Wilson, M.R.; O'Donovan, B.D.; Gelfand, J.M.; Sample, H.A.; Chow, F.C.; Betjemann, J.P.; Shah, M.P.; Richie, M.B.; Gorman, M.P.; Hajj-Ali, R.A.; et al. Chronic Meningitis Investigated via Metagenomic Next-Generation Sequencing. *JAMA Neurol.* **2018**, *75*, 947–955. [CrossRef]
110. Wang, J.; Zhou, W.; Ling, H.; Dong, X.; Zhang, Y.; Li, J.; Zhang, Y.; Song, J.; Liu, W.J.; Li, Y.; et al. Identification of *Histoplasma* causing an unexplained disease cluster in Matthews Ridge, Guyana. *Biosaf. Health* **2019**, *1*, 150–154. [CrossRef]
111. Chen, J.; Li, Y.; Li, Z.; Chen, G.; Liu, X.; Ding, L. Metagenomic next-generation sequencing identified *Histoplasma capsulatum* in the lung and epiglottis of a Chinese patient: A case report. *Int. J. Infect. Dis.* **2020**, *101*, 33–37. [CrossRef]
112. Muldoon, J.L.; Wysozan, T.R.; Toubin, Y.; Relich, R.F.; Davis, T.E.; Zhang, C.; Alomari, A.K. An unusual presentation of cutaneous histoplasmosis as a recurrent solitary and spontaneously healing lesion in an immunocompetent patient. *Access Microbiol.* **2020**, *2*, acmi000156. [CrossRef]
113. Wang, N.; Zhao, C.; Tang, C.; Wang, L. Case Report and Literature Review: Disseminated Histoplasmosis Infection Diagnosed by Metagenomic Next-Generation Sequencing. *Infect. Drug. Resist.* **2022**, *15*, 4507–4514. [CrossRef] [PubMed]
114. Bansal, S.; Yadav, M.; Singhania, N.; Samal, S.; Singhania, G. Blastomycosis Detected by Microbial Cell-Free DNA in Renal Transplant Recipient. *Am. J. Med.* **2020**, *133*, e599–e600. [CrossRef]
115. Wang, N.; Luo, Z.; Deng, S.; Li, Q. A young male with chronic nonproductive cough diagnosed with blastomycosis in China: A case report. *BMC Pulm. Med.* **2020**, *20*, 189. [CrossRef] [PubMed]
116. Zhang, H.C.; Zhang, Q.R.; Ai, J.W.; Cui, P.; Wu, H.L.; Zhang, W.H.; Wang, T. The role of bone marrow metagenomics next-generation sequencing to differential diagnosis among visceral leishmaniasis, histoplasmosis, and talaromycosis marneffei. *Int. J. Lab. Hematol.* **2020**, *42*, e52–e54. [CrossRef]
117. Larkin, P.M.K.; Lawson, K.L.; Contreras, D.A.; Le, C.Q.; Trejo, M.; Realegeno, S.; Hilt, E.E.; Chandrasekaran, S.; Garner, O.B.; Fishbein, G.A.; et al. Amplicon-Based Next-Generation Sequencing for Detection of Fungi in Formalin-Fixed, Paraffin-Embedded Tissues: Correlation with Histopathology and Clinical Applications. *J. Mol. Diagn.* **2020**, *22*, 1287–1293. [CrossRef] [PubMed]
118. Bernal-Martinez, L.; Herrera, L.; Valero, C.; de la Cruz, P.; Ghimpu, L.; Mesa-Arango, A.C.; Santoni, G.; Goterris, L.; Millan, R.; Buitrago, M.J. Differential Diagnosis of Fungal Pneumonias vs. Tuberculosis in AIDS Patients by Using Two New Molecular Methods. *J. Fungi* **2021**, *7*, 336. [CrossRef]

Journal of
Fungi

MDPI

Review

Chest Imaging in Systemic Endemic Mycoses

Célia Sousa [1], Edson Marchiori [2], Ali Youssef [3], Tan-Lucien Mohammed [3], Pratik Patel [3], Klaus Irion [3], Romulo Pasini [3], Alexandre Mançano [3], Arthur Souza [4], Alessandro C. Pasqualotto [5] and Bruno Hochhegger [3,*]

1 Radiology Department, Centro Hospitalar Universitário de São João, 4200-319 Porto, Portugal
2 Radiology Department, Universidade Federal de Rio de Janeiro, Rio de Janeiro 21941-901, Brazil
3 Radiology Department, University of Florida Health Shands Hospital, Gainesville, FL 32608, USA
4 Radiology Department, Faculdade de Medicina de São José do Rio Preto, São José do Rio Preto 15090-000, Brazil
5 Radiology Department, Santa Casa de Misericórdia de Porto Alegre, Porto Alegre 90020-090, Brazil
* Correspondence: brunohochhegger@gmail.com

Abstract: Endemic fungal infections are responsible for high rates of morbidity and mortality in certain regions of the world. The diagnosis and management remain a challenge, and the reason could be explained by the lack of disease awareness, variability of symptoms, and insidious and often overlooked clinical presentation. Imaging findings are nonspecific and frequently misinterpreted as other more common infectious or malignant diseases. Patient demographics and clinical and travel history are important clues that may lead to a proper diagnosis. The purpose of this paper is to review the presentation and differential diagnosis of endemic mycoses based on the most common chest imaging findings.

Keywords: endemic mycoses; histoplasmosis; coccidioidomycosis; *cryptococcosis*; blastomycosis; paracoccidioidomycosis; computed tomography

Citation: Sousa, C.; Marchiori, E.; Youssef, A.; Mohammed, T.-L.; Patel, P.; Irion, K.; Pasini, R.; Mançano, A.; Souza, A.; Pasqualotto, A.C.; et al. Chest Imaging in Systemic Endemic Mycoses. *J. Fungi* **2022**, *8*, 1132. https://doi.org/10.3390/jof8111132

Academic Editor: David S. Perlin

Received: 29 September 2022
Accepted: 24 October 2022
Published: 27 October 2022

1. Introduction

Fungi may have a widespread or endemic geographic distribution. Immunocompromised patients are most affected by ubiquitous fungi, such as *Aspergillus*, *Candida*, *Mucor*, *Cryptococcoci neoformans*, and *Pneumocystis jiroveci*. Immunocompetent individuals or those with chronic conditions may be susceptible to endemic mycoses. Overall, patients with reduced immunity are more vulnerable to complications and disseminated disease from endemic and ubiquitous fungal infections. Additionally, endemic mycoses can be divided into two groups, such as implantation or subcutaneous mycoses in which the pathogen generally infects transcutaneous wounds and systemic mycoses in which the respiratory tract is the primary route of transmission into the human host, i.e., aerogenic and dust inhalation (construction, farming, landscaping, excavation) [1,2]. Systemic endemic mycoses are a group of dimorphic fungi prevalent in specific geographical locations. *Histoplasma capsulatum*, *Coccidioides* spp., *Blastomyces dermatitidis*, *Cryptococcosis gattii, and Paracoccidioides brasiliensis* are the primary pulmonary fungal pathogens of otherwise healthy people. *Histoplasma capsulatum* is found worldwide, but particularly in North, Central, and South America. Pulmonary coccidioidomycosis or valley fever is caused by the dimorphic fungi *C. immitis* and *C. posadasii*. It is endemic in the southwestern parts of the USA (California, Arizona, Utah, New Mexico, and Nevada) and parts of Central and South America (Brazil, Argentina, Mexico). Blastomycosis is mainly reported in North America and Africa. Cryptococcosis (*C. gattii*) is endemic in the Pacific Northwest, Central and South America, Asia, and India. Paracoccidioidomycosis (*P. brasiliensis*) is endemic in Latin America, particularly Brazil, Colombia, Venezuela, and Argentina [3–5]. Additionally, talaromycosis and emergomycosis are two other important endemic systemic mycoses; however, they are more frequently reported as opportunistic infections in immunocompromised patients, particularly in advanced HIV disease. Talaromycosis, caused by *Talaromyces marneffei*, is endemic to north-eastern India, southeast Asia, and southern China. Emergomycosis are endemic to Africa (*Emergomyces pasteurianus, Es africanus*), Europe (*Es pasteurianus, Es europaeus*), North America (*Es canadensis*), and Asia (*Es pasteurianus, Es orientalis*) [6,7].

The majority of infected people remain asymptomatic or develop self-limiting respiratory symptoms (up to 6 weeks). The clinical presentation of these granulomatous diseases can vary from asymptomatic to disseminated infection and depends on the amount of environmental exposure, virulent strain, host's immune status, and extremes of age. Symptoms are often subacute but may have an acute presentation. Cough, fever, chills, anorexia, weight loss, fatigue, headache, and chest pain are all possible clinical symptoms. Patients may also manifest dermatological symptoms, such as erythema nodosum or multiforme, and rheumatological manifestations. The severe disease form may spread hematogenously to bones, skin, joints, and central nervous system [1,4].

The diagnosis of systemic endemic mycosis is challenging and frequently misinterpreted for other diseases (e.g., bacterial or viral pneumonia, tuberculosis, sarcoidosis, lung cancer, metastases). Both clinicians and radiologists may be unfamiliar with the disease manifestations, especially those from non-endemic areas. Familiarization with these diseases by health professionals is of crucial importance as they are becoming increasingly relevant worldwide due to traveling and immigration [4,8].

Imaging findings are nonspecific and overlap with both other non-fungal diseases and among the numerous fungal pathogens. The aim of this paper is to review the most common imaging presentation and differential diagnoses of systemic endemic mycoses.

2. Imaging Findings

2.1. Lung Nodule or Mass

Acute or chronic fungal infections may manifest as solitary or multiple lung nodules or masses. The most frequent pathogens are histoplasmosis, coccidioidomycosis, *cryptococcosis*, and blastomycosis [4]. The nodules have a nonspecific appearance, may be ill or well-defined, have regular or spiculated borders, and may also demonstrate cavitation or ground glass halo (Figure 1). Additionally, a presenting dominant nodule or mass may be associated with other satellite nodules or bronchovascular beading [9]. Three-in-bud opacities are less common but may be seen in cases of bronchiolitis. More severe infections may present with multiple solid or cavitary nodules with a tendency toward confluence (Figure 2) [4,10]. In histoplasmosis, cryptococcosis, and coccidioidomycosis, the size of the nodules frequently ranges from less than 10 to 30 mm and they have a lower lobe predominance. The nodules in histoplasmosis and cryptococcosis are also predominantly peripherally located (Figure 3) [4,5,10–13]. Lung masses are the second most common finding in blastomycosis (after lobar and segmental consolidation), and they may measure up to 10 cm in diameter [4,14]. Pulmonary granulomata resulting from histoplasmosis may continue to enlarge over time (average of 1.7 mm per year) due to proliferation of fibrosis at the periphery of the nodule, secondary to an abnormal host response [9].

Hilar and mediastinal lymphadenopathies are typically visualized in histoplasmosis and coccidioidomycosis, which may be bulky, especially in the latter [4]. In acute infection, Fluorine 18 (18F)–fluorodeoxyglucose (FDG) PET/CT shows avid uptake in both the pulmonary infection and lymph nodes but with a greater degree of pulmonary uptake. During the subacute phase of infection, the pulmonary nodules or masses demonstrate a decrease in FDG activity more quickly than the adenopathy (Figures 4 and 5). This is the opposite of lung cancer in which the malignancy usually retains higher FDG avidity. This finding, also known as the flip-flop fungus sign, discloses the benign granulomatous disease nature, particularly associated with histoplasmosis [15]. As the infection heals, the infected mediastinal lymph nodes and pulmonary granulomata can calcify, showing central (target) or diffuse calcification, a typical finding in histoplasmosis [4,8,9]. Hilar and mediastinal lymphadenopathy and pulmonary calcified nodules are not typical findings in coccidioidomycosis, cryptococcosis and blastomycosis [12–14].

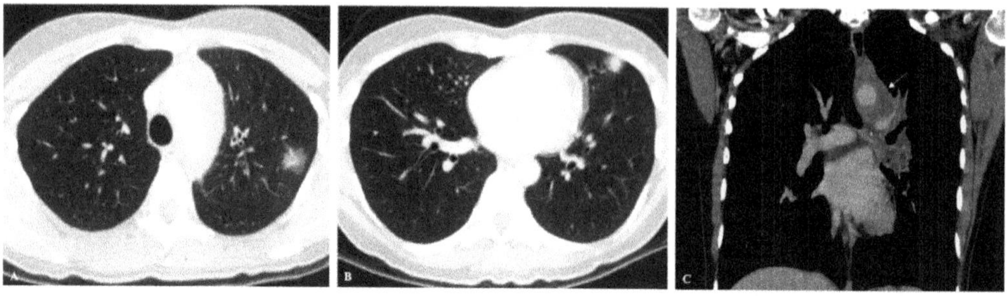

Figure 1. (**A**,**B**) Axial CT images showing peripherally located solid nodules with ground-glass halo in the upper left lobe, in a patient with acute histoplasmosis infection. (**C**) Coronal CT image depicts left mediastinal and hilar enlarged lymph nodes (arrow).

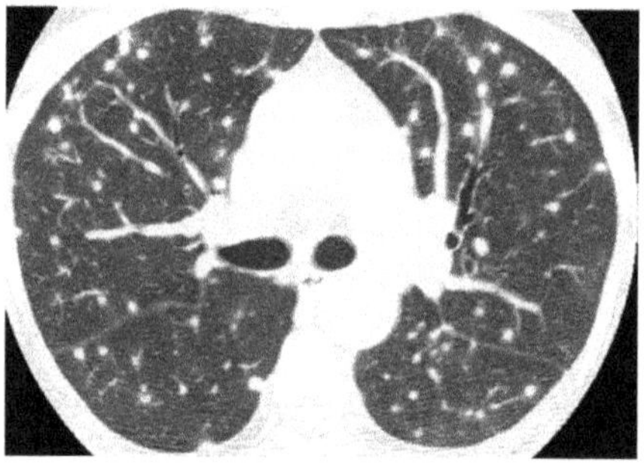

Figure 2. Axial CT image in a patient with histoplasmosis shows numerous bilateral and randomly distributed solid nodules, some of them with peripheral ground-glass halos.

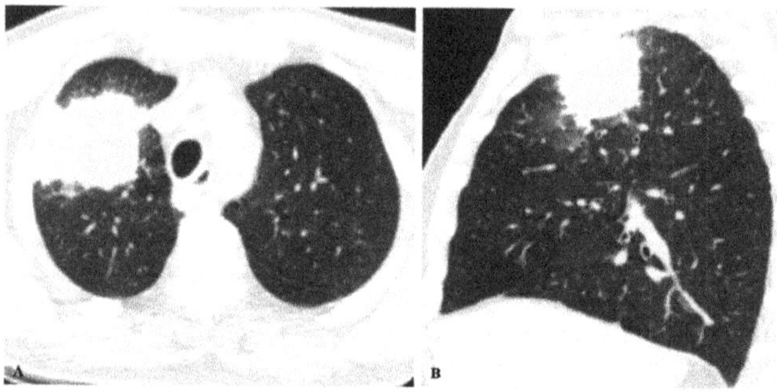

Figure 3. Cryptococcosis in a 63-year-old man with clinical history of colon cancer, heavy smoking and cough for 3 months. (**A**) Axial and sagittal (**B**) images show an irregular solid mass in the right upper lobe adjacent to the pleura and ground-glass halo.

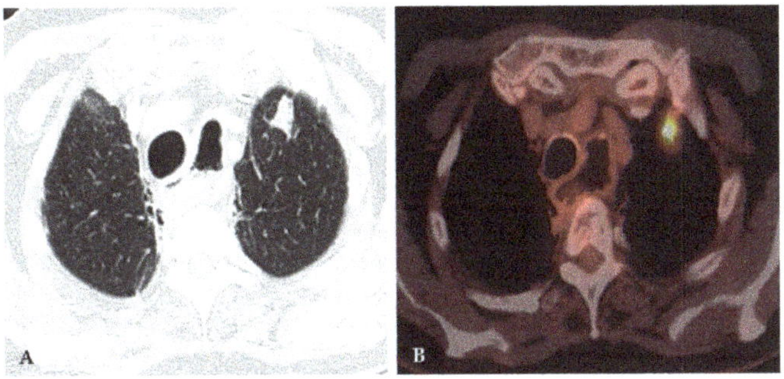

Figure 4. Acute histoplasmosis infection. (**A**) Axial CT image depicts a left upper lobe nodule. (**B**) FDG PET/CT shows avid uptake in the pulmonary infection.

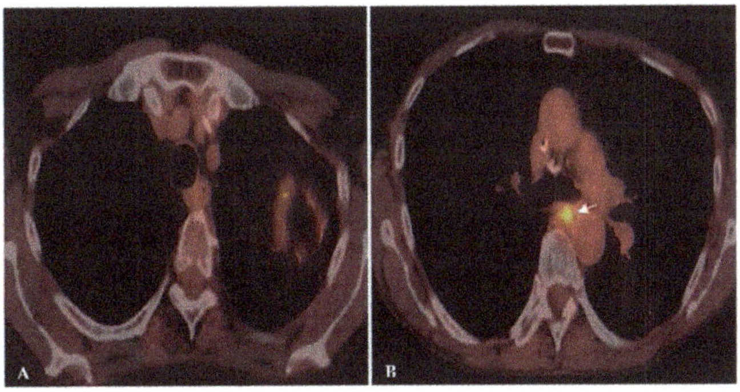

Figure 5. Chronic histoplasmosis infection. FDG PET/CT images show higher FDG activity in a mediastinal draining lymph node (arrow) in (**B**) than in the left upper lobe cavitary mass (**A**). The flip-flop fungus sign discloses the benign granulomatous nature of the disease.

Although nodules or masses can predominate on imaging findings, mixed patterns and additional findings are frequently visualized: consolidations, ground-glass opacities, bronchovascular and septal thickening, and airway disease. Pleural effusions are not a common finding but may be present. More rarely, other pulmonary complications, such as pneumothorax, bronchopleural fistula, and lung abscess, can be found [4,5,8,11].

The main concern in differential diagnosis is the exclusion of malignancy (primary lung cancer and metastases). Other differentials include tuberculosis, nontuberculous mycobacteria, sarcoidosis, vasculitis, and further infectious fungi, such as semi-invasive or invasive aspergillosis, mucormycosis, and candidiasis, especially in immunocompromised patients [4,8,11,16].

2.2. Non-Resolving Pneumonia

Acute fungal infections can present clinically with flu-like symptoms or as community acquired bacterial pneumonia with fever, cough, pleuritic chest pain, myalgia, and headache. Lobar, segmental or patchy multifocal consolidations involving several lobes concurrent with hilar and mediastinal lymphadenopathy are common findings of coccidioidomycosis and histoplasmosis (Figure 6). A tree-in-bud pattern may also be present. Progressive cavitation of the consolidations can occur as the disease progresses. The consolidations in coccidioidomycosis tend to be unilateral, basilar, or perihilar [11,17,18]. Additionally, nodules or masses may evolve at sites of prior consolidation, and eventually can form residual granuloma [4,9].

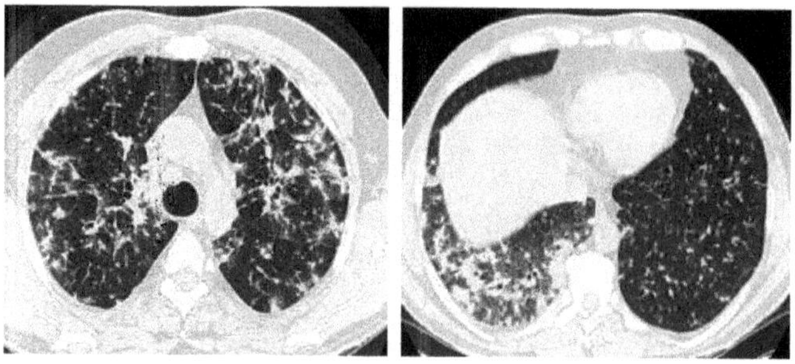

Figure 6. Acute histoplasmosis infection. Axial CT images from upper (**left image**) and lower levels (**right image**) showing bilateral peribronchovascular consolidations and groundglass opacities, and ill-defined centriacinar nodules.

The most prevalent pattern of blastomycosis is patchy, ill-defined consolidations with air bronchograms. The disease may be unilateral or bilateral, and areas of confluence can become quite extensive. These findings may also be visualized alongside the presence of large nodules or masses [19,20].

Segmental or lobar consolidations are also recognized features of cryptococcosis, usually coexisting with parenchymal nodules and masses (Figure 7) [21,22].

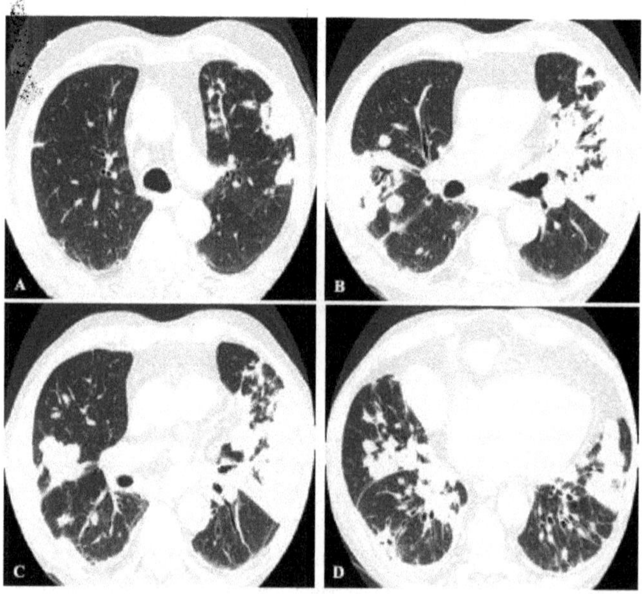

Figure 7. 72-year-old male with acute cryptococcosis. Axial CT images from upper to lower levels (**A–D**) displaying bilateral consolidations with air bronchograms and multiple nodules.

Associated hilar and mediastinal lymphadenopathies are frequent in histoplasmosis and coccidioidomycosis. Pleural or pericardial effusions are not a common finding but may be present. When this complication is present, it occurs most frequently secondary to a hypersensitivity reaction to the fungus antigens. It develops mainly in younger patients and is self-limited in the vast majority of cases. Less frequently, pleural or pericardial effusions can also occur during immune reconstitution syndrome, due to an excessive inflammatory response against the pathogen after a rapidly restored immune system. This finding has been especially reported in histoplasmosis infection in immunocompromised patients (HIV, immunomodulatory therapy) [23,24]. Additionally, empyema may complicate some cases of pleural effusions [4].

The differential diagnoses include community-acquired pneumonia, malignancy, tuberculosis, aspiration, airway invasive aspergillosis, and organizing pneumonia [8].

2.3. Chronic Cavitating Disease

Chronic infections, mainly associated with histoplasmosis and coccidioidomycosis, may develop over time into scarring or cavitating disease. The findings are similar to tuberculosis and most frequently appear in the upper lobes, especially in the apical and posterior segments (Figure 8). Imaging depicts chronic consolidation with progressive cavitation. The cavity may present wall thickening, air-fluid level, internal fungus balls, and may enlarge or collapse, ultimately resulting in volume loss [9,11]. Pleural thickening adjacent to cavitary lesions is also common. Additionally, coccidioidomycosis may evolve thin-walled cavities, a finding termed grape-skin cavities [25]. Calcified lymph nodes may be present in histoplasmosis, but lymphadenopathy is usually absent. Conversely, lymphadenopathy is common in coccidioidomycosis, especially in the acute phase. Other pulmonary findings are recurrently present, such as patchy consolidations, ground-glass opacities, nodules, bronchiectasis, fibrosis, and architectural disorganization [4].

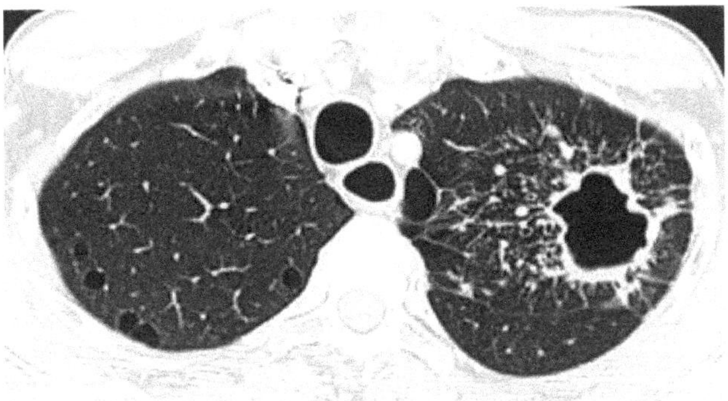

Figure 8. Chronic histoplasmosis infection. CT axial image depicts a spiculated cavity with thick walls in the left upper lobe.

Complications are uncommon and include pneumothorax, fistula formation, broncholiths, empyema, hemoptysis, atelectasis, and pleural and chest wall invasion [4,8]. Involvement of the airways (trachea and bronchi) or larynx can also occur, especially in chronic coccidioidomycosis and paracoccidioidomycosis [11,26].

Chronic pulmonary aspergillosis (CPA) is an important differential diagnosis and usually occurs in non-immunocompromised patients with prior or current lung disease. Chronic cavitating PA is the most common form of CPA, and if untreated it may progress to chronic fibrosing PA (CFPA). Subacute invasive aspergillosis is another form of CPA, occurring in mildly immunocompromised patients (such as those with a history of alcoholism, diabetes mellitus, connective tissue disorders) and usually has a more rapid progression. Imaging findings include necrotic nodules or masses, cavities with thin or thick walls, perhaps containing intraluminal debris and aspergilloma. Pericavitary infiltrates, consolidation, and fibrosis may also be present (Figure 9). Severe fibrotic destruction involving at least two lobes is a hallmark of CFPA [27].

Additional differential diagnoses include postprimary pattern of tuberculosis, lung cancer, nontuberculous mycobacterial infection, Actinomyces, mucormycosis, and Nocardia [8].

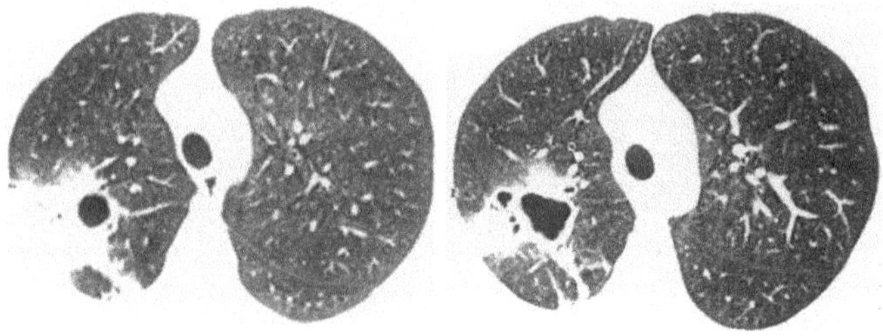

Figure 9. Subacute invasive aspergillosis in a 35-year-old male with HIV infection. CT axial images from an upper and lower levels (left to right) show an irregular cavitary mass in the right upper and lower lobes transgressing the major fissure.

2.4. Disseminated Infection

Endemic mycosis is a self-limited disease in the vast majority of immunocompetent individuals. Immunosuppressed patients are at risk for disseminated disease: HIV, transplant recipients, biologic modifying agents, hematological malignancies, corticosteroid therapy, and extremes of age. Symptoms may include fever, malaise, weight loss, septicemia, severe respiratory distress, coagulopathy, and multiorgan failure. This form of disease presentation is most frequently associated with histoplasmosis and coccidioidomycosis. Chest imaging may depict diffuse air-space opacities, multiple consolidations, acute respiratory distress syndrome (ARDS), and diffuse miliary micronodules (Figure 10). The severe chest imaging findings represent the spread of infection and the difficulty of controlling fungal proliferation in immunosuppressed patients. Extrathoracic dissemination may affect any organ, including skin, lymph nodes, heart, adrenal glands, gastrointestinal tract, bones, joints, and central nervous system [5,9,28,29].

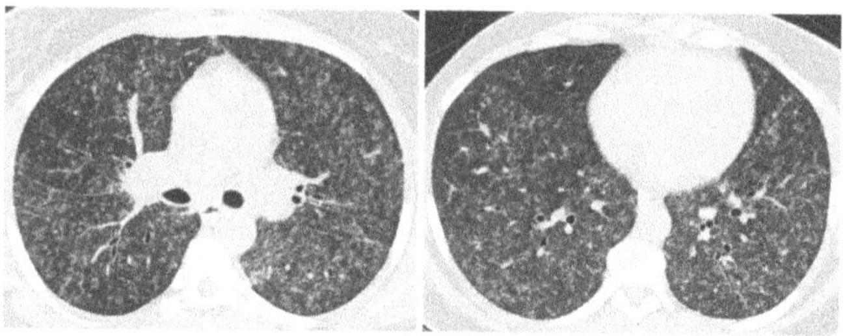

Figure 10. Diffuse miliary nodules in an immunosuppressed patient with histoplasmosis. Axial CT images from upper and lower levels (from **left** to **right**) depict diffuse micronodules randomly distributed in both lungs.

Talaromycosis and emergomycosis are endemic mycoses that occur predominantly in immunosuppressed patients and frequently manifest as multiorgan disseminated diseases. Talaromycosis can involve the upper and lower respiratory tract. Although rare, the infection of the pharynx, larynx, trachea, and bronchi is a distinctive manifestation, and frequently accompanied by cervical lymphadenopathy. Chest imaging findings are diverse and may include patchy consolidations, ground-glass changes, multiple nodules, cavitary disease, pleural effusions, and lymphadenopathy [6]. Patients with emergomycosis commonly manifest widespread polymorphic skin lesions at clinical presentation (umbilicated papules, verrucous lesions, nodules, erythema, and hyperkeratotic plaques). Chest imaging findings are also highly nonspecific and can include diffuse and focal infiltrates, consolidation, lobar atelectasis, pleural effusions, and hilar lymphadenopathy [7].

Differential diagnoses include miliary tuberculosis, hematogenous metastases, and pneumocystis pneumonia (Figure 11).

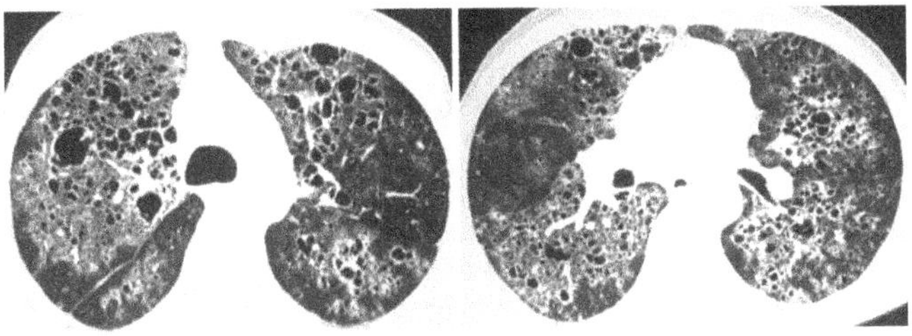

Figure 11. Pneumocystis pneumonia in a patient with acquired immunodeficiency syndrome (AIDS). Axial CT images from upper and lower levels (from **left** to **right**) reveal diffuse bilateral ground-glass opacities with some peripheral sparing. Multiple bilateral pneumatoceles of varying size, shape and wall thickness are also visualized.

2.5. Mixed Pattern

Paracoccidioidomycosis is endemic in Latin America, with the greatest number of cases originating in Brazil. However, several cases have been reported in Europe and North America, particularly among travelers and immigrants. The initial infection is similar to the primary complex of tuberculosis and may have a self-limited path controlled by the host immune response or may progress to symptomatic disease. The two leading clinical forms of paracoccidioidomycosis are the acute form or juvenile type and the chronic form or adult type. The adult form may manifest clinically several years after the inhalation of the infectious particles and accounts for the vast majority of the disease presentation (90% of cases). The disease installation is typically insidious and classically characterized by pulmonary involvement in more than 90% of patients and chronic development of mucocutaneous lesions in approximately 50% of patients. The dissociation between inoculation and proliferation of the fungus and the initial clinical symptoms is responsible for the frequent extensive imagiological findings and severe lung damage in the first exams. Pulmonary fibrosis may be present from the time of diagnosis in 32% of patients, even with indirect signs of pulmonary hypertension and *cor pulmonale* [30–32]. Imaging findings are nonspecific and may include: patchy ground-glass opacities, consolidation, nodules of variable sizes, interlobular septal thickening, intralobular lines, cavitation, and manifestations of fibrosis, such as traction bronchiectasis, paracicatricial emphysema, architectural distortion (Figures 12 and 13). A miliary pattern and lung opacities with "halo" and "reversed halo" signs are also described. These changes are often combined and tend to be bilateral, symmetrical, and involve at least one-third of the lung parenchyma. The disease regularly affects all lung zones; however, a tendency toward the middle zones configuring a butterfly wing pattern may also suggest the diagnosis. Residual fibrotic changes may persist in 60% of patients. Enlarged mediastinal or hilar lymph nodes and pleural effusions are uncommon in the chronic form of the disease. Patients may also manifest extrapulmonary findings, such as tracheal involvement, pneumothorax, joint and osseous lesions. Tracheal infection may manifest as irregular circumferential wall thickening with submucosal nodules, and it may result from direct contact with infected sputum, lymphatic drainage, or hematogenous dissemination. Paracoccidioidomycosis is uncommon in immunocompromised patients, such as those with hematologic malignancies, HIV infection, or transplant recipients. Differential diagnoses include community-acquired pneumonia, other fungal diseases, tuberculosis, and malignancy [5,33,34].

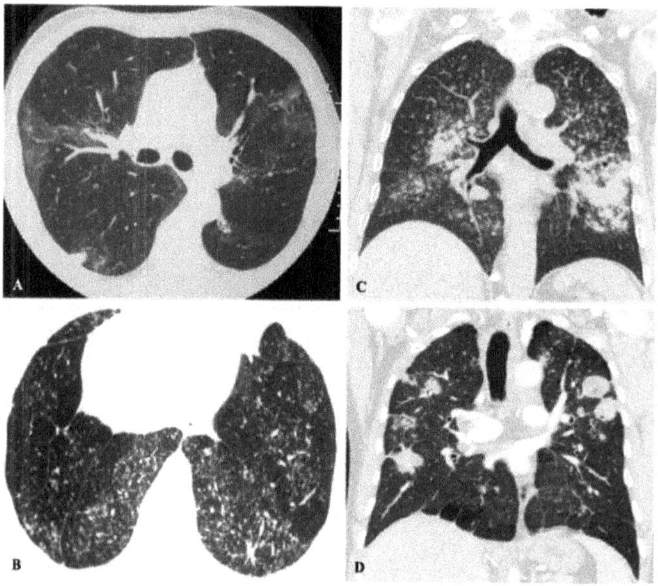

Figure 12. Imaging findings in different patients with paracoccidioidomycosis. The changes are often bilateral, affecting both central and peripheral lung areas and may predominate in the middle zones. The changes are often combined; however, some patterns may prevail. Imaging findings may include patchy bilateral predominant ground-glass opacities (**A**), multiple bilateral small nodules (**B**), bilateral consolidations with associated nodules (**C**), and a dominant pattern of large bilateral nodules (**D**).

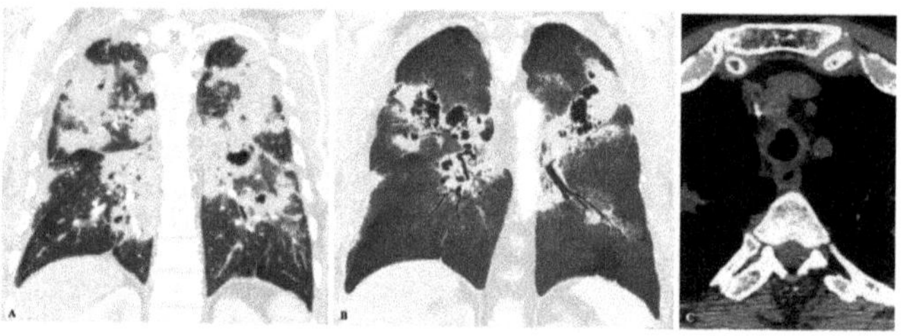

Figure 13. 52-year-old man with paracoccidioidomycosis, presenting with a history of 1 year of shortened of breath that had worsened in the last 20 days. (**A**) coronal CT and (**B**) coronal minimum-intensity projection images depict bilateral and symmetrical consolidations and cavitations in a "butterfly wing" pattern, nodules, and ground-glass opacities. CT also reveals tracheal infection, with irregular circumferential thickening of the wall (**C**). Diagnosis was established by transbronchial biopsy.

2.6. Central Bronchiectasis and Asthma

Allergic bronchopulmonary aspergillosis (ABPA) is an uncommon hypersensitivity reaction to fungal infection that affects immunocompetent individuals. It has a ubiquitous distribution and does not belong to the family of endemic mycosis; however, this disease is worth mentioning. ABPA is classically portrayed in asthmatic patients but also occurs in individuals with cystic fibrosis, lung transplantation, or Kartagener's syndrome. Asthma-induced mucosal damage of the proximal airways enables the proliferation of the fungus, creating an additional mucosal injury, mucus production, and bronchiectasis [8]. Imaging findings can be either transient or permanent and include predominant central bronchiectasis, mucous plugging, consolidation or non-homogeneous infiltrations frequently surrounding the secretion-filled bronchiectasis, centrilobular nodules, parenchymal scarring, fibrosis, mosaic perfusion, and air trapping on expiration. Plain films show tubular and branching opacities with peri-hilar predominance, related to bronchiectasis with mucoid impaction (finger in glove sign). High-attenuation mucus plugging on CT is a pathognomonic feature of ABPA, noticed in 28% of patients, and represents metallic ions and calcium produced by the fungus (Figure 14). Lobar or segmental collapse is not uncommon [8,35].

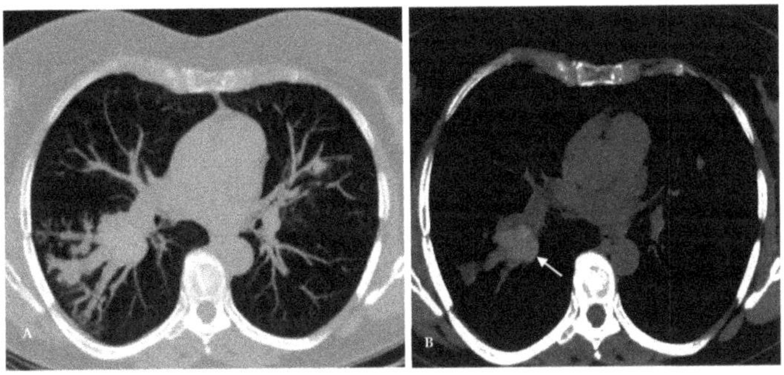

Figure 14. 39-year-old female with ABPA and asthma. (**A**) Axial CT image depicts right lower lobe varicose bronchiectasis with mucoid impaction. (**B**) High-attenuation mucus plugging on CT (arrow): a pathognomonic feature of ABPA.

2.7. Additional Intra and Extrathoracic Findings

Fibrosing mediastinitis is a rare, delayed and potentially life-threatening fibroinflammatory process associated most commonly with *H capsulatum* infection (usually years earlier). Most cases are seen in young patients and are thought to be caused by an abnormal reaction to fungus antigens in genetically susceptible individuals. On CT, it appears as an infiltrative mediastinal fibrous mass, replacing the fat and encasing the adjacent airways and vascular structures, usually with calcified lymph nodes. This complication commonly causes compression of the superior vena cava (with resulting vena cava syndrome), pulmonary arteries and veins, esophagus, and bronchi. The main causes of morbidity and mortality are related to pulmonary arterial hypertension and *cor pulmonale* [36].

Signs of previous histoplasmosis infection: calcified pulmonary nodules (dense or target calcification), broncholithiasis, hepatic and splenic granulomata [4].

Involvement of the larynx and airways: laryngeal, tracheal, and bronchial disease is a chronic disease presentation, especially in coccidioidomycosis and paracoccidioidomycosis, usually associated with parenchymal disease and with disseminated disease, and most commonly results from direct invasion. Additionally, upper and lower respiratory tract infection is also a distinctive manifestation of talaromycosis. Imaging findings include wall thickening, endoluminal nodules, stenosis, and obstruction. Lymph nodes may also compress and erode into the airways, causing obstruction. Complications of airway obstruction include postobstructive atelectasis or pneumonia, fistula formation, hemoptysis, and lithoptysis (expectoration of broncholiths) [6,33,37].

3. Conclusions

We described the most frequent chest imaging findings of endemic mycoses, such as the presence of lung nodules or masses, non-resolving pneumonia, chronic cavitating disease, and disseminated infection, with discussion of the differential diagnoses (Table 1). However, a mixed pattern or a combination of multiple findings are frequently observed and may be focal, multifocal, or diffuse. Immunocompromised patients often present with mixed, diffuse, and severe chest findings. As the geographic distribution of mycoses is spreading, familiarity with the imaging findings, along with a proper clinical, occupational/recreational, and travel history are essential to make an appropriate diagnosis.

Table 1. Chest imaging findings of systemic endemic mycoses.

Lung Nodule or Mass		Non-Resolving Pneumonia		Chronic Cavitating Disease		Disseminated Infection		Bronchiectasis & Asthma	
Imaging Clue	Dx	Imaging Clue	Dx	Imaging Clue	Dx	Imaging Clue	Dx	Imaging Clue	Dx
Adenopathy	Coccidioidomycosis Histoplasmosis	Consolidation + large nodules/masses	Blastomycosis Cryptococcosis Paracoccidioidomycosis	Grape-skin cavities + Lymphadenopathy	Coccidioidomycosis	Miliary ARDS Extrathoracic	++ Histoplasmosis Coccidioidomycosis	High-attenuation mucus plugging Finger in glove	ABPA
Lung Mass	Cryptococcosis Blastomycosis	Adenopathy	Coccidioidomycosis Histoplasmosis	Calcified nodes	Histoplasmosis				
Flip-flop node SUVmax > lung mass	Granulomatous Infection								

SUV, standardized uptake value; ARDS, Acute respiratory distress syndrome; ABPA, Allergic bronchopulmonary aspergillosis; ++, most frequent agents.

Author Contributions: Conceptualization, C.S. and B.H.; writing—original draft preparation, C.S. and B.H.; writing—review and editing, C.S., E.M., A.Y., T.-L.M., P.P., K.I., R.P., A.M., A.S., A.C.P. and B.H. All authors have read and agreed to the published version of the manuscript.

Funding: This research received no external funding.

Conflicts of Interest: The authors declare no conflict of interest.

References

1. Salzer, H.J.; Burchard, G.; Cornely, O.A.; Lange, C.; Rolling, T.; Schmiedel, S.; Libman, M.; Capone, D.; Le, T.; Dalcolmo, M.P.; et al. Diagnosis and Management of Systemic Endemic Mycoses Causing Pulmonary Disease. *Respiration* **2018**, *96*, 283–301. [CrossRef] [PubMed]
2. Denning, D.W.; Chakrabarti, A. Pulmonary and sinus fungal diseases in non-immunocompromised patients. *Lancet Infect. Dis.* **2017**, *17*, e357–e366. [CrossRef]
3. Borchardt, R.A. Endemic Fungal Infections in the United States. *Physician Assist. Clin.* **2017**, *2*, 297–312. [CrossRef]
4. Kunin, J.R.; Flors, L.; Hamid, A.; Fuss, C.; Sauer, D.; Walker, C.M. Thoracic Endemic Fungi in the United States: Importance of Patient Location [published correction appears in Radiographics. *Radiographics* **2021**, *41*, 380–398. [CrossRef] [PubMed]
5. Di Mango, A.L.; Zanetti, G.; Penha, D.; Menna Barreto, M.; Marchiori, E. Endemic pulmonary fungal diseases in immunocompetent patients: An emphasis on thoracic imaging. *Expert Rev. Respir. Med.* **2019**, *13*, 263–277. [CrossRef]
6. Narayanasamy, S.; Dougherty, J.; van Doorn, H.R.; Le, T. Pulmonary talaromycosis: A window into the immunopathogenesis of an endemic mycosis. *Mycopathologia* **2021**, *186*, 707–715. [CrossRef]
7. Reddy, D.L.; Nel, J.; Govender, N.P. Emergomycosis. *J. Med. Mycol.* **2022**, 101313. [CrossRef]
8. Chong, S.; Lee, K.S.; Yi, C.A.; Chung, M.J.; Kim, T.S.; Han, J. Pulmonary fungal infection: Imaging findings in immunocompetent and immunocompromised patients. *Eur. J. Radiol.* **2006**, *59*, 371–383. [CrossRef]
9. Semionov, A.; Rossi, A.; Perillo, M.; Sayegh, K.; Pressacco, J.; Kosiuk, J. Many Faces of Thoracic Histoplasmosis-Pictorial Essay. *Can. Assoc. Radiol. J.* **2019**, *70*, 273–281. [CrossRef] [PubMed]
10. Torres, P.P.T.E.S.; Rabahi, M.F.; Moreira, M.A.C.; Santana, P.R.P.; Gomes, A.C.P.; Marchiori, E. Tomographic assessment of thoracic fungal diseases: A pattern and signs approach. *Radiol. Bras.* **2018**, *51*, 313–321. [CrossRef] [PubMed]
11. Jude, C.M.; Nayak, N.B.; Patel, M.K.; Deshmukh, M.; Batra, P. Pulmonary coccidioidomycosis: Pictorial review of chest radiographic and CT findings. *Radiographics* **2014**, *34*, 912–925. [CrossRef] [PubMed]
12. Fox, D.L.; Müller, N.L. Pulmonary cryptococcosis in immunocompetent patients: CT findings in 12 patients. *AJR Am. J. Roentgenol.* **2005**, *185*, 622–626. [CrossRef] [PubMed]
13. Wang, D.-X.; Zhang, Q.; Wen, Q.-T.; Ding, G.-X.; Wang, Y.-G.; Du, F.-X.; Zhang, T.-Y.; Zheng, X.-Y.; Cong, H.-Y.; Du, Y.-L.; et al. Comparison of CT findings and histopathological characteristics of pulmonary cryptococcosis in immunocompetent and immunocompromised patients. *Sci. Rep.* **2022**, *12*, 5712. [CrossRef] [PubMed]
14. Maini, R.; Ranjha, S.; Tandan, N.; Regmi, M.R.; Ullah, S.; Stone, T.; Nayeemuddin, F.; Pervin, N. Pulmonary Blastomycosis: A case series and review of unique radiological findings. *Med. Mycol. Case Rep.* **2020**, *28*, 49–54. [CrossRef] [PubMed]
15. Nagelschneider, A.; Broski, S.M.; Holland, W.P.; E Midthun, D.; Sykes, A.-M.; Lowe, V.J.; Peller, P.J.; Johnson, G.B. The flip-flop fungus sign: An FDG PET/CT sign of benignity. *Am. J. Nucl. Med. Mol. Imaging* **2017**, *7*, 212–217. [PubMed]
16. Kshatriya, R.; Patel, V.; Chaudhari, S.; Patel, P.; Prajapati, D.; Khara, N.; Paliwal, R.; Patel, S. Cannon ball appearance on radiology in a middle-aged diabetic female. *Lung India* **2016**, *33*, 562–568. [CrossRef]
17. Gabe, L.M.; Malo, J.; Knox, K.S. Diagnosis and Management of Coccidioidomycosis. *Clin. Chest Med.* **2017**, *38*, 417–433. [CrossRef]
18. Capone, D.; Marchiori, E.; Wanke, B.; E Dantas, K.; Cavalcanti, M.A.S.; Filho, A.D.; Escuissato, D.L.; Warszawiak, D. Acute pulmonary coccidioidomycosis: CT findings from 15 patients. *Br. J. Radiol.* **2008**, *81*, 721–724. [CrossRef]
19. Fang, W.; Washington, L.; Kumar, N. Imaging manifestations of blastomycosis: A pulmonary infection with potential dissemination. *Radiographics* **2007**, *27*, 641–655. [CrossRef]
20. Ronald, S.; Strzelczyk, J.; Moore, S.; Trepman, E.; Cheang, M.; Limerick, B.; Wiebe, L.; Sarsfield, P.; Macdonald, K.; Meyers, M.; et al. Computed tomographic scan evaluation of pulmonary blastomycosis. *Can. J. Infect. Dis. Med. Microbiol.* **2009**, *20*, 112–116. [CrossRef]
21. Liu, K.; Ding, H.; Xu, B.; You, R.; Xing, Z.; Chen, J.; Lin, Q.; Qu, J. Clinical analysis of non-AIDS patients pathologically diagnosed with pulmonary cryptococcosis. *J. Thorac. Dis.* **2016**, *8*, 2813–2821. [CrossRef] [PubMed]
22. Qu, Y.; Liu, G.; Ghimire, P.; Liao, M.; Shi, H.; Yang, G.; Xu, L.; Wang, G. Primary pulmonary cryptococcosis: Evaluation of CT characteristics in 26 immunocompetent Chinese patients. *Acta Radiol.* **2012**, *53*, 668–674. [CrossRef] [PubMed]
23. Melzani, A.; Michel, R.D.R.D.S.; Ntab, B.; Djossou, F.; Epelboin, L.; Nacher, M.; Blanchet, D.; Demar, M.P.; Couppie, P.; Adenis, A. Incidence and trends in immune reconstitution inflammatory syndrome associated with histoplasma capsulatum among people living with human immunodeficiency virus: A 20-year case series and literature review. *Clin. Infect. Dis.* **2020**, *70*, 643–652. [CrossRef] [PubMed]
24. Wright, T.; Coruh, B.; Fredricks, D.; Kim, N. Immune reconstitution inflammatory syndrome associated with disseminated histoplasmosis and TNF-alpha inhibition. *Med. Mycol. Case Rep.* **2018**, *23*, 62–64. [CrossRef]
25. Batra, P. Pulmonary coccidioidomycosis. *J. Thorac. Imaging* **1992**, *7*, 29–38. [CrossRef]

26. Abreu e Silva, M.À.; Salum, F.G.; Figueiredo, M.A.; Cherubini, K. Important aspects of oral paracoccidioidomycosis–a literature review. *Mycoses* **2013**, *56*, 189–199. [CrossRef]
27. Denning, D.W.; Cadranel, J.; Beigelman-Aubry, C.; Ader, F.; Chakrabarti, A.; Blot, S.; Ullmann, A.J.; Dimopoulos, G.; Lange, C.; Dimopoulos, Christoph Lange on behalf of the European Society for Clinical Microbiology and Infectious Diseases and European Respiratory Society. Chronic pulmonary aspergillosis: Rationale and clinical guidelines for diagnosis and management. *Eur. Respir. J.* **2016**, *47*, 45–68. [CrossRef]
28. Wheat, L.J.; Connolly-Stringfield, P.A.; Baker, R.L.; Curfman, M.F.; Eads, M.E.; Israel, K.S.; Norris, S.A.; Webb, D.H.; Zeckel, M.L. Disseminated histoplasmosis in the acquired immune deficiency syndrome: Clinical findings, diagnosis and treatment, and review of the literature. *Medicine (Baltimore)* **1990**, *69*, 361–374. [CrossRef]
29. Stockamp, N.W.; Thompson GR 3rd. Coccidioidomycosis. *Infect. Dis. Clin. N. Am.* **2016**, *30*, 229–246. [CrossRef]
30. Restrepo, A.; Benard, G.; de Castro, C.C.; Agudelo, C.A.; Tobón, A.M. Pulmonary paracoccidioidomycosis. *Semin. Respir. Crit. Care Med.* **2008**, *29*, 182–197. [CrossRef]
31. Mendes, R.P.; Cavalcante, R.D.S.; Marques, S.A.; Marques, M.E.A.; Venturini, J.; Sylvestre, T.F.; Paniago, A.M.M.; Pereira, A.C.; Silva, J.D.F.D.; Fabro, A.T.; et al. Paracoccidioidomycosis: Current perspectives from Brazil. *Open Microbiol. J.* **2017**, *11*, 224–282. [CrossRef] [PubMed]
32. Tobon, A.M.; Agudelo, C.A.; Osorio, M.L.; Alvarez, D.L.; Arango, M.; Cano, L.E.; Restrepo, A. Residual pulmonary abnormalities in adult patients with chronic paracoccidioidomycosis: Prolonged follow-up after itraconazole therapy. *Clin. Infect. Dis.* **2003**, *37*, 898–904. [CrossRef] [PubMed]
33. Barreto, M.M.; Marchiori, E.; Amorim, V.B.; Zanetti, G.; Takayassu, T.C.; Escuissato, D.L.; Souza, A.S.; Rodrigues, R.S. Thoracic paracoccidioidomycosis: Radiographic and CT findings [published correction appears in Radiographics. *Radiographics* **2012**, *32*, 71–84. [CrossRef] [PubMed]
34. Marchiori, E.; Valiante, P.M.; Mano, C.M.; Zanetti, G.; Escuissato, D.L.; Souza, A.S.; Capone, D. Paracoccidioidomycosis: High-resolution computed tomography-pathologic correlation. *Eur. J. Radiol.* **2011**, *77*, 80–84. [CrossRef]
35. Shah, A.; Panjabi, C. Allergic aspergillosis of the respiratory tract. *Eur. Respir. Rev.* **2014**, *23*, 8–29. [CrossRef]
36. Garrana, S.H.; Buckley, J.R.; Rosado-de-Christenson, M.L.; Martínez-Jiménez, S.; Muñoz, P.; Borsa, J.J. Multimodality Imaging of Focal and Diffuse Fibrosing Mediastinitis. *Radiographics* **2019**, *39*, 651–667. [CrossRef]
37. Polesky, A.P.; Kirsch, C.M.; Synder, L.S.; LoBue, P.; Kagawa, F.T.; Dykstra, B.J.; Wehner, J.H.; Catanzaro, A.; Ampel, N.M.; Stevens, D.A. Airway coccidioidomycosis-report of cases and review. *Clin. Infect. Dis.* **1999**, *28*, 1273–1280. [CrossRef]

Journal of
Fungi

MDPI

Review

Immunologic Diagnosis of Endemic Mycoses

Rodrigo Almeida-Paes, Andrea Reis Bernardes-Engemann, Beatriz da Silva Motta, Claudia Vera Pizzini, Marcos de Abreu Almeida, Mauro de Medeiros Muniz, Renata Alves Barcelos Dias and Rosely Maria Zancopé-Oliveira *

Laboratório de Micologia, Instituto Nacional de Infectologia Evandro Chagas, Fundação Oswaldo Cruz, Rio de Janeiro 21040-900, Brazil
* Correspondence: rosely.zancope@ini.fiocruz.br

Abstract: The endemic mycoses blastomycosis, coccidioidomycosis, histoplasmosis, paracoccidioidomycosis, cryptococcosis, sporotrichosis, talaromycosis, adiaspiromycosis, and emergomycosis are mostly caused by geographically limited thermally dimorphic fungi (except for cryptococcosis), and their diagnoses can be challenging. Usual laboratory methods involved in endemic mycoses diagnosis include microscopic examination and culture of biological samples; however, serologic, histopathologic, and molecular techniques have been implemented in the last few years for the diagnosis of these mycoses since the recovery and identification of their etiologic agents is time-consuming and lacks in sensitivity. In this review, we focus on the immunologic diagnostic methods related to antibody and antigen detection since their evidence is presumptive diagnosis, and in some mycoses, such as cryptococcosis, it is definitive diagnosis.

Keywords: antibody; antigen; blastomycosis; coccidioidomycosis; histoplasmosis; paracoccidioidomycosis; cryptococcosis; sporotrichosis; talaromycosis; emergomycosis

Citation: Almeida-Paes, R.; Bernardes-Engemann, A.R.; da Silva Motta, B.; Pizzini, C.V.; de Abreu Almeida, M.; de Medeiros Muniz, M.; Dias, R.A.B.; Zancopé-Oliveira, R.M. Immunologic Diagnosis of Endemic Mycoses. *J. Fungi* **2022**, *8*, 993. https://doi.org/10.3390/jof8100993

Academic Editor: Alessandro C. Pasqualotto

Received: 29 July 2022
Accepted: 13 September 2022
Published: 22 September 2022

1. Introduction

Endemic mycoses are mostly caused by thermally dimorphic fungi that present a limited geographic distribution, occupying specific ecologic niches in the environment, and can cause both primary or opportunistic diseases [1]. In addition, endemic mycoses are recognized as significant causes of morbidity and mortality predominantly in HIV/AIDS and other immunosuppressive conditions, including immunosuppressant drugs [2]. The most common endemic mycoses are blastomycosis, coccidioidomycosis, histoplasmosis, paracoccidioidomycosis, cryptococcosis, sporotrichosis, and, more recently, talaromycosis, adiaspiromycosis, and emergomycosis, considered emerging endemic mycoses [3]. In recent years, the number of endemic mycoses cases has risen worldwide [1]. In addition, there are significant variations in their geography, clinical presentation, roentgen manifestations, analytic diagnostic methods, and therapeutics. Their proper control involves recognition of risk factors (e.g., putative environmental sources of fungal exposure in endemic areas), correct diagnostic procedures, and therapeutic management [4].

The diagnosis of endemic mycoses is difficult to achieve. Precise laboratory data evaluation is necessary to guarantee appropriate therapy for patients. Although the manifestations of endemic mycoses are well defined, their diagnosis cannot be centered solely on patient's clinical data, since the signs and symptoms of endemic mycoses overlap among them and with other infectious diseases [3].

The association of clinical, epidemiological, and laboratorial data typically diagnoses endemic mycoses. To corroborate the diagnosis, laboratorial tests must be performed. The usual laboratory tests involved in endemic mycoses diagnosis comprise the microscopic examination and culture of several types of biological samples. The microscopic aspect of the agents is often indicative in the case of endemic mycosis, but considerable laboratory expertise is necessary and sensitivity of these methods is variable. Culture from possibly involved sites remains the diagnostic gold standard method, despite longtime fungal growth (up to six weeks in some cases) and the need for biosafety level 3 facilities for handling some agents in the laboratory [5].

Currently, there are further diagnostic tools available for diagnosis of endemic mycoses to complement culture and direct examination [6]. These complementary methods have fast turnaround time and satisfactory efficiency. Different immunologic techniques concerning antibody and antigen detection have been developed to aid in the diagnosis of endemic mycoses (Table 1). Several antigenic preparations have been used in these tests, from crude to purified antigens, as well as recombinant proteins and synthetic peptides. However, the latter are not used in the validated assays for routine mycology laboratories. As mentioned before, serologic evidence of these infections is valuable due to the time-consuming nature and low sensitivity of gold-standard methods. In addition to antigen and antibody detection methods, intradermal skin tests were largely employed in the last century [7–9], but their current use for diagnostic purposes in medical mycology is severely limited, due to the lack of standardized antigens, advances in antibody and antigen detection methods, and biosafety requirements to perform the skin tests. Molecular tools of dimorphic fungal DNA detection in biological samples are also being standardized and validated in numerous laboratories to simplify diagnosis. Unfortunately, although promising and useful, non-culture diagnostic tools are not accessible in most low-income countries.

Table 1. Immunologic methods used for antigen or antibody detection for the diagnosis of the major endemic mycoses.

Method	BLM	CDM	HPM	PCM	CRY	SPT	TLM
Complement fixation	Ab	Ab	Ab	Ab	-	-	-
Immunodiffusion	Ab	Ab/Ag	Ab	Ab	-	Ab	Ab/Ag
Counterimmunoelectrophoresis	-	-	-	Ab/Ag	-	Ab	-
Tube precipitin	-	Ab	-	-	-	Ab	-
Latex agglutination	-	Ab	Ab	Ab	Ag	Ab	-
Lateral flow assay	-	Ab	Ag	-	Ag	Ab	Ag
ELISA	Ab/Ag	Ab/Ag	Ab/Ag	Ab/Ag	Ab/Ag	Ab	Ab/Ag
Western blot	-	Ab	Ab	Ab/Ag	-	Ab	-
Radioimmunoassay	Ab	-	Ag	Ag	-	-	-

BLM: blastomycosis; CDM: coccidioidomycosis; HPM: histoplasmosis; PCM: paracoccidioidomycosis; CRY: cryptococcosis; SPT: sporotrichosis; TLM: talaromycosis; Ab: antibody; Ag: antigen; ELISA: enzyme-linked immunosorbent assay.

The next sections will focus on the immunologic diagnostic applications of the endemic mycoses. A number of well-established tools will be discussed and reviewed. Furthermore, we will summarize the progress in the development of new serologic tests and their relative advantages.

2. Blastomycosis

Blastomycosis is a fungal infection of humans and mammals caused by dimorphic fungi of the genus *Blastomyces*. Its major etiologic agents include *Blastomyces dermatitidis*, *Blastomyces gilchristii*, and *Blastomyces percursus*. Blastomycosis frequently affects immunocompetent individuals; however, immunocompromised patients are more likely to present severe disease [10].

Direct visualization of *Blastomyces* spp. in biological samples can provide fast diagnosis, making it possible to initiate proper antifungal therapy. Correct visualization of fungal elements is sometimes difficult to achieve by hematoxylin-eosin (H&E) staining, thus the periodic acid–Schiff or methenamine silver stains are recommended. Potassium hydroxyde or calcofluor white direct examination are valuable for specimens from the respiratory tract [3]. As for other endemic mycoses, culture is the gold standard diagnostic method. Sabouraud dextrose agar cultures usually demonstrate mold *Blastomyces* sp. colonies within weeks to months.

2.1. Antibody Detection

The complement fixation reaction was used as the first immunological test for blastomycosis diagnosis using the yeast-form derived antigen, but its sensitivity (57%) and specificity (30%) were low. After the introduction of the immunodiffusion test in 1973 using *B. dermatitidis* specific "A" antigen, a culture filtrate [11], the efficiency of the test was enhanced [12,13]. The evaluation of the purified "A" antigen, instead of the crude yeast antigen, and its application in the complement fixation test resulted in high specificity, even though sensitivity was 62% [14]. Afterwards, the value of purified "A" antigen was evaluated in an enzyme-linked immunosorbent assay (ELISA), revealing 92% sensitivity and 84% specificity when comparing its diagnostic efficacy to blastomycosis with complement fixation and immunodiffusion tests as gold standards [15].

The reagents commercially available for *Blastomyces* antibody detection have been used for several years; nevertheless, they are currently judged unsatisfactory for blastomycosis diagnosis. Immunodiffusion performed with the purified *B. dermatitidis* "A" antigen was shown to be more efficient than complement fixation. Immunodiffusion precipitation bands are specific for blastomycosis, but their absence does not rule out diagnosis, since the test sensitivity ranges from 65% to 80% [14,16]. Undoubtedly, EIA for diagnosis of blastomycosis developed for antibody detection is more sensitive than immunodiffusion; however, it is less specific [15]. The EIA provided a noteworthy advance in immunologic testing for blastomycosis and could be performed during outbreaks as an epidemiological tool to detect acute *B. dermatitidis* infection; titers higher than or equal to 1:32 powerfully support blastomycosis diagnosis, while titers of 1:8 or 1:16 are only indicative of blastomycosis [15].

Linder and Kauffman (2020) summarize the main points of standard immunodiffusion and complement fixation assays, which are valuable for histoplasmosis diagnosis, but have not showed satisfactory sensitivity and specificity to support blastomycosis diagnosis [17]. Further improvements aiming to measure antibodies to the WI-1 (BAD-1) antigen, an important fungal adhesin, look like they have better sensitivity [18]. An EIA directed to the WI-1 antigen presented superior efficiency when compared to previous tests [19]. The report of sensitivity and specificity was 88% and 99%, respectively.

Klein and Jones (1990) developed a radioimmunoassay (RIA) to detect antibodies anti-WI-1 (also known as BAD-1) antigen, demonstrating positivity in 85% of blastomycosis patients and just 3% of patients with other mycoses, proving to be superior to the EIA using the "A" antigen (58% positivity) [20]. Several studies validate the initial results [21–23], but up to now, no commercially available kit for clinical testing using this method has been manufactured.

2.2. Antigen Detection

The quantitative Sandwich EIA is produced by the MVista®, and is applied for the antigen detection test for the diagnosis of blastomycosis. A galactomannan of the cell wall of *B. dermatitidis* is the target of the MVista® enzyme immunoassay. The test is useful for diagnosis and monitoring of disease and patient management. Cross-reactions are seen with histoplasmosis, paracoccidioidomycosis, talaromycosis, infrequently in coccidioidomycosis, uncommonly in aspergillosis, and probably in patients with sporotrichosis [24,25]. This assay can be performed with urine, serum, bronchoalveolar lavage (BAL), and cerebrospinal fluid (CSF) samples [26–28]. The majority of the data about efficiency are stated for urine, in which sensitivity varies from 76% to 90% in several studies [24,26,27,29]. Sensitivity of the sandwich EIA is lower in serum, ranging from 56% to 82% [25,27,28]. Pretreatment with ethylenediaminetetraacetic acid and boiling to 104 °C to dissociate immune complexes improved antigen detection in serum samples, increasing antigenemia from 35.7% to 57.1% [25]. Sensitivity of antigen detection in BAL and CSF is unknown, but has been informed to help diagnosis of particular cases [26].

Several fungi share galactomannan antigens, and therefore, the specificity for a particular genus is usually not high enough. For instance, the cross-reactivity of *Histoplasma* and *Blastomyces* antigens has ranged between 93 and 96% using this EIA assay [24,25]. Although several patients with blastomycosis present false positive results for *Aspergillus* galactomannan, no patient with aspergillosis presented a positive antigen assay for *B. dermatitidis* [30].

Some research groups have stated that the antigen follow-up in urine can be helpful for checking the resolution or progression of blastomycosis [28,31,32]. The *Blastomyces* urine antigen detection could also be valuable to follow up the therapeutic response, since clearance of antigen correlates well with the patient's recovery [28]. In addition, this test appears to diagnose blastomycosis regardless of its etiological agent [33].

3. Coccidioidomycosis

The two cryptic and dimorphic fungi *Coccidioides immitis* and *Coccidioides posadasii* cause coccidioidomycosis. The former species occurs in the Central Valley of California (San Joaquin Valley), but has now been detected as far north as east of Washington [34]. The latter species is regularly found in Arizona, Texas, Utah, Mexico, and Central and South America [35]. The at-risk individuals to be infected with these fungi include archeologists, laboratory staff handling the coccidioidomycosis agents, and visitors to endemic areas [34]. Coccidioidomycosis is often asymptomatic or occurs as a respiratory syndrome with undistinguishable, self-limiting symptoms. On the other hand, depending on the patient's immunity, symptomatic, disseminated, and severe infections may occur [36].

Microscopic examination is fast and effective for coccidioidomycosis diagnosis, and culture confirms the species. Spherules of 20 to 70 µm in diameter, or even larger, with a double membrane containing endospores (2–5 µm), are observed in biological samples by microscopy [4]. Histopathologic tests show tuberculoid and mixed granulomas with spherules of different sizes [37]. The intradermal coccidioidin skin test and antibody detection by complement fixation are the most used immunological diagnostic tools for coccidioidomycosis. Both tests are useful for prognosis [4].

Among the immunological methods on hand for the endemic mycoses, those for coccidioidomycosis have been the most trustworthy. The following tests have been used for diagnosis: complement fixation and precipitin reactions in numerous versions, i.e., tube precipitin, immunodiffusion tube precipitin, immunodiffusion complement fixation, and quantitative immunodiffusion complement fixation, agar gel precipitin-inhibition test, and counterimmunoelectrophoresis; latex particle agglutination; fluorescent antibody; RIA, ELISA, and mycoarray.

3.1. Antibody Detection

Smith and collaborators established the tube precipitin and complement fixation tests [38,39]. They observed positive tube precipitin reactivity within weeks of infection. On the other hand, complement fixation positivity occurred later, usually within 2 to 3 months after fungal exposure. Moreover, complement fixation titers might increase if coccidioidomycosis was not under control.

The appearance of immunodiffusion complement fixation and sporadic immunodiffusion tube precipitin bands between a serum sample and the *Coccidioides* antigen is probable evidence of coccidioidomycosis, active or recently acquired, and a negative test does not exclude mycosis [40].

Latex agglutination and complement fixation assays may offer relevant additional data about the patient status [41]. The complement fixation assay has high sensitivity; however, its performance is complex and laborious. In addition, the complement fixation assay has low specificity due to cross-reactivity that may occur with antibodies recognizing common fungal carbohydrate moieties. The immunodiffusion assay is more specific, and complement fixation is more sensitive [42].

The use of complement fixation and immunodiffusion tests for coccidioidomycosis diagnosis is well established with crude antigen samples, known as coccidioidins, which include the reactive complement fixation antigen, as well as several important molecules. The production of purified antigens could improve the effectiveness of immunological tests. The *C. posadasii* Silveira antigenic preparation has a protein with a 110 kDa molecular weight that migrated at 48 kDa when fractionated under heated and reducing SDS-PAGE conditions. The use of recombinant chitinase antigen in conventional complement fixation and immunodiffusion complement fixation has been reported [43].

Several tests for antibody detection in coccidioidomycosis have been reported, demonstrating very relevant outcomes [44]. In an effort to improve sensitivity of immunologic diagnosis, Meridian Diagnostics (Cincinnati, OH, USA) established an ELISA (Premier Coccidioides EIA kit) for IgM and IgG antibody detection against *Coccidioides* spp. as well as for the detection of antibodies against a 33 kDa cell-wall purified antigenic molecule from immature *C. immitis* spherules [45].

The method "mycoarray" is composed of three antigen extracts (histoplasmin, coccidioidin, and *Coccidioides* "TP") for antibody detection. Microarray slides are probed with coccidioidomycosis and histoplasmosis serum samples from patients or from healthy individuals and the detection of immunocomplexes is carried out by indirect immunofluorescence. In concordance with clinical and mycological diagnosis, the "mycoarray" could distinguish between these two mycoses and clearly discriminate between IgM and IgG antibody reactivity. After a proper validation and with its employ as a large-scale array, the "mycoarray" could be applied to help clinicians provide coccidioidomycosis diagnosis [46].

The immunologic response of an in-house antigen preparation, obtained from a *C. posadasii* strain isolated in Ceará, northeastern Brazil, was evaluated by immunodiffusion and Western blot. In addition, its biochemical characterization was performed. Two immunoreactive proteins were characterized as a β-glucosidase and a glutamine synthetase after analyses of their respective N-terminal sites. This in-house *Coccidioides* preparation could be promising as a fast and low-cost diagnostic method [47]. This study, however, does not contain conclusions on immunologic data.

Other studies directed to antigenic fractions recognized by anti-*Coccidioides* antibodies in serum samples from coccidioidomycosis patients were carried out more recently, and the obtained proteins were analyzed by homology to species-specific *Coccidioides* peptides. A *C. immitis* specific peptide was selected from the "GPI anchored serine-threonine rich protein OS" that recognized both *C. immitis* and *C. posadasii*. These peptides can be employed in diagnostic reagents, immunobiologicals, and antifungal drugs [48].

3.2. Antigen Detection

Antibody detection has been used as the principal coccidioidomycosis diagnostic method, but it presents some weaknesses. Kassis and collaborators evaluated in retrospect the efficiency of antigen and antibody detection in 158 coccidioidomycosis cases and 487 controls. The sensitivity of combining antigen and immunodiffusion antibody detection was 93.0%. The sensitivity of antigen detection in urine and serum samples was 55% in proven coccidioidomycosis and 59% in probable coccidioidomycosis, 79% in disseminated coccidioidomycosis, 42% in pulmonary cases, 75% in immunocompromised individuals, and 40% in immunocompetent individuals. Specificity was 99% for antigen detection and 96% for antibody detection using the immunodiffusion method. Accuracy was determined as 95% for immunodiffusion antibody and antigen detection, 94% for immunodiffusion antibody alone, and 89% for pathology or culture [49]. These findings supported the detection of antibodies and antigens to diagnose progressive coccidioidomycosis. An incorrect diagnosis would occur if antigen detection was not carried out.

An inhibition ELISA was developed to detect and quantify *Coccidioides* chitinase-1 (CTS1) in human sera using a monoclonal antibody reactive for this protein. CTS1 was quantified in commercial antigenic reagents using recombinant CTS1 as the standard. The amounts of CTS1 in diagnostic commercial antigens from distinct suppliers varied. CTS1 antigenemia was observed in 87% of patients with proven or probable coccidioidomycosis. Specificity was determined to be 97% using sera from Phoenix, Arizona residents who did not have coccidioidomycosis. Levels of CTS1 could be associated with low- and high-titer serology from individuals with proven coccidioidomycosis diagnosis [50]. The CTS1 antigen detection assay has the possibility of similar or better performance than other immunologic assays as well as the distinct advantage of a direct measurement of fungal antigen concentrations in blood. Even though further studies are necessary to specify the real role of this assay in mycology laboratories, it could be used as a convenient instrument for difficult-to-diagnose cases.

4. Histoplasmosis

Histoplasmosis is a systemic mycosis caused by the dimorphic fungus *H. capsulatum* and is the major endemic mycosis in the United States and in a large part of Latin America [51,52]. In Africa, in addition to classical histoplasmosis, African histoplasmosis, caused by *Histoplasma duboisii*, is also endemic [53]. *H. capsulatum* is a primary fungus, and can cause serious infection in immunocompetent patients, and depending on the patient's immunity, symptomatic, disseminated, and severe infections may occur.

Histoplasmosis diagnosis is a challenge and often requires a multifactorial approach. Identification of *H. capsulatum* in biological samples by direct microscopy and/or culture is still the gold standard for diagnosis [54,55]. However, these tests still have some limitations: (i) the low sensitivity, which varies according to the clinical form of HPM; (ii) the lengthy cultivation of the fungus, taking 4 to 6 weeks and still requiring conversion to the yeast-like form; (iii) the need for a biosafety level 3 facility for handling *H. capsulatum* [56]. Thus, immunologic methods of antibody and antigen detection are options for the presumptive diagnosis of histoplasmosis using serum, plasma, CSF, and urine as clinical specimens [55].

4.1. Antibody Detection

The time required for anti-*H. capsulatum* antibody development is two to six weeks after fungal exposure [54]. Some of the available serological tests for detecting anti-*Histoplasma* antibodies are immunodiffusion, complement fixation, latex agglutination, ELISA, and Western blot. The two most used methods until recently for antibody detection in biological samples are immunodiffusion and complement fixation, usually performed in reference laboratories due to the convenience, availability, and precision of these assays [57,58].

Immunodiffusion using the histoplasmin antigen (HMIN), an antigenic preparation obtained from the mycelium-form cultures of *H. capsulatum*, detects the presence of antibodies through the appearance of H and M precipitins. The H precipitin usually co-exists with the M precipitin; however, the latter often occurs alone. Anti-M antibodies are triggered in acute or chronic histoplasmosis and in some individuals who have undertaken the histoplasmin skin test. In addition, the M precipitin can persist for years [59]. H precipitin usually appears after the M precipitin and is suggestive of chronic or severe histoplasmosis. Anti-H antibodies are rarely observed in the routine diagnosis (20%), but, when detected, corroborate with a histoplasmosis diagnosis [55]. Although the specificity of the test is 100%, the sensitivity varies from 70 to 95%, according to the histoplasmosis clinical form [58]. The detection of both precipitins (H and M) is thought to be decisive for the histoplasmosis diagnosis, although the mycosis condition requires an evaluation of the patient [57].

The complement fixation test detects antibodies against the yeast and mycelial phase histoplasmin. Although often more sensitive (72–95%) than immunodiffusion, depending on the antigen used, complement fixation is less specific and may present cross-reactivity with serum samples from patients with *B. dermatitides*, *C. immitis*, *Paracoccidioides brasiliensis*, and *Candida* sp. infections [58,60]. As for the interpretation of the results, titers equal or higher than 1:32 or four times increase in antibody titers of acute and convalescent disease indicate active infection. Titers of 1:8 generally suggest prior *H. capsulatum* exposure [61].

Latex agglutination tests were developed for the diagnosis of histoplasmosis, and despite some reports that this test is more sensitive than complement fixation using histoplasmin as the antigen, the specificity of the test was compromised [62]. False positive results may occur in patients with another infectious disease, e.g., tuberculosis [63], and with inflammatory diseases such as rheumatoid arthritis [64].

It has already been demonstrated that immunoassays such as ELISA [65] and Western blot [66] have higher sensitivity than immunodiffusion and complement fixation in the detection of antibodies. Several ELISA protocols for detecting antibodies anti-*Histoplasma* using different antigenic preparations have been described; however, most of them are developed for an *in house* use and present varied degrees of sensitivity and specificity [57,58]. For instance, an ELISA assay with an *H. capsulatum* yeast cell antigenic preparation showed an 86% sensitivity and a specificity of 91% in patients with acute pulmonary histoplasmosis detecting human IgG, but when detecting IgM, the sensitivity decreased to 66% and the specificity rose to 100% [67]. The ELISA test with a proprietary MVista® *Histoplasma* antigen used for evaluating the acute pulmonary form of histoplasmosis detected IgG antibodies in 87%, IgM antibodies in 67%, and IgG and/or IgM antibodies in 89% of patients with this clinical form of histoplasmosis [68]. Another indirect ELISA using purified and deglycosylated histoplasmin was 92% sensitive and 96% specific [65]. The same assay was evaluated for different clinical forms of histoplasmosis, yielding positive results in 100% of acute patients, 90% of chronic patients, 89% of disseminated infection in individuals without HIV infection, 86% of disseminated disease in people living with HIV/AIDS (PLWHA), and 100% of mediastinal histoplasmosis patients [69]. More recently, an ELISA using a similar antigen, deglycosylated extracellular released antigen, showed 72% and 98% sensitivity and specificity, respectively. In this study, 100% from the patients with acute form, 50% with chronic form, and 66.67% with disseminated form, respectively, were positive [70].

A Western blot test using purified and deglycosylated histoplasmin was developed, evaluated, and validated, showing sensitivity of 94.9% and specificity of 94.1%. In addition to being simpler and faster, strips sensitized with the purified and deglycosylated histoplasmin antigen were shown to be viable for use for at least five years [66,71,72], and can also be applied with high sensitivity even in PLWHA [73].

4.2. Antigen Detection

Antigen detection tests are particularly valuable in the diagnosis of disseminated histoplasmosis in PLWHA whose antibody levels are low or inexistent. They provide high sensitivity for the diagnosis of histoplasmosis, and are now incorporated in the World Health Organization (WHO) Essential Diagnostics List [74]. During histoplasmosis, the antigen can be liberated from fungal cells and detected in biological samples such as serum, urine, CSF, BAL, and pleural fluid [58]. Antigen detection assays can also be applied in the histoplasmosis follow-up. However, a limitation to these tests is the substantial cross-reactivity with other mycoses, including paracoccidioidomycosis, blastomycosis, talaromycosis, coccidioidomycosis, and aspergillosis [55].

The RIA method was the first test described for the recognition of *H. capsulatum* antigens. Based on the detection of *H. capsulatum* polysaccharide antigen in urine and serum of patients, it proved to be effective to diagnose this infection, especially in individuals with disseminated histoplasmosis. Since its development in 1986, and with the improvement of the technique, there has been an increase in detection levels of antigens, demonstrating a sensitivity of 96.7% in urine and 78.7% in sera from PLWHA and disseminated histoplasmosis patients [75,76]. False positive results may occur in individuals with blastomycosis or paracoccidioidomycosis [77] and this test has been performed in an EIA format to avoid exposing workers to radioactivity [78].

ELISA, in its various protocols, is another method using for *Histoplasma* antigen detection [79–82]. A quantitative ELISA assay was developed, and the concentrations of *H. capsulatum* galactomannan antigen were established by comparing them to a standard curve constructed with a purified galactomannan from the *H. capsulatum* yeast-like form. Serum and urine samples were tested, showing a sensitivity of 92.3% in serum samples and 100% in urine from the disseminated histoplasmosis cases. Cross-reactions were detected in 70% of patients with other endemic mycoses (blastomycosis, paracoccidioidomycosis, coccidioidomycosis, and talaromycosis) [83]. The same test, MVista *Histoplasma* antigen enzyme assay, was changed to allow the quantitative determination of antigen in BAL and this method was compared to culture and cytopathology. Antigen was detected in BAL in 93% of patients with histoplasmosis, and culture and cytopathology both showed 48% sensitivity. Combining antigen detection and cytopathology in BAL, both rapid diagnostic tools, the sensitivity was 96.8%. Thus, BAL antigen detection complements antigenemia and antigenuria as a diagnostic tool for histoplasmosis. However, cross-reactivity is observed in patients with blastomycosis (80%) [84].

A multicenter study described by Hage and collaborators [85] evaluated the sensitivity and specificity of the ELISA for the detection of MVista® *Histoplasma* antigen (MiraVista Dianostics) in different clinical forms. A sensitivity of 91.8% was found in urine from individuals with disseminated histoplasmosis, 83.3% with acute histoplasmosis, 30.4% with the subacute form, and 87.5% with the chronic pulmonary form. In serum samples, the test showed a sensitivity of 100% in cases of disseminated histoplasmosis. Specificity was 99% between individuals with non-fungal infections and healthy individuals; however, 90% of patients with blastomycosis presented cross-reactivity.

Another study evaluated two commercial kits for histoplasmosis diagnosis in immunocompromised individuals. The FDA-cleared in vitro diagnostic assay kit (Alpha *Histoplasma* Antigen EIA) uses a rabbit polyclonal antibody or a monoclonal anti-*Histoplasma* galactomannan antibody (Immuno Mycologics – IMMY, Norman, OK, USA). The assay using the monoclonal antibody presented higher sensitivity (90.5%) and specificity (96.3%) than the test performed with the polyclonal antibody (61.9 and 79.3%) [86]. More recently, an ELISA for the detection of *Histoplasma* antigenuria, developed by Optimum Imaging Diagnostics, was studied, presenting 92% sensitivity. However, false positive results occurred in 68% of samples tested [87].

In the search for rapid tests for the diagnosis of histoplasmosis, a lateral flow assay to detect *Histoplasma* antigenemia settled by MiraVista Diagnostics was studied in three populations: PLWHA with proven histoplasmosis, PLWHA with other infectious diseases, and people without HIV. The test sensitivity was 96% when read visually and 92% when an automated reader was used and the specificities were 90% and 94% for the same conditions [88]. Afterwards, a validation study was carried out on the MVista® Diagnostics *Histoplasma* urine antigen lateral flow assay for antigen detection, and was associated with the MVista® *Histoplasma* Ag quantitative ELISA. The sensitivity of both tests was 96%; however, the specificity was 96% and 77% for LFA and ELISA, respectively [89].

4.3. *African Histoplasmosis*

Most cases of African histoplasmosis reported up to now were diagnosed by culture and histology. Serological tests to diagnose histoplasmosis have been applied in just a few African countries (Benin, Chad Republic, Egypt, Republic of Congo, South Africa, Tanzania, and Uganda). In four cases, the serological test was carried out outside Africa [90–92].

Diagnosis of histoplasmosis in Africa is currently attainable using traditional mycological methods (culture or histopathology). Antigen detection, although very sensitive, is not available in most of Africa [91]. According to Cipriano and collaborators, antigen detection in serum and urine has only been developed for *H. capsulatum* [93].

Immunodiffusion tests using histoplasmin produced with *H. duboisii* and *H. capsulatum* strains were used to investigate the presence of antibodies in inhabitants around a natural focus of *H. duboisii* [94]. In a case report of suspected disseminated histoplasmosis by *H. duboisii* in a child from the Chad Republic, cultures could not be performed due to the lack of laboratory infrastructure, but immunodiffusion tests with soluble antigens of *H. capsulatum* and *H. duboisii* were performed and precipitins were observed against both antigens [90].

A recent literature review about African histoplasmosis in the Republic of Congo, with the majority of cases of histoplasmosis reported in Africa, demonstrated fifty-four cases of African histoplasmosis, and only one of them was diagnosed by *Histoplasma* antigen, which was tested in France, in an HIV-positive woman originating from the Republic of Congo [92].

5. Paracoccidioidomycosis

Paracoccidioidomycosis, a neglected tropical disease recognized by the WHO [95], has the fungi from the genus *Paracoccidioides* as etiologic agents, *Paracoccidioides brasiliensis* being the most common species, followed by *Paracoccidioides lutzii*. The other cryptic species *P. americana*, *P. restrepiensis*, and *P. venezuelensis* have also been reported. All species are widespread in Latin America, from Mexico to Argentina, being more frequent in Brazil [96].

Paracoccidioidomycosis diagnosis is typically achieved by the association of clinical, epidemiological, and laboratorial information [97]. Substantial progress has occurred in non-culture-based methods employed in the diagnosis of this systemic mycosis, with the development of a diversity of techniques for antibody, antigen, or nucleic acid detection. Immunologic techniques are typically simpler than mycological traditional methods, e.g., culture, and are extremely helpful in the diagnosis and follow-up of patients infected with *Paracoccidioides* spp. These methods underwent substantial advances in the past years as a result of the development of original detection methods and identification of pertinent *Paracoccidioides* antigens [96].

5.1. Antibody Detection

Over the years, several groups have evaluated serological techniques for the diagnosis of paracoccidioidomycosis, which provides a presumptive diagnosis and therapeutic follow-up. Despite the variety of assays proposed during years of study, the immunodiffusion test, described by Ouchterlony [98], is employed as the gold standard in the serological diagnosis of paracoccidioidomycosis. Currently, immunodiffusion, counter-immunoelectrophoresis, ELISA, and Western blot are the immunologic tests offered by different reference laboratories [96,97]. The tests employ similar and adequate antigens, presenting sensitivity values ranging from 80% to 95% [96]. The detection of seric anti-*Paracoccidioides* spp. antibodies involves a fungal antigenic preparation, which needs to have satisfactory reactivity in the chosen immunologic test format. This reactivity needs to involve patients infected with either *P. brasiliensis* or *P. lutzii*, in addition to being low in serum samples from patients with other mycoses [99]. The glycoprotein gp43 is the major antigen of *Paracoccidioides* spp., but its production is variable among different fungal species, especially in *P. lutzii*, whose gp43 is released in lower quantities and may present with an altered molecular organization [100].

Negative results in serological testing of patients with confirmed mycological paracoccidioidomycosis are described. Failure in the detection of anti-*Paracoccidioides* antibodies has been associated with issues either related to the methods or to the immunologic condition of the patient. In the last few years, this has been clarified by the perception that *P. lutzii* and *P. brasiliensis* have different antigenic profiles and, consequently, may drive different humoral immune responses in the host [101].

The literature about the immunologic diagnosis of paracoccidioidomycosis is wide and diverse. A variety of serological methods have proven suitable for proper diagnosis in adequate time [102]. For a detailed summary of works in this subject up to 2016, the review of Silva is recommended [103]. Here, we will prioritize the most used tests in the paracoccidioidomycosis presumptive diagnosis.

Double immunodiffusion has been the most used method for the primary diagnosis of individuals with paracoccidioidomycosis [97]. This test has high efficiency, which may range from 65 to 100%, depending on the antigenic preparation employed [104]. The test performance is not affected by HIV-driven immunosuppression [105]. For many years, different antigens have been used for paracoccidioidomycosis immunologic diagnosis, and various antigens lack a standardized preparation from one laboratory to another. The glycoprotein gp43 is considered the most important *P. brasiliensis* antigen. Patients affected by severe paracoccidioidomycosis present high amounts of anti-gp43 antibodies, therefore a strong and long-lasting humoral response to this antigen is usually seen in *P. brasiliensis* infected patients [96]. Actually, around 90% of paracoccidioidomycosis patients can be easily diagnosed by gp43-based immunodiffusion [106]. Some laboratories noticed that false negative results can occur in some cases [96,97,107,108]. Two possibilities could be associated with this situation: (i) immunosuppression of the patient, with insufficient precipitating antibodies in immunodiffusion; (ii) the presence of IgG asymmetric antibodies with a structure predominantly based on the mannose-rich oligosaccharide part connected to the Fc moiety of just one of the Fab arms of the antibody, which are functionally univalent and, therefore, non-precipitating [107]. Additionally, the lack of reactivity on serum samples from patients with *P. brasiliensis* may be associated with the production of low-avidity IgG2 antibodies that bind carbohydrate epitopes [109]. During efficacious treatment, the serum antibody levels detected by immunodiffusion decrease progressively until becoming negative [96], thereby constituting the most useful test for paracoccidioidomycosis cure control [97].

An additional method used in the early diagnosis of paracoccidioidomycosis is counterimmunoelectrophoresis. The time to obtain results of immunodiffusion and counterimmunoelectrophoresis is virtually the same [107]. In addition, counterimmunoelectrophoresis has a sensitivity equal to or slightly higher than immunodiffusion [110]. However, this test is more expensive and is not accessible in several laboratories from endemic areas [107]. Other precipitation techniques, such as immunoelectrophoresis and immunoelectrophoresis-immunodiffusion, are less specific than immunodiffusion, and are also more expensive [111,112]. Therefore, they are usually applied more in research studies than diagnostic tests [113].

Several latex agglutination tests have been described [103]. They have lower specificity and sensitivity, but are faster and are simple to be carried out. For instance, Santos and collaborators described a latex agglutination test with high sensitivity and specificity to detect the anti-gp43 antibody and gp43 antigen when employing latex particles linked to the purified gp43 and anti-gp43 monoclonal antibody [114].

ELISA has been extensively employed for the detection of antibodies to *Paracoccidioides* spp. in biological samples. Different research groups use a variety of antigens in ELISA tests because partially purified crude antigens and purified proteins, such as gp43, normally present high sensitivity but do not have high specificity with certainty [103]. Recombinant antigens, such as r*Pb*27 and r*Pb*40, can be used as well [115]. After some years, a standardization of a yeast filtrate as an antigen provided an increase of sensitivity and specificity [116]. Capture ELISA uses adsorbed monoclonal antibodies directed to gp43 on the plate, and represents progress in the detection of specific antibodies [107]. The standard ELISA is an excellent method for the detection of humoral immune responses in paracoccidioidomycosis patients for laboratories with medium infrastructure [103]. In addition, antibody responses evaluated by this method differ according to the clinical form of the disease. Patients with acute paracoccidioidomycosis present higher IgG titers, while patients with chronic paracoccidioidomycosis have higher IgA production [117].

Western blot has been used to recognize *Paracoccidioides* antigens that react with antibodies in serum samples. Mendes-Giannini and collaborators were pioneers in the development of this methodology for paracoccidioidomycosis diagnosis [118,119], being followed by several other groups [120–124]. The benefit of Western blot in relation to routine immunologic tests is a fast diagnosis of some patients, before complement fixation and immunodiffusion detect seroconversion, in addition to high efficiency [97].

Dot-ELISA has been recognized as a quick, versatile, and effortless test based on the principle of EIA, for the detection of many protozoan, virus, and fungal infections [125]. Dot-ELISA use in paracoccidioidomycosis diagnosis was hitherto presented by three research groups, without fundamental differences in efficiency [126–128]. In fact, Dot-ELISA was a particularly innovative tool as an immunologic screening technique, due to its high sensitivity (91%) and specificity (95%) [102]. Furthermore, Dot-ELISA could be performed by laboratories with little infrastructure or even in field work.

5.2. Antigen Detection

Although antigen detection would have crucial benefits over antibody detection in paracoccidioidomycosis diagnosis, especially in immunocompromised patients [129], a large scope for enhancement in antigen detection remains, as the last relevant contribution in this field was provided in 2011. Different assays of antigen detection were described, involving different ELISA formats, antigenic targets, and clinical samples, e.g., serum, urine, BAL, and CSF [130–132]. The majority of them, however, showed low sensitivity [107].

6. Cryptococcosis

Cryptococcosis is a mycosis of worldwide significance, involving both immunocompromised and immunocompetent patients. Traditionally, *Cryptococcus neoformans* and *Cryptococcus gattii* have been the major agents of cryptococcosis [133]. These two fungi share numerous similarities, but diverge in endemic areas, epidemiology, and clinical presentation [134]. *C. neoformans* occurs worldwide. *C. gattii* has been identified for several years, especially in tropical parts of Australia, Asia, Africa, and the Americas, but after the outbreak in Vancouver Island, *C. gattii* has gained prominence, with cases of *Cryptococcus gattii* in mammals from other areas of southwestern Canada and the northwestern United States [135].

The diagnosis of cryptococcosis, regardless of the causative species (*C. neoformans* or *C. gattii*), is classically performed using microscopy, immunologic methods, or microbiologic methods. India ink direct examination is a low-cost and fast method to detect *Cryptococcus* spp. in CSF and other body fluids. The stain fills the background field, but is not taken up by the *Cryptococcus* capsule, forming a bright light halo under regular microscopy. Though highly specific, the sensitivity of India ink microscopy (around 86%) is user-dependent and notably inferior in early infection, when the fungal burden is minor [136]. Therefore, its use is less frequent, particularly in the setting of broadly available rapid cryptococcal antigen tests.

The majority of the *Cryptococcus* capsule mass is constituted of glucoronoxylomannan and is generally known as cryptococcal antigen or CrAg [137]. Glucoronoxylomannan is produced by all *Cryptococcus* species. *Cryptococcus* spp. differ in the architecture on their capsular polysaccharides, and this may have consequences for development of diagnostic tests, possibly affecting glucoronoxylomannan detection.

There are many immunologic techniques for the diagnosis of cryptococcosis and innovative approaches have significantly diminished complexity and time of the test until results. Immunochemically based methods such as EIA, reverse passive latex agglutination, and immunochromatography are broadly used due to present advantages such as efficiency, easiness, and speed [138]. The main tools for the immunologic diagnosis of cryptococcosis are the tests aiming at CrAg in biological samples. The serum CrAg test is the most used noninvasive method to detect cryptococcal infection [139]. All these tests are able to diagnose disease caused by either *C. neoformans* or *C. gattii*.

6.1. Antibody Detection

Anti-*Cryptococcus* antibodies are usually not detectable during active cryptococcosis; therefore, antibody detection is not applied to the diagnosis of cryptococcosis. Tests based on antibody detection present highly variable results, depending on the type of assay. A study demonstrated that patients infected with *C. gattii* had a significantly higher prevalence of IgA and a non-significant higher prevalence of IgG compared to immunocompetent patients with *C. neoformans* infection [140].

6.2. Antigen Detection

An EIA, the PREMIER Cryptococcal Antigen Assay (Meridian Diagnostics, Inc.), was developed to detect cryptococcal capsular polysaccharide molecules in either serum or CSF specimens. This EIA does not need specimen pretreatment. The sensitivity with the serum samples is 100% and the specificity is 99%. Only the genus *Trichosporon*, which also produces glucoronoxylomannan-like molecules [141], caused a false positive reaction with this test [142].

Latex agglutination has been used for routine serological diagnosis, and some commercial kits are globally available to improve the diagnosis of cryptococcosis: CALAS Cryptococcal Antigen LA System (Meridian Bioscience Inc., Cincinnati, OH, USA), The Murex Cryptococcus Test (Remel), Crypto-LA test (Wampole Laboratories, Waltham, MA, USA) IMMY Latex-Crypto, and Pastorex Crypto Plus [143]. Usually, latex particles are linked to specific hyperimmune rabbit immunoglobulins, and are combined with different dilutions of clinical samples (CSF, sera, and urine) from cryptococcosis patients. A positive result at a 1:4 dilution clearly suggests *Cryptococcus* infection. Titers higher than 8 generally denote active disease. False positive results may be related to the presence of rheumatoid factor, which can be abolished after the biological sample treatment with pronase, dithiothreitol, or with boiling in EDTA. False positive results seldom occur when a glucoronoxylomannan-similar antigen is present in the clinical sample, for instance the polysaccharide of *Trichosporon* spp. The prozone-like effects due the excessive concentration of antigen or immune complexes can be abolished after clinical sample dilutions or treatment with pronase, respectively. The sensitivity of the latex agglutination is 94–100% and the specificity is 86–97% [144]. Performing the latex agglutination test requires a laboratory facility, and skilled laboratory workers, heat inactivation, and refrigeration of reagents [145].

A significant advance in testing for cryptococcal antigen was the development of an immunochromatographic assay, in a lateral flow assay format, the CrAg test (IMMY, Norman, OK, USA). This test identifies free capsular carbohydrates that have been released by *Cryptococcus* spp. into body fluids and addresses all pathogenic *Cryptococcus* species. The test is inexpensive and its sensitivity is equal to or higher than latex agglutination. The lateral flow assay detects the same antigen that is demonstrable by the commonly used latex agglutination and ELISA tests, and consequently, it has similar specificity. However, it is faster than the latex agglutination and ELISA, and produces clear results within 10 min. The test does not require electricity or advanced laboratory infrastructure, presents rapid turnaround time, and little technical knowledge is needed for development [146,147]. The CrAg lateral flow assay offers both qualitative and semi-quantitative results and can be used as a point-of-care assay for cryptococcosis diagnosis [148,149].

WHO published a guideline to diagnose and treat PLWHA infected by *Cryptococcus* spp. [150]. This guideline is centered on the fact that early diagnosis and treatment of cryptococcosis are essential to reduce mortality. The guideline emphasizes the value of a rapid CrAg test and pointed out that the CrAg test has high efficiency, is easy to perform, and is less dependent on laboratory expertise. This test would detect patients at high risk for cryptococcal disease and permit preemptive treatment with antifungals to avoid disease. This tactic is based on the fact that *Cryptococcus* antigenemia is noticeable around three weeks preceding the onset of clinical signs [151,152].

A number of other serologic procedures have been developed and are reported to improve cryptococcosis diagnosis. The alternative assays are based upon recombinant multi-epitope proteins, specific monoclonal antibodies, and the fungal heat shock protein 70 [153–155]. These tests have great potential to be inserted into the array of diagnostic tests.

7. Sporotrichosis

Sporotrichosis is a subcutaneous disease caused by human pathogenic fungi belonging to the genus *Sporothrix*, especially *Sporothrix schenckii*, *Sporothrix brasiliensis*, and *Sporothrix globosa*. The current major endemic areas of sporotrichosis include South America, especially Brazil; Asia, especially China and India; and Australia, but cases are also described in Europe and North America [4]. Since the 1970s, certain attempts were made to use immunologic tests as a tool for sporotrichosis diagnosis. For this mycosis, immunological tests are considered to be of inestimable diagnostic value, especially for extracutaneous and atypical forms of the disease [156].

7.1. Antibody Detection

Initially, immunoelectrophoresis, tube or latex agglutination, and immunodiffusion, using *Sporothrix* spp. antigens, were proposed for the serodiagnosis of sporotrichosis, but the efficacy of these methods was considered low, especially in cutaneous forms of the disease, which account for most cases of the mycosis [157–159]. Despite the low sensitivity, the immunodiffusion test is specific, without cross-reactions with cutaneous leishmaniasis or chromoblastomycosis, two infectious diseases with similar manifestations. Immunoelectrophoresis presents better sensitivity, with the presence of an anodic precipitation arc, called S arc, in reactive samples [157]. The sensitivity of agglutination tests was higher than that of precipitation tests, but only latex agglutination yielded satisfactory specificity results [158]. Due to this good specificity and sensitivity, latex agglutination is currently commercially available for sporotrichosis diagnosis (LA-*Sporothrix* antibody system—IMMY, Norman, OK, USA).

Because of the high endemicity of sporotrichosis in Brazil, noticed since the late 1990s, tests that are more sensitive for the cutaneous forms of the disease have been developed [160–162]. However, there is no consensus on the antigens used. ELISA has been performed with antigens obtained from the crude extract of yeast-like or filamentous forms of *Sporothrix*, besides purified antigens. The first described ELISA protocol for sporotrichosis diagnosis used a soluble antigen preparation from the *S. schenckii* yeast form. This antigen presented proteins ranging from 22 to 70 kDa and the whole performance of the assay showed 100% sensitivity and 90.5% specificity [163]. Later, the ELISA with a *S. schenckii* concanavalin-A binding antigenic fraction, isolated from the yeast cell wall, proved efficient in the detection of IgG antibodies, with a sensitivity of 90%, specificity of 80% and 86% global efficacy [160]. This ELISA can give results in a few hours and is very useful for therapeutic follow-up [164]. In addition, an ELISA with exoantigens produced by the filamentous form of *S. brasiliensis* was developed. The detection of IgG antibodies against these exoantigens resulted in 97% sensitivity and 89% specificity, with similar reactivity among patients with different clinical forms of the disease [161]. Moreover, IgA and IgM antibodies can also be evaluated using this method and a combination of IgG and IgA detection improves immunologic diagnosis, while the combination of IgG and IgM reactivities is suitable for therapeutic follow-up [165]. This ELISA protocol was also validated for diagnosis of sporotrichosis in cats, with better efficiency than the *S. schenckii* concanavalin-A binding antigenic fraction, as seen with human sera [166]. In addition, it was used to investigate an area supposedly without sporotrichosis endemicity, but with around 31% of positive samples from cats living in urban areas of the city [167]. Alvarado and collaborators developed an ELISA using a similar antigen produced by *S. schenckii*. Samples with immunologic reactivity were evaluated by immunodiffusion and counterimmunoelectrophoresis. They observed 100% of specificity and sensitivity superior to 98% with immunodiffusion, counterimmunoelectrophoresis, and ELISA [168]. Finally, an ELISA protocol using yeast cellular lysate proteins from *S. schenckii* was successfully used for a seroepidemiological survey in an endemic area in Brazil [169].

Western blot is less studied in the context of sporotrichosis diagnosis. The first protocol used the same soluble antigen from an *S. schenckii* strain described for the ELISA. All sera from patients with sporotrichosis presented reactivity against this antigen. Moreover, serum samples from individuals with extracutaneous manifestations of the disease reacted to more proteins than those from patients with cutaneous SPT: 15 to 20 and 8 to 10, respectively [163]. The other protocol uses a cell-free antigen preparation with the yeast-like form of *S. brasiliensis*, which presents up to 13 immunologic reactive bands, from 40 to 186 kDa. This Western blot showed 100% sensitivity, but just 50% specificity when an individual band was considered. Conversely, if only sera reactive to at least two distinct proteins are considered positive, sensitivity slightly decreases to 92.9% but specificity rises to 80% [170].

More recently, a lateral flow assay was developed to aid in the diagnosis of this mycosis. The test showed an accuracy of 82%, with sensitivity values dependent on sporotrichosis clinical forms. The sensitivity was greater for extracutaneous disease (92% sensitivity for ocular sporotrichosis) and lower for fixed-cutaneous sporotrichosis (78% sensitivity) [171].

An advance in the applicability of immunologic tests in sporotrichosis diagnosis is the possibility of analyzing different biological samples in addition to blood, such as CSF and synovial fluid. Furthermore, the serology is associated with an efficient clinical-serological correlation and cure control and can provide diagnosis even in immunocompromised patients [160,172–174].

7.2. Antigen Detection

The human pathogenic *Sporothrix* species produces an important antigen, rhamnomannan [164], that is not shared with other common mycoses agents, which usually produce galactomannan or glucuronoxylomannan. This would make antigen detection an interesting tool for sporotrichosis diagnosis. However, to the best of our knowledge, there are no immunological tests based on antigen detection for sporotrichosis diagnosis.

8. Talaromycosis

Talaromyces marneffei (formerly *Penicillium marneffei*) is a thermodimorphic fungal pathogen endemic in several countries of Southeast Asia, where it is a major threat to PLWHA. It was also reported in individuals traveling from several countries to an endemic area. There is a single report in Ghana of a patient who certainly did not travel to Southeast Asia [3]. Talaromycosis, the disease caused by *T. marneffei*, is a life-threatening infection with unspecific symptoms, which makes diagnosis difficult, especially in cases of imported disease [175].

The talaromycosis gold standard method is the culture of its agent from bone marrow, blood, sputum, and skin samples. The fungal filamentous form will develop at 25 to 30 °C, consisting of a white mycelium that turns green after sporulation with red diffusible pigment production. At 37 °C, yeast cerebriform or smooth colonies formed by cells that divide by binary fission will grow. Since rapid diagnosis is necessary, immunologic tests centered on antibody and antigen detection were developed to aid in this task [176].

8.1. Antibody Detection

The first immunologic test used for talaromycosis diagnosis was an immunodiffusion test using a concentrated filamentous secretome of the fungus grown for six weeks at 25 °C. Two or three precipitin bands are observed in this test against a rabbit hyperimmune serum [177]. However, this test presented low sensitivity and a single precipitin in positive samples when used for serum antibody detection in culture-proven talaromycosis patients, probably due to their immunosuppression [178].

To overcome this problem, an ELISA was developed using a recombinant 90 kDa mannoprotein of the fungus, Mp1p, expressed in *Escherichia coli*. This test was developed to be used in both immunocompetent and immunosuppressed patients, reaching an overall 82% sensitivity and 100% specificity, with around 80% positivity among PLWHA [179]. Later, the test with this mannoprotein was remodeled, now using *Pichia pastoris* to express the recombinant protein and a double-antigen sandwich ELISA format to detect anti-Mp1p antibodies. Again, 100% specificity was observed, but the sensitivity lowered to 13.3% [180].

The major limitation in employing antibody detection assays for talaromycosis diagnosis is their low sensitivity. In fact, these methods are highly specific, but culture, although time-consuming, has higher sensitivity [181].

8.2. Antigen Detection

As occurs with other endemic mycoses, antigen detection is particularly useful to diagnose talaromycosis in PLWHA that fail to produce specific *T. marneffei* antibodies. Special attention must be given to possible serologic cross-reactions that may occur when patients with talaromycosis are tested for *Histoplasma*, *Blastomyces*, or *Aspergillus* antigens [129,182].

Initially, antigen detection was performed using immunodiffusion, with better sensitivity values than antibody detection. In fact, a study with eight patients with proven talaromycosis revealed seven positive patients for antigen and two for antibodies (one patient was positive for both) using this method [178]. Afterwards, several methods for antigen detection were described, most of them in an ELISA format using polyclonal or monoclonal antibodies reactive to the Mp1p antigenemia or antigenuria. Their pooled sensitivity and specificity values, which have enrolled 320 *T. marneffei* infected patients and 1873 control individuals, are 82% and 99%, respectively [183]. The ELISA for Mp1p antigen detection is faster and more sensitive than traditional culture-based methods of diagnosis [184].

A dot-blot ELISA with a polyclonal anti-*T. marneffei* antibody coupled to FITC and an anti-FITC amplification system to detect antigen in urine samples presented 94.6% sensitivity and 97.3% specificity; however, these values were lower than those observed with the traditional ELISA format. The same reagents were used in a latex agglutination format and, with this method, 100% sensitivity was reached, with 99.3% specificity [185].

Lately, a lateral flow assay was created for talaromyces point-of-care diagnostics. This assay uses a monoclonal antibody reactive to a 50–180 kDa mannoprotein with a broad high molecular mass pattern conjugated with nanoparticles of colloidal gold for specific *T. marneffei* antigenuria detection. The detection limit is 3.12 μg/mL for *T. marneffei* antigen and sensitivity and specificity were 87.87% and 100%, respectively [176].

Some authors also report the combined detection of specific antigen and antibodies in patients with talaromyces. For instance, a study that used two ELISA formats to detect Mp1p, one with monoclonal antibody and the other with polyclonal antibody, and an ELISA to detect IgG anti-Mp1p presented 55%, 75%, and 30% sensitivity, respectively. However, the combined results of these tests yielded 100% sensitivity and 98% specificity [186]. Another study found 93.3% sensitivity when combining antigen and antibody detection results using a *P. pastoris* recombinant Mp1p in a sandwich ELISA [180].

Despite the high sensitivity and excellent specificity of the tests to detect *T. marneffei* antigens, no validation studies to endorse their use in the clinical setting exist. Moreover, there is a paucity of commercially available diagnostic kits for the serodiagnosis of talaromycosis [129].

9. Endemic Mycoses without Specific Immunologic Tests

9.1. *Lacaziosis*

The diagnosis of lacaziosis, a deep fungal infection caused by *Lacazia loboi*, is confirmed by clinical and histopathological methods. One study demonstrated that individuals with lacaziosis possess antibodies reactive to the gp43 antigen of *P. brasiliensis*, and also to a 193 kDa major *L. loboi* antigen through WB. The cross-reactivity occurs because, as supported by molecular studies, *L. loboi* and *P. brasiliensis* share a similar ancestor [187]. In contrast to prior reports, this study proposes that, during infection, *L. loboi* presents antigens that are distinct from that presented during paracoccidioidomycosis [188,189]. The molecular report of the 193 kDa molecule could generate precious data to comprehend the immunology of lacaziosis and its diagnostic applications, probably aiding in the management of infections caused by this resilient fungus.

9.2. *Adiaspiromycosis*

Adiaspiromycosis is a pulmonary infection of rodents, fossorial mammals, and their predators, infrequently occurring in humans, and is caused by the *Emmonsia crescens* and *Emmonsia parva*. During adiaspiromycosis, inhaled conidia enlarge to form non-replicating adiaspores. The infection usually involves the lungs, with rare cases of infection at other sites. Diagnosis is usually made by histopathology and the etiologic agent is identified based on the size of adiaspores [190]. There are a few immunologic tests to aid in adiaspiromycosis diagnosis; they are all designed to diagnose this mycosis in wild animals.

Immunodiffusion and complement fixation showed good correlation with histopathology, with valuable sensitivity and specificity [191].

9.3. Emergomycosis

In the last few years, the emergence of emergomycosis, a rare, cosmopolitan fungal infection caused by the unusual dimorphic fungus *Emergomyces* spp. has been noticed among immunocompromised patients [192]. This mycosis also affects wild mammals. The main agents are *Emergomyces pasteuriana* (formerly *Emmosia pasteuriana*), *Emergomyces africanus*, *Emergomyces orientalis*, and *Emergomyces canadiensis* [193]. Multifocal pneumonia and cutaneous forms including papule-crusted injuries, nodules, wart-like lesions, or ulcerated plaques on the face, trunk, and extremities are the most common symptoms.

Diagnosis of emergomycosis remains challenging. Differential diagnosis includes HPM, BLM, tuberculosis, *Listeria* sp., and TLM. Among the endemic mycoses, emergomycosis should be considered in the histoplasmosis differential diagnosis since there is substantial clinical and histopathological findings overlapping between the two diseases [194]. The gold standard laboratory diagnosis is culture, with the growth, around 20–30 days, of white to beige, fastidious, and mycelial fungal colonies. Microscopic examination presents hyaline and thin hyphae with microconidia. However, potassium hydroxide slides of sputum or secretions are also helpful, where several small yeast cells around 1–3 µm in size are typically noticed. Histopathology displays an inflammatory, granulomatous process, with intra- and extra-cellular, 2–5 µm, round or ovoid yeast-like cells, alike in *H. capsulatum*, but *Emergomyces* have smaller and halo-less yeast cells [195].

There is no immunologic test or biomarkers with sufficient efficiency for emergomycosis diagnosis. Nevertheless, cross-reactivity has been seen with other dimorphic fungi. A retrospective case series was reported with two emergomycosis patients with a positive *Histoplasma* antigenuria, and one with positive 1,3-ß-D-glucan antigen detection [196]. Other reports also revealed cross-reactivity of the *Histoplasma* galactomannan in urine samples of patients infected with *E. africanus*. A commercial *Histoplasma* EIA had suitable accuracy to diagnose proven histoplasmosis, but cross-reactions were seen in urine samples from individuals with invasive infections due to *E. africanus* and in culture filtrates of this species and other related fungal pathogens [197]. *Emergomyces* spp. may present cross-reactivity with *Histoplasma* antigenuria assays, but a negative result cannot reject diagnosis [198]. Clinical research main concerns must incorporate the validation of existing and new diagnostic tests to improve comprehension of emergomycosis epidemiology, to aid in diagnosis, and to feasibly identify individuals who may benefit from preemptive therapeutics.

10. Conclusions

Endemic mycoses result from infection mainly due to dimorphic fungi, and continue to cause substantial morbidity and mortality, especially in selected regions. The diagnosis of endemic mycoses is typically achieved by an association of clinical, epidemiological, and laboratory information. Substantial progress has been made in non-culture-based methods to diagnose these mycoses with the development of a variety of techniques for the detection of antibodies (Table 2), antigens (Table 3), and nucleic acids. The serological methods described for the diagnosis of endemic mycoses have their strengths and weaknesses and demand critical evaluation by mycologists and medical doctors. Nevertheless, not all tests herein described are entirely available across the world, which complicates the competence to diagnose and treat patients with endemic mycoses. Moreover, the immunological status of the patient and manifestation of these diseases influence the efficacy of the diagnostic test. Continuing efforts to improve or develop diagnostic tests will facilitate our diagnostic capacity. However, such assays will require validation in populations from diverse regions of the world prior to their general application in routine diagnosis. Results obtained from a panel of serologic diagnostic tests play an important role in the diagnosis of endemic mycoses, allowing more rapid and precise diagnosis, which would lead to earlier treatment. However, the gold standard for diagnosis continues to be the culture, and the correlation between molecular data and phenotypic characteristics is crucial in identifying the etiological agents of endemic mycoses.

Table 2. Summary of sensitivity and specificity values of immunological tests used for diagnosis of endemic mycoses by antibody detection.

Disease	Test	Sensitivity	Specificity	References
Histoplasmosis	ID	75–95	100	[58]
	CF	72–95	70–80	[58]
	EIA	66–97	54–100	[65,67–70]
	Western blot	95	94	[72,73]
Paracoccidioidomycosis	ID	17–100	43.3–100	[99,103,104,110,122]
	EIA	75–100	100	[103,115,116]
	Western blot	77.3–100	73.3–100	[114,118–120,123]
	CIE	95–100	100	[99,103,110]
Blastomycosis	ID	28	100	[6]
	CF	9	100	[6]
	EIA	77	92	[6]
Coccidioidomycosis	ID	50–90		[6,44]
	CF	67–75		[6]
	EIA	54–92	97	[6,44]
Sporotrichosis	ID	98	100	[168]
	CIE	98	100	[168]
	EIA	90–100	80–100	[160,161,163,168]
	Western blot	93–100	50–80	[163,170]
Talaromycosis	ID	25	NE	[177,178]
	EIA	13–82	81–100	[179,180]

ID: immunodiffusion; CF: complement fixation; EIA: enzyme immunoassay; CIE: counterimmunoelectrophoresis.

Table 3. Summary of sensitivity and specificity values of immunological tests used for diagnosis of endemic mycoses by antigen detection.

Disease	Test	Target	Specimen	Sensitivity	Specificity	References
Histoplasmosis	RIA	100 kDa (HPA)	Urine	96.7	100	[76]
			Serum	78.7	100	
	EIA	69–70 kDa	Serum	71.4	85.4	[80]
	EIA	Galactomannan	Urine	61.9–100	32–99.8	[82–87]
			Serum	92.3	99	[83]
			BAL	93.5	97.8	[84]
	EIA	Cell wall antigen	Serum	81	95	[79]
	EIA	100 kDa (HPA)	Urine	86	94	[81]
	LFA	Galactomannan	Urine	96	96	[88,89]
			Serum	92	94	
Blastomycosis	EIA	Galactomannan	Urine	76–90	NE	[24]
			Serum	52–82	NE	
Coccidioidomycosis	EIA	Chitinase-1	Serum	87	97	[50]
Paracoccidioidomycosis	EIA	gP43 glycoprotein	Serum	95.1	97.5	[131]
	EIA	*P. brasiliensis* total and filtrate antigen	Urine	75	100	[132]
	EIA	87 kDa	Serum	80.4	81.4	[130]

Table 3. *Cont.*

Disease	Test	Target	Specimen	Sensitivity	Specificity	References
Talaromycosis	ID	*T. marneffei* yeast secretome	Serum	58.8	100	[176]
		T. marneffei filamentous secretome	Serum	87.5	NE	[176,178]
	LA	*T. marneffei* yeast secretome	Serum	76.5	100	[176]
		Whole-fission-form yeast of *T. marneffei*	Urine	100	99.3	[176,185]
	Dot-blot	Whole-fission-form yeast of *T. marneffei*	Urine	94.6	97.3	[176,185]
	EIA	Whole-fission-form yeast of *T. marneffei*	Urine	97.3	98	[176,185]
	EIA	Yeast and mycelial antigens	Serum	72–92.5	97.5–100	[176]
	EIA	Mp1p	Serum	55–75	99.4–99.6	[176,179,186]
	EIA	TM cytoplasmic yeast antigen	Serum	100	100	[176]
	LFA	TM cytoplasmic yeast antigen	Urine	87.9	100	[176]
Cryptococcosis	LA	Cryptococcal capsular antigen	Serum, CSF	94–100	86–97	[144]
	EIA	Cryptococcal capsular antigen	Serum, CSF	99	97	[142]
	LFA	Cryptococcal capsular antigen	Serum, CSF	100	97–99	[149]

RIA: radioimmunoassay; EIA: enzyme immunoassay; LFA: lateral flow assay; LA: latex agglutination; ID: immunodiffusion; CSF: cerebrospinal fluid.

Author Contributions: Conceptualization, R.M.Z.-O.; methodology, R.A.-P., A.R.B.-E., B.d.S.M., C.V.P., M.d.A.A., M.d.M.M. and R.A.B.D.; data curation, M.d.A.A.; writing—original draft preparation, R.A.-P., A.R.B.-E., B.d.S.M., C.V.P., M.d.A.A., M.d.M.M. and R.A.B.D.; writing—review and editing, R.A.-P. and R.M.Z.-O.; supervision, R.M.Z.-O.; project administration, R.M.Z.-O.; funding acquisition, R.M.Z.-O. All authors have read and agreed to the published version of the manuscript.

Funding: This work was supported in part by Conselho Nacional de Desenvolvimento Científico e Tecnológico [CNPq 302796/2017-7] and Fundação Carlos Chagas Filho de Amparo à Pesquisa do Estado do Rio de Janeiro [FAPERJ E-26/202.527/2019]. The APC was funded by Programa Jovens Pesquisadores (INI/Fiocruz).

Institutional Review Board Statement: Not applicable.

Informed Consent Statement: Not applicable.

Data Availability Statement: Not applicable.

Conflicts of Interest: The authors declare no conflict of interest. The funders had no role in the design of the study; in the collection, analyses, or interpretation of data; in the writing of the manuscript; or in the decision to publish the results.

References

1. Lockhart, S.R.; Toda, M.; Benedict, K.; Caceres, D.H.; Litvintseva, A.P. Endemic and Other Dimorphic Mycoses in The Americas. *J. Fungi* **2021**, *7*, 151. [CrossRef]
2. Ashraf, N.; Kubat, R.C.; Poplin, V.; Adenis, A.A.; Denning, D.W.; Wright, L.; McCotter, O.; Schwartz, I.S.; Jackson, B.R.; Chiller, T.; et al. Re-Drawing the Maps for Endemic Mycoses. *Mycopathologia* **2020**, *185*, 843–865. [CrossRef] [PubMed]
3. Tirado-Sánchez, A.; González, G.M.; Bonifaz, A. Endemic Mycoses: Epidemiology and Diagnostic Strategies. *Expert Rev. Anti Infect. Ther.* **2020**, *18*, 1105–1117. [CrossRef] [PubMed]
4. Thompson, G.R.; Le, T.; Chindamporn, A.; Kauffman, C.A.; Alastruey-Izquierdo, A.; Ampel, N.M.; Andes, D.R.; Armstrong-James, D.; Ayanlowo, O.; Baddley, J.W.; et al. Global Guideline for the Diagnosis and Management of the Endemic Mycoses: An Initiative of the European Confederation of Medical Mycology in Cooperation with the International Society for Human and Animal Mycology. *Lancet Infect. Dis.* **2021**, *21*, e364–e374. [CrossRef]

5. Thompson, G.R.; Pasqualotto, A.C. Endemic Mycoses: Expansion of Traditional Geographic Ranges and Pitfalls in Management. *Mycoses* **2021**, *64*, 989–992. [CrossRef] [PubMed]
6. Wheat, L.J. Approach to the Diagnosis of the Endemic Mycoses. *Clin. Chest Med.* **2009**, *30*, 379–389. [CrossRef]
7. Gifford, J.; Catanzaro, A. A Comparison of Coccidioidin and Spherulin Skin Testing in the Diagnosis of Coccidioidomycosis. *Am. Rev. Respir. Dis.* **1981**, *124*, 440–444. [CrossRef]
8. Dominguez-Soto, L.; Hojyo-Tomoka, M.T. The Intradermal Sporotrichin Test and the Diagnosis of Sporotrichosis. *Int. J. Dermatol.* **1983**, *22*, 520. [CrossRef]
9. Torres-Rodríguez, J.M.; Ribas-Forcadell, E.; Gascón, J.; López-Jodra, O.; Espasa, M. Utilidad diagnóstica de la prueba intradérmica con histoplasmina, en áreas no endémicas de histoplasmosis. *Rev. Iberoam. Micol.* **2000**, *17*, 97–101.
10. Schwartz, I.S.; Kauffman, C.A. Blastomycosis. *Semin. Respir. Crit. Care Med.* **2020**, *41*, 31–41. [CrossRef]
11. Kaufman, L.; McLaughlin, D.W.; Clark, M.J.; Blumer, S. Specific Immunodiffusion Test for Blastomycosis. *Appl. Microbiol.* **1973**, *26*, 244–247. [CrossRef] [PubMed]
12. Green, J.H.; Harrell, W.K.; Johnson, J.E.; Benson, R. Preparation of Reference Antisera for Laboratory Diagnosis of Blastomycosis. *J. Clin. Microbiol.* **1979**, *10*, 1–7. [CrossRef] [PubMed]
13. Williams, J.E.; Murphy, R.; Standard, P.G.; Phair, J.P. Serologic Response in Blastomycosis: Diagnostic Value of Double Immunodiffusion Assay. *Am. Rev. Respir. Dis.* **1981**, *123*, 209–212. [CrossRef]
14. Klein, B.S.; Kuritsky, J.N.; Chappell, W.A.; Kaufman, L.; Green, J.; Davies, S.F.; Williams, J.E.; Sarosi, G.A. Comparison of the Enzyme Immunoassay, Immunodiffusion, and Complement Fixation Tests in Detecting Antibody in Human Serum to the A Antigen of *Blastomyces dermatitidis*. *Am. Rev. Respir. Dis.* **1986**, *133*, 144–148. [CrossRef]
15. Klein, B.S.; Vergeront, J.M.; Kaufman, L.; Bradsher, R.W.; Kumar, U.N.; Mathai, G.; Varkey, B.; Davis, J.P. Serological Tests for Blastomycosis: Assessments during a Large Point-Source Outbreak in Wisconsin. *J. Infect. Dis.* **1987**, *155*, 262–268. [CrossRef]
16. Turner, S.; Kaufman, L.; Jalbert, M. Diagnostic Assessment of an Enzyme-Linked Immunosorbent Assay for Human and Canine Blastomycosis. *J. Clin. Microbiol.* **1986**, *23*, 294–297. [CrossRef] [PubMed]
17. Linder, K.A.; Kauffman, C.A. Current and New Perspectives in the Diagnosis of Blastomycosis and Histoplasmosis. *J. Fungi* **2020**, *7*, 12. [CrossRef] [PubMed]
18. Klein, B.S.; Bradsher, R.W.; Vergeront, J.M.; Davis, J.P. Development of Long-Term Specific Cellular Immunity after Acute *Blastomyces dermatitidis* Infection: Assessments Following a Large Point-Source Outbreak in Wisconsin. *J. Infect. Dis.* **1990**, *161*, 97–101. [CrossRef]
19. Richer, S.M.; Smedema, M.L.; Durkin, M.M.; Brandhorst, T.T.; Hage, C.A.; Connolly, P.A.; Leland, D.S.; Davis, T.E.; Klein, B.S.; Wheat, L.J. Development of a Highly Sensitive and Specific Blastomycosis Antibody Enzyme Immunoassay Using *Blastomyces dermatitidis* Surface Protein BAD-1. *Clin. Vaccine Immunol. CVI* **2014**, *21*, 143–146. [CrossRef]
20. Klein, B.S.; Jones, J.M. Isolation, Purification, and Radiolabeling of a Novel 120-KD Surface Protein on *Blastomyces dermatitidis* Yeasts to Detect Antibody in Infected Patients. *J. Clin. Investig.* **1990**, *85*, 152–161. [CrossRef]
21. Klein, B.S.; Hogan, L.H.; Jones, J.M. Immunologic Recognition of a 25-Amino Acid Repeat Arrayed in Tandem on a Major Antigen of *Blastomyces dermatitidis*. *J. Clin. Investig.* **1993**, *92*, 330–337. [CrossRef] [PubMed]
22. Klein, B.S.; Jones, J.M. Purification and Characterization of the Major Antigen WI-1 from *Blastomyces dermatitidis* Yeasts and Immunological Comparison with A Antigen. *Infect. Immun.* **1994**, *62*, 3890–3900. [CrossRef]
23. Soufleris, A.J.; Klein, B.S.; Courtney, B.T.; Proctor, M.E.; Jones, J.M. Utility of Anti-WI-1 Serological Testing in the Diagnosis of Blastomycosis in Wisconsin Residents. *Clin. Infect. Dis. Off. Publ. Infect. Dis. Soc. Am.* **1994**, *19*, 87–92. [CrossRef] [PubMed]
24. Durkin, M.; Witt, J.; Lemonte, A.; Wheat, B.; Connolly, P. Antigen Assay with the Potential to Aid in Diagnosis of Blastomycosis. *J. Clin. Microbiol.* **2004**, *42*, 4873–4875. [CrossRef] [PubMed]
25. Connolly, P.; Hage, C.A.; Bariola, J.R.; Bensadoun, E.; Rodgers, M.; Bradsher, R.W.; Wheat, L.J. *Blastomyces dermatitidis* Antigen Detection by Quantitative Enzyme Immunoassay. *Clin. Vaccine Immunol. CVI* **2012**, *19*, 53–56. [CrossRef] [PubMed]
26. Bariola, J.R.; Hage, C.A.; Durkin, M.; Bensadoun, E.; Gubbins, P.O.; Wheat, L.J.; Bradsher, R.W. Detection of *Blastomyces dermatitidis* Antigen in Patients with Newly Diagnosed Blastomycosis. *Diagn. Microbiol. Infect. Dis.* **2011**, *69*, 187–191. [CrossRef] [PubMed]
27. Bariola, J.R.; Perry, P.; Pappas, P.G.; Proia, L.; Shealey, W.; Wright, P.W.; Sizemore, J.M.; Robinson, M.; Bradsher, R.W. Blastomycosis of the Central Nervous System: A Multicenter Review of Diagnosis and Treatment in the Modern Era. *Clin. Infect. Dis. Off. Publ. Infect. Dis. Soc. Am.* **2010**, *50*, 797–804. [CrossRef]
28. Frost, H.M.; Novicki, T.J. Blastomyces Antigen Detection for Diagnosis and Management of Blastomycosis. *J. Clin. Microbiol.* **2015**, *53*, 3660–3662. [CrossRef]
29. Baumgardner, D.J. Use of Urine Antigen Testing for *Blastomyces* in an Integrated Health System. *J. Patient-Cent. Res. Rev.* **2018**, *5*, 176–182. [CrossRef]
30. Van Der Veer, J.; Lewis, R.J.; Emtiazjoo, A.M.; Allen, S.D.; Wheat, L.J.; Hage, C.A. Cross-Reactivity in the Platelia™ *Aspergillus* Enzyme Immunoassay Caused by Blastomycosis. *Med. Mycol.* **2012**, *50*, 396–398. [CrossRef]
31. Mongkolrattanothai, K.; Peev, M.; Wheat, L.J.; Marcinak, J. Urine Antigen Detection of Blastomycosis in Pediatric Patients. *Pediatr. Infect. Dis. J.* **2006**, *25*, 1076–1078. [CrossRef]
32. Tarr, M.; Marcinak, J.; Mongkolrattanothai, K.; Burns, J.L.; Wheat, L.J.; Durkin, M.; Ismail, M. Blastomyces Antigen Detection for Monitoring Progression of Blastomycosis in a Pregnant Adolescent. *Infect. Dis. Obstet. Gynecol.* **2007**, *2007*, 89059. [CrossRef] [PubMed]

33. Laux, K.L.; Anderson, J.L.; Bentivenga, S.P.; McBride, J.A.; Sterkel, A.; Matkovic, E.; Gauthier, G.M.; Meece, J.K. Urine Antigen Testing Is Equally Sensitive to *B. dermatitidis* and *B. gilchristii* Infections. *Clin. Med. Res.* **2020**, *18*, 133–139. [CrossRef] [PubMed]
34. Crum, N.F. Coccidioidomycosis: A Contemporary Review. *Infect. Dis. Ther.* **2022**, *11*, 713–742. [CrossRef]
35. Kollath, D.R.; Miller, K.J.; Barker, B.M. The Mysterious Desert Dwellers: *Coccidioides immitis* and *Coccidioides posadasii*, Causative Fungal Agents of Coccidioidomycosis. *Virulence* **2019**, *10*, 222–233. [CrossRef] [PubMed]
36. Kirkland, T.N.; Fierer, J. *Coccidioides immitis* and *Posadasii*; A Review of Their Biology, Genomics, Pathogenesis, and Host Immunity. *Virulence* **2018**, *9*, 1426–1435. [CrossRef]
37. Saubolle, M.A. Laboratory Aspects in the Diagnosis of Coccidioidomycosis. *Ann. N. Y. Acad. Sci.* **2007**, *1111*, 301–314. [CrossRef]
38. Smith, C.E. Diagnosis of Pulmonary Coccidioidal Infections. *Calif. Med.* **1951**, *75*, 385–394.
39. Smith, C.E.; Saito, M.T.; Beard, R.R.; Kepp, R.M.; Clark, R.W.; Eddie, B.U. Serological Tests in the Diagnosis and Prognosis of Coccidioidomycosis. *Am. J. Hyg.* **1950**, *52*, 1–21. [CrossRef]
40. Kaufman, L.; Clark, M.J. Value of the Concomitant Use of Complement Fixation and Immunodiffusion Tests in the Diagnosis of Coccidioidomycosis. *Appl. Microbiol.* **1974**, *28*, 641–643. [CrossRef]
41. Huppert, M.; Bailey, J.W. The Use of Immunodiffusion Tests in Coccidioidomycosis. I. The Accuracy and Reproducibility of the Immunodiffusion Test Which Correlates with Complement Fixation. *Tech. Bull. Regist. Med. Technol. Am. Soc. Clin. Pathol. Regist. Med. Technol.* **1965**, *35*, 150–154.
42. Reiss, E.; Kaufman, L.; Kovacs, J.; Lindslay, M. Clinical Immunomycology. In *Manual of Clinical Laboratory Immunology*; Rose, N., Hamilton, R., Detrick, B., Eds.; ASM Press: Washington, DC, USA, 2002; pp. 559–583.
43. Pishko, E.J.; Kirkland, T.N.; Cole, G.T. Isolation and Characterization of Two Chitinase-Encoding Genes (Cts1, Cts2) from the Fungus *Coccidioides immitis*. *Gene* **1995**, *167*, 173–177. [CrossRef]
44. de Aguiar Cordeiro, R.; Patoilo, K.R.N.; Praciano, S.B.; Medrano, D.J.A.; de Farias Marques, F.J.; Martins, L.M.S.; Eulalio, K.D.; de Deus Filho, A.; Cavalvanti, M.D.A.S.; Fechine, M.A.B.; et al. Antigens of *Coccidioides posadasii* as an Important Tool for the Immunodiagnosis of Coccidioidomycosis. *Mycopathologia* **2013**, *175*, 25–32. [CrossRef] [PubMed]
45. Wieden, M.A.; Lundergan, L.L.; Blum, J.; Delgado, K.L.; Coolbaugh, R.; Howard, R.; Peng, T.; Pugh, E.; Reis, N.; Theis, J.; et al. Detection of Coccidioidal Antibodies by 33-KDa Spherule Antigen, *Coccidioides* EIA, and Standard Serologic Tests in Sera from Patients Evaluated for Coccidioidomycosis. *J. Infect. Dis.* **1996**, *173*, 1273–1277. [CrossRef]
46. Ardizzoni, A.; Baschieri, M.C.; Manca, L.; Orsi, C.F.; Venturelli, C.; Meacci, M.; Peppoloni, S.; Farina, C.; Blasi, E. The Protein "Mycoarray": A Novel Serological Assay for the Laboratory Diagnosis of Primitive Endemic Mycoses. *New Microbiol.* **2011**, *34*, 307–316.
47. Filho, R.E.M.; Bandeira, S.P.; Brillhante, R.S.N.; Rocha, M.F.G.; Vasconcelos, I.M.; Pereira, M.L.; de Souza Collares Maia Castelo-Branco, D.; Rocha, F.A.C.; de Camargo, Z.P.; Ramos, M.V.; et al. Biochemical Characterization of an In-House *Coccidioides* Antigen: Perspectives for the Immunodiagnosis of Coccidioidomycosis. *Molecules* **2012**, *17*, 7854–7863. [CrossRef]
48. Duarte Escalante, E.; De Frías León, M.G.; Martínez García, L.G.; Herrera, J.; Acosta Altamirano, G.; Cabello, C.; Palma, G.; Reyes Montes, M.D.R. Selection of Specific Peptides for *Coccidioides* spp. Obtained from Antigenic Fractions through SDS-PAGE and Western Blot Methods by the Recognition of Sera from Patients with Coccidioidomycosis. *Molecules* **2018**, *23*, 3145. [CrossRef]
49. Kassis, C.; Durkin, M.; Holbrook, E.; Myers, R.; Wheat, L. Advances in Diagnosis of Progressive Pulmonary and Disseminated Coccidioidomycosis. *Clin. Infect. Dis. Off. Publ. Infect. Dis. Soc. Am.* **2021**, *72*, 968–975. [CrossRef]
50. Grill, F.J.; Grys, T.E.; Grill, M.F.; Roeder, A.; Blair, J.E.; Lake, D.F. Development of a Quantitative Antigen Assay to Detect Coccidioidal Chitinase-1 (CTS1) in Human Serum. *Open Forum Infect. Dis.* **2021**, *8*, ofab344. [CrossRef]
51. Couppié, P.; Aznar, C.; Carme, B.; Nacher, M. American Histoplasmosis in Developing Countries with a Special Focus on Patients with HIV: Diagnosis, Treatment, and Prognosis. *Curr. Opin. Infect. Dis.* **2006**, *19*, 443–449. [CrossRef]
52. Nacher, M.; Couppié, P.; Epelboin, L.; Djossou, F.; Demar, M.; Adenis, A. Disseminated Histoplasmosis: Fighting a Neglected Killer of Patients with Advanced HIV Disease in Latin America. *PLoS Pathog.* **2020**, *16*, e1008449. [CrossRef] [PubMed]
53. Valero, C.; Gago, S.; Monteiro, M.C.; Alastruey-Izquierdo, A.; Buitrago, M.J. African Histoplasmosis: New Clinical and Microbiological Insights. *Med. Mycol.* **2018**, *56*, 51–59. [CrossRef]
54. Leimann, B.C.Q.; Pizzini, C.V.; Muniz, M.M.; Albuquerque, P.C.; Monteiro, P.C.F.; Reis, R.S.; Almeida-Paes, R.; Lazera, M.S.; Wanke, B.; Pérez, M.A.; et al. Histoplasmosis in a Brazilian Center: Clinical Forms and Laboratory Tests. *Rev. Iberoam. Micol.* **2005**, *22*, 141–146. [CrossRef]
55. Azar, M.M.; Hage, C.A. Laboratory Diagnostics for Histoplasmosis. *J. Clin. Microbiol.* **2017**, *55*, 1612–1620. [CrossRef] [PubMed]
56. Azar, M.M.; Hage, C.A. Clinical Perspectives in the Diagnosis and Management of Histoplasmosis. *Clin. Chest Med.* **2017**, *38*, 403–415. [CrossRef] [PubMed]
57. Guimarães, A.J.; Nosanchuk, J.D.; Zancopé-Oliveira, R.M. Diagnosis of Histoplasmosis. *Braz. J. Microbiol. Publ. Braz. Soc. Microbiol.* **2006**, *37*, 1–13. [CrossRef]
58. Scheel, C.M.; Gómez, B.L. Diagnostic Methods for Histoplasmosis: Focus on Endemic Countries with Variable Infrastructure Levels. *Curr. Trop. Med. Rep.* **2014**, *1*, 129–137. [CrossRef]
59. Kauffman, C.A. Histoplasmosis: A Clinical and Laboratory Update. *Clin. Microbiol. Rev.* **2007**, *20*, 115–132. [CrossRef]
60. Wheat, J.; French, M.L.; Kamel, S.; Tewari, R.P. Evaluation of Cross-Reactions in *Histoplasma capsulatum* Serologic Tests. *J. Clin. Microbiol.* **1986**, *23*, 493–499. [CrossRef]

61. Wheat, J.; French, M.L.; Kohler, R.B.; Zimmerman, S.E.; Smith, W.R.; Norton, J.A.; Eitzen, H.E.; Smith, C.D.; Slama, T.G. The Diagnostic Laboratory Tests for Histoplasmosis: Analysis of Experience in a Large Urban Outbreak. *Ann. Intern. Med.* **1982**, *97*, 680–685. [CrossRef]
62. Gerber, J.D.; Riley, R.E.; Jones, R. Evaluation of a Microtiter Latex Agglutination Test for Histoplasmosis. *Appl. Microbiol.* **1972**, *24*, 191–197. [CrossRef] [PubMed]
63. DiSalvo, A.F.; Corbett, D.S. Apparent False Positive Histoplasmin Latex Agglutination Tests in Patients with Tuberculosis. *J. Clin. Microbiol.* **1976**, *3*, 306–308. [CrossRef] [PubMed]
64. Oxenhandler, R.W.; Adelstein, E.H.; Rogers, W.A. Rheumatoid Factor: A Cause of Fals Positive Histoplasmin Latex Agglutination. *J. Clin. Microbiol.* **1977**, *5*, 31–33. [CrossRef]
65. Guimarães, A.J.; Pizzini, C.V.; de Matos Guedes, H.L.; Albuquerque, P.C.; Peralta, J.M.; Hamilton, A.J.; Zancopé-Oliveira, R.M. ELISA for Early Diagnosis of Histoplasmosis. *J. Med. Microbiol.* **2004**, *53*, 509–514. [CrossRef] [PubMed]
66. Pizzini, C.V.; Zancopé-Oliveira, R.M.; Reiss, E.; Hajjeh, R.; Kaufman, L.; Peralta, J.M. Evaluation of a Western Blot Test in an Outbreak of Acute Pulmonary Histoplasmosis. *Clin. Diagn. Lab. Immunol.* **1999**, *6*, 20–23. [CrossRef] [PubMed]
67. Torres, M.; Díaz, H.; Herrera, T.; Sada, E. Evaluation of Enzyme Linked Immunosorbent-Assay and Western Blot for Diagnosis of Histoplasmosis. *Rev. Investig. Clin. Organo. Hosp. Enferm. Nutr.* **1993**, *45*, 155–160.
68. Richer, S.M.; Smedema, M.L.; Durkin, M.M.; Herman, K.M.; Hage, C.A.; Fuller, D.; Wheat, L.J. Improved Diagnosis of Acute Pulmonary Histoplasmosis by Combining Antigen and Antibody Detection. *Clin. Infect. Dis. Off. Publ. Infect. Dis. Soc. Am.* **2016**, *62*, 896–902. [CrossRef]
69. Guimarães, A.J.; Pizzini, C.V.; De Abreu Almeida, M.; Peralta, J.M.; Nosanchuk, J.D.; Zancopé-Oliveira, R.M. Evaluation of an Enzyme-Linked Immunosorbent Assay Using Purified, Deglycosylated Histoplasmin for Different Clinical Manifestations of Histoplasmosis. *Microbiol. Res.* **2010**, *1*, e2. [CrossRef]
70. Alvarado, P.; Pérez-Rojas, Y.; Zambrano, E.A.; Gonzatti, M.I.; Roschman-González, A. Improved Serodiagnosis of Histoplasmosis by Use of Deglycosylated Extracellular Released Antigens of *Histoplasma capsulatum*. *J. Microbiol. Methods* **2020**, *175*, 105981. [CrossRef]
71. Zancopé-Oliveira, R.M.; Bragg, S.L.; Reiss, E.; Wanke, B.; Peralta, J.M. Effects of Histoplasmin M Antigen Chemical and Enzymatic Deglycosylation on Cross-Reactivity in the Enzyme-Linked Immunoelectrotransfer Blot Method. *Clin. Diagn. Lab. Immunol.* **1994**, *1*, 390–393. [CrossRef]
72. Almeida, M.D.A.; Pizzini, C.V.; Damasceno, L.S.; Muniz, M.D.M.; Almeida-Paes, R.; Peralta, R.H.S.; Peralta, J.M.; Oliveira, R.D.V.C.; Vizzoni, A.G.; de Andrade, C.L.T.; et al. Validation of Western Blot for *Histoplasma capsulatum* Antibody Detection Assay. *BMC Infect. Dis.* **2016**, *16*, 87. [CrossRef] [PubMed]
73. Almeida, M.A.; Damasceno, L.S.; Pizzini, C.V.; de Muniz, M.M.; Almeida-Paes, R.; Zancopé-Oliveira, R.M. Role of Western Blot Assay for the Diagnosis of Histoplasmosis in AIDS Patients from a National Institute of Infectious Diseases in Rio de Janeiro, Brazil. *Mycoses* **2019**, *62*, 261–267. [CrossRef] [PubMed]
74. World Health Organization. *Selection of Essential in Vitro Diagnostics at Country Level: Using the WHO Model List of Essential in Vitro Diagnostics to Develop and Update a National List of Essential in Vitro Diagnostics*; World Health Organization: Geneva, Switzerland, 2021; ISBN 978-92-4-003092-3.
75. Wheat, L.J.; Garringer, T.; Brizendine, E.; Connolly, P. Diagnosis of Histoplasmosis by Antigen Detection Based upon Experience at the Histoplasmosis Reference Laboratory. *Diagn. Microbiol. Infect. Dis.* **2002**, *43*, 29–37. [CrossRef]
76. Wheat, L.J.; Connolly-Stringfield, P.; Kohler, R.B.; Frame, P.T.; Gupta, M.R. *Histoplasma capsulatum* Polysaccharide Antigen Detection in Diagnosis and Management of Disseminated Histoplasmosis in Patients with Acquired Immunodeficiency Syndrome. *Am. J. Med.* **1989**, *87*, 396–400. [CrossRef]
77. Garner, J.A.; Kernodle, D. False-Positive *Histoplasma* Antigen Test in a Patient with Pulmonary Blastomycosis. *Clin. Infect. Dis. Off. Publ. Infect. Dis. Soc. Am.* **1995**, *21*, 1054. [CrossRef] [PubMed]
78. Zimmerman, S.E.; Stringfield, P.C.; Wheat, L.J.; French, M.L.; Kohler, R.B. Comparison of Sandwich Solid-Phase Radioimmunoassay and Two Enzyme-Linked Immunosorbent Assays for Detection of *Histoplasma capsulatum* Polysaccharide Antigen. *J. Infect. Dis.* **1989**, *160*, 678–685. [CrossRef]
79. Scheel, C.M.; Samayoa, B.; Herrera, A.; Lindsley, M.D.; Benjamin, L.; Reed, Y.; Hart, J.; Lima, S.; Rivera, B.E.; Raxcaco, G.; et al. Development and Evaluation of an Enzyme-Linked Immunosorbent Assay to Detect *Histoplasma capsulatum* Antigenuria in Immunocompromised Patients. *Clin. Vaccine Immunol. CVI* **2009**, *16*, 852–858. [CrossRef]
80. Gomez, B.L.; Figueroa, J.I.; Hamilton, A.J.; Ortiz, B.L.; Robledo, M.A.; Restrepo, A.; Hay, R.J. Development of a Novel Antigen Detection Test for Histoplasmosis. *J. Clin. Microbiol.* **1997**, *35*, 2618–2622. [CrossRef]
81. Caceres, D.H.; Scheel, C.M.; Tobón, A.M.; Ahlquist Cleveland, A.; Restrepo, A.; Brandt, M.E.; Chiller, T.; Gómez, B.L. Validation of an Enzyme-Linked Immunosorbent Assay That Detects *Histoplasma capsulatum* Antigenuria in Colombian Patients with AIDS for Diagnosis and Follow-up during Therapy. *Clin. Vaccine Immunol. CVI* **2014**, *21*, 1364–1368. [CrossRef]
82. Theel, E.S.; Jespersen, D.J.; Harring, J.; Mandrekar, J.; Binnicker, M.J. Evaluation of an Enzyme Immunoassay for Detection of *Histoplasma capsulatum* Antigen from Urine Specimens. *J. Clin. Microbiol.* **2013**, *51*, 3555–3559. [CrossRef]
83. Connolly, P.A.; Durkin, M.M.; Lemonte, A.M.; Hackett, E.J.; Wheat, L.J. Detection of *Histoplasma* Antigen by a Quantitative Enzyme Immunoassay. *Clin. Vaccine Immunol. CVI* **2007**, *14*, 1587–1591. [CrossRef] [PubMed]

84. Hage, C.A.; Davis, T.E.; Fuller, D.; Egan, L.; Witt, J.R.; Wheat, L.J.; Knox, K.S. Diagnosis of Histoplasmosis by Antigen Detection in BAL Fluid. *Chest* **2010**, *137*, 623–628. [CrossRef] [PubMed]
85. Hage, C.A.; Ribes, J.A.; Wengenack, N.L.; Baddour, L.M.; Assi, M.; McKinsey, D.S.; Hammoud, K.; Alapat, D.; Babady, N.E.; Parker, M.; et al. A Multicenter Evaluation of Tests for Diagnosis of Histoplasmosis. *Clin. Infect. Dis. Off. Publ. Infect. Dis. Soc. Am.* **2011**, *53*, 448–454. [CrossRef] [PubMed]
86. Zhang, X.; Gibson, B.; Daly, T.M. Evaluation of Commercially Available Reagents for Diagnosis of Histoplasmosis Infection in Immunocompromised Patients. *J. Clin. Microbiol.* **2013**, *51*, 4095–4101. [CrossRef]
87. Cáceres, D.H.; Gómez, B.L.; Tobón, Á.M.; Chiller, T.M.; Lindsley, M.D. Evaluation of OIDx *Histoplasma* Urinary Antigen EIA. *Mycopathologia* **2022**, *187*, 129–131. [CrossRef]
88. Cáceres, D.H.; Gómez, B.L.; Tobón, A.M.; Chiller, T.M.; Lindsley, M.D. Evaluation of a *Histoplasma* Antigen Lateral Flow Assay for the Rapid Diagnosis of Progressive Disseminated Histoplasmosis in Colombian Patients with AIDS. *Mycoses* **2020**, *63*, 139–144. [CrossRef]
89. Cáceres, D.H.; Gómez, B.L.; Tobón, Á.M.; Minderman, M.; Bridges, N.; Chiller, T.; Lindsley, M.D. Validation and Concordance Analysis of a New Lateral Flow Assay for Detection of *Histoplasma* Antigen in Urine. *J. Fungi.* **2021**, *7*, 799. [CrossRef]
90. Garcia-Guiñon, A.; Torres-Rodríguez, J.M.; Ndidongarte, D.T.; Cortadellas, F.; Labrín, L. Disseminated Histoplasmosis by *Histoplasma capsulatum* var. *Duboisii* in a Paediatric Patient from the Chad Republic, Africa. *Eur. J. Clin. Microbiol. Infect. Dis. Off. Publ. Eur. Soc. Clin. Microbiol.* **2009**, *28*, 697–699. [CrossRef]
91. Oladele, R.O.; Ayanlowo, O.O.; Richardson, M.D.; Denning, D.W. Histoplasmosis in Africa: An Emerging or a Neglected Disease? *PLoS Negl. Trop. Dis.* **2018**, *12*, e0006046. [CrossRef]
92. Amona, F.M.; Denning, D.W.; Moukassa, D.; Develoux, M.; Hennequin, C. Histoplasmosis in the Republic of Congo Dominated by African Histoplasmosis, *Histoplasma capsulatum* var. *Duboisii*. *PLoS Negl. Trop. Dis.* **2021**, *15*, e0009318. [CrossRef]
93. Cipriano, A.; Neves-Maia, J.; Lopes, V.; Fleming, C.E.; Ferreira, M.A.; Bathay, J. African Histoplasmosis in a Guinea Bissau Patient with HIV-2: Case Report and Review. *J. Mycol. Med.* **2020**, *30*, 100904. [CrossRef]
94. Muotoe-Okafor, F.A.; Gugnani, H.C.; Gugnani, A. Skin and Serum Reactivity among Humans to Histoplasmin in the Vicinity of a Natural Focus of *Histoplasma capsulatum* var. *Duboisii*. *Mycopathologia* **1996**, *134*, 71–74. [CrossRef] [PubMed]
95. Griffiths, J.; Lopes Colombo, A.; Denning, D.W. The Case for Paracoccidioidomycosis to Be Accepted as a Neglected Tropical (Fungal) Disease. *PLoS Negl. Trop. Dis.* **2019**, *13*, e0007195. [CrossRef] [PubMed]
96. Shikanai-Yasuda, M.A.; Mendes, R.P.; Colombo, A.L.; Queiroz-Telles, F.D.; Kono, A.S.G.; Paniago, A.M.M.; Nathan, A.; Valle, A.C.F.D.; Bagagli, E.; Benard, G.; et al. Brazilian Guidelines for the Clinical Management of Paracoccidioidomycosis. *Rev. Soc. Bras. Med. Trop.* **2017**, *50*, 715–740. [CrossRef] [PubMed]
97. Zancope-Oliveira, R.M.; Pizzini, C.V.; de Medeiros Muniz, M.; do Valle, A.C.F.; Almeida-Paes, R. Diagnostic Aspects of Paracoccidioidomycosis. *Curr. Trop. Med. Rep.* **2014**, *1*, 111–118. [CrossRef]
98. Ouchterlony, O. Antigen-Antibody Reactions in Gels. *Acta. Pathol. Microbiol. Scand.* **1949**, *26*, 507–515. [CrossRef]
99. Cocio, T.A.; Martinez, R. Serological Diagnosis of Paracoccidioidomycosis Using a *Paracoccidioides* spp. Comercial Antigen and the Counterimmunoelectrophoresis Method. *Braz. J. Infect. Dis. Off. Publ. Braz. Soc. Infect. Dis.* **2021**, *25*, 101607. [CrossRef]
100. Leitão, N.P.; Vallejo, M.C.; Conceição, P.M.; Camargo, Z.P.; Hahn, R.; Puccia, R. Paracoccidioides Lutzii Plp43 Is an Active Glucanase with Partial Antigenic Identity with P. Brasiliensis Gp43. *PLoS Negl. Trop. Dis.* **2014**, *8*, e3111. [CrossRef]
101. Buccheri, R.; Morais, V.D.S.; Kamikawa, C.M.; Vidal, M.S.M.; Naves, G.; Del Negro, G.M.B.; Vicentini, A.P.; Benard, G. Case Report: Misleading Serological Diagnosis of Paracoccidioidomycosis in a Young Patient with the Acute Form Disease: *Paracoccidioides Brasiliensis* or *Paracoccidioides Lutzii*? *Am. J. Trop. Med. Hyg.* **2018**, *98*, 1082–1085. [CrossRef]
102. Kamikawa, C.M.; Mendes, R.P.; Vicentini, A.P. Standardization and Validation of Dot-ELISA Assay for *Paracoccidioides Brasiliensis* Antibody Detection. *J. Venom. Anim. Toxins Trop. Dis.* **2017**, *23*, 11. [CrossRef]
103. de Fátima da Silva, J.; de Oliveira, H.C.; Marcos, C.M.; Assato, P.A.; Fusco-Almeida, A.M.; Mendes-Giannini, M.J.S. Advances and Challenges in Paracoccidioidomycosis Serology Caused by *Paracoccidioides* Species Complex: An Update. *Diagn. Microbiol. Infect. Dis.* **2016**, *84*, 87–94. [CrossRef] [PubMed]
104. Cano, L.E.; Restrepo, A. Predictive Value of Serologic Tests in the Diagnosis and Follow-up of Patients with Paracoccidioidomycosis. *Rev. Inst. Med. Trop. Sao Paulo* **1987**, *29*, 276–283. [CrossRef] [PubMed]
105. Falcão, E.M.; de Macedo, P.M.; Freitas, D.F.S.; Freitas, A.D.A.; Grinsztejn, B.; Veloso, V.G.; Almeida-Paes, R.; do Valle, A.C.F. Paracoccidioidomycosis in People Living with HIV/AIDS: A Historical Retrospective Cohort Study in a National Reference Center for Infectious Diseases, Rio de Janeiro, Brazil. *PLoS Negl. Trop. Dis.* **2022**, *16*, e0010529. [CrossRef] [PubMed]
106. Blotta, M.H.; Camargo, Z.P. Immunological Response to Cell-Free Antigens of *Paracoccidioides Brasiliensis*: Relationship with Clinical Forms of Paracoccidioidomycosis. *J. Clin. Microbiol.* **1993**, *31*, 671–676. [CrossRef] [PubMed]
107. de Camargo, Z.P. Serology of Paracoccidioidomycosis. *Mycopathologia* **2008**, *165*, 289–302. [CrossRef]
108. de Macedo, P.M.; Teixeira, M.D.M.; Barker, B.M.; Zancopé-Oliveira, R.M.; Almeida-Paes, R.; Francesconi do Valle, A.C. Clinical Features and Genetic Background of the Sympatric Species *Paracoccidioides Brasiliensis* and *Paracoccidioides Americana*. *PLoS Negl. Trop. Dis.* **2019**, *13*, e0007309. [CrossRef]
109. Neves, A.R.; Mamoni, R.L.; Rossi, C.L.; de Camargo, Z.P.; Blotta, M.H.S.L. Negative Immunodiffusion Test Results Obtained with Sera of Paracoccidioidomycosis Patients May Be Related to Low-Avidity Immunoglobulin G2 Antibodies Directed against Carbohydrate Epitopes. *Clin. Diagn. Lab. Immunol.* **2003**, *10*, 802–807. [CrossRef]

110. Del Negro, G.M.; Garcia, N.M.; Rodrigues, E.G.; Cano, M.I.; de Aguiar, M.S.; de Lírio, V.S.; da Lacaz, C.S. The Sensitivity, Specificity and Efficiency Values of Some Serological Tests Used in the Diagnosis of Paracoccidioidomycosis. *Rev. Inst. Med. Trop. Sao Paulo* **1991**, *33*, 277–280. [CrossRef]
111. Conti-Diaz, I.A.; Mackinnon, J.E.; Calegari, L.; Casserone, S. Estudio comparativo de la inmunoelectroforesis (IEF) y de la inmunoelectroosmoforesis-inmunodifusion (IEOF-ID) aplicadas al diagnostico de la paracoccidioidomicosis. *Mycopathologia* **1978**, *63*, 161–165. [CrossRef]
112. Ferreira-da-Cruz, M.F.; Galvão-Castro, B.; Wanke, B. Produção e padronização dos antígenos de *Paracoccidioides brasiliensis* (Pb), *Histoplasma capsulatum* (Hc) e *Aspergillus fumigatus* (Af) para uso no imunodiagnóstico: Comparação entre as técnicas de imunodifusão e imunoeletroosmoforese. *Mem. Inst. Oswaldo Cruz* **1985**, *80*, 301–305. [CrossRef]
113. Brummer, E.; Castaneda, E.; Restrepo, A. Paracoccidioidomycosis: An Update. *Clin. Microbiol. Rev.* **1993**, *6*, 89–117. [CrossRef] [PubMed]
114. Dos Santos, P.O.; Rodrigues, A.M.; Fernandes, G.F.; da Silva, S.H.M.; Burger, E.; de Camargo, Z.P. Immunodiagnosis of Paracoccidioidomycosis Due to *Paracoccidioides Brasiliensis* Using a Latex Test: Detection of Specific Antibody Anti-Gp43 and Specific Antigen Gp43. *PLoS Negl. Trop. Dis.* **2015**, *9*, e0003516. [CrossRef] [PubMed]
115. Fernandes, V.C.; Coitinho, J.B.; Veloso, J.M.R.; Araújo, S.A.; Pedroso, E.P.; Goes, A.M. Combined Use of *Paracoccidioides Brasiliensis* Recombinant RPb27 and RPb40 Antigens in an Enzyme-Linked Immunosorbent Assay for Immunodiagnosis of Paracoccidioidomycosis. *J. Immunol. Methods* **2011**, *367*, 78–84. [CrossRef] [PubMed]
116. Mendes-Giannini, M.J.; Camargo, M.E.; Lacaz, C.S.; Ferreira, A.W. Immunoenzymatic Absorption Test for Serodiagnosis of Paracoccidioidomycosis. *J. Clin. Microbiol.* **1984**, *20*, 103–108. [CrossRef] [PubMed]
117. Baida, H.; Biselli, P.J.; Juvenale, M.; Del Negro, G.M.; Mendes-Giannini, M.J.; Duarte, A.J.; Benard, G. Differential Antibody Isotype Expression to the Major *Paracoccidioides Brasiliensis* Antigen in Juvenile and Adult Form Paracoccidioidomycosis. *Microbes Infect.* **1999**, *1*, 273–278. [CrossRef]
118. Mendes-Giannini, M.J.; Bueno, J.P.; Shikanai-Yasuda, M.A.; Ferreira, A.W.; Masuda, A. Detection of the 43,000-Molecular-Weight Glycoprotein in Sera of Patients with Paracoccidioidomycosis. *J. Clin. Microbiol.* **1989**, *27*, 2842–2845. [CrossRef]
119. Giannini, M.J.; Bueno, J.P.; Shikanai-Yasuda, M.A.; Stolf, A.M.; Masuda, A.; Amato Neto, V.; Ferreira, A.W. Antibody Response to the 43 KDa Glycoprotein of *Paracoccidioides Brasiliensis* as a Marker for the Evaluation of Patients under Treatment. *Am. J. Trop. Med. Hyg.* **1990**, *43*, 200–206. [CrossRef]
120. Do Valle, A.C.; Costa, R.L.; Fialho Monteiro, P.C.; Von Helder, J.; Muniz, M.M.; Zancopé-Oliveira, R.M. Interpretation and Clinical Correlation of Serological Tests in Paracoccidioidomycosis. *Med. Mycol.* **2001**, *39*, 373–377. [CrossRef]
121. Bertoni, T.A.; Perenha-Viana, M.C.Z.; Patussi, E.V.; Cardoso, R.F.; Svidzinski, T.I.E. Western Blotting Is an Efficient Tool for Differential Diagnosis of Paracoccidioidomycosis and Pulmonary Tuberculosis. *Clin. Vaccine Immunol. CVI* **2012**, *19*, 1887–1888. [CrossRef]
122. Gegembauer, G.; Araujo, L.M.; Pereira, E.F.; Rodrigues, A.M.; Paniago, A.M.M.; Hahn, R.C.; de Camargo, Z.P. Serology of Paracoccidioidomycosis Due to *Paracoccidioides Lutzii*. *PLoS Negl. Trop. Dis.* **2014**, *8*, e2986. [CrossRef]
123. Peron, G.; Fernandes, F.F.; Landgraf, T.N.; Martinez, R.; Panunto-Castelo, A. Recombinant 60-KDa Heat Shock Protein from *Paracoccidioides Brasiliensis*: Is It a Good Antigen for Serological Diagnosis of Paracoccidioidomycosis? *Braz. J. Med. Biol. Res. Rev. Bras. Pesqui. Medicas E Biol.* **2017**, *50*, e5928. [CrossRef] [PubMed]
124. Sylvestre, T.F.; Franciscone Silva, L.R.; de Cavalcante, R.S.; Moris, D.V.; Venturini, J.; Vicentini, A.P.; de Carvalho, L.R.; Mendes, R.P. Prevalence and Serological Diagnosis of Relapse in Paracoccidioidomycosis Patients. *PLoS Negl. Trop. Dis.* **2014**, *8*, e2834. [CrossRef] [PubMed]
125. Pappas, M.G. Recent Applications of the Dot-ELISA in Immunoparasitology. *Vet. Parasitol.* **1988**, *29*, 105–129. [CrossRef]
126. Correa, M.M.; Bedoya, A.M.; Guerrero, M.P.; Méndez, J.; Restrepo, A.; McEwen, J.G. Diagnosis of Paracoccidioidomycosis by a Dot Blot Assay Using a Recombinant *Paracoccidioides Brasiliensis* P27 Protein. *Mycoses* **2007**, *50*, 41–47. [CrossRef]
127. Taborda, C.P.; Camargo, Z.P. Diagnosis of Paracoccidioidomycosis by Dot Immunobinding Assay for Antibody Detection Using the Purified and Specific Antigen Gp43. *J. Clin. Microbiol.* **1994**, *32*, 554–556. [CrossRef]
128. Martins, R.; Marques, S.; Alves, M.; Fecchio, D.; de Franco, M.F. Serological Follow-up of Patients with Paracoccidioidomycosis Treated with Itraconazole Using Dot-Blot, ELISA and Western-Blot. *Rev. Inst. Med. Trop. Sao Paulo* **1997**, *39*, 261–269. [CrossRef]
129. Wheat, L.J. Antigen Detection, Serology, and Molecular Diagnosis of Invasive Mycoses in the Immunocompromised Host. *Transpl. Infect. Dis. Off. J. Transplant. Soc.* **2006**, *8*, 128–139. [CrossRef]
130. Gómez, B.L.; Figueroa, J.I.; Hamilton, A.J.; Diez, S.; Rojas, M.; Tobón, A.M.; Hay, R.J.; Restrepo, A. Antigenemia in Patients with Paracoccidioidomycosis: Detection of the 87-Kilodalton Determinant during and after Antifungal Therapy. *J. Clin. Microbiol.* **1998**, *36*, 3309–3316. [CrossRef]
131. Marques da Silva, S.H.; Colombo, A.L.; Blotta, M.H.S.L.; Lopes, J.D.; Queiroz-Telles, F.; de Pires Camargo, Z. Detection of Circulating Gp43 Antigen in Serum, Cerebrospinal Fluid, and Bronchoalveolar Lavage Fluid of Patients with Paracoccidioidomycosis. *J. Clin. Microbiol.* **2003**, *41*, 3675–3680. [CrossRef]
132. Salina, M.A.; Shikanai-Yasuda, M.A.; Mendes, R.P.; Barraviera, B.; Mendes Giannini, M.J. Detection of Circulating *Paracoccidioides Brasiliensis* Antigen in Urine of Paracoccidioidomycosis Patients before and during Treatment. *J. Clin. Microbiol.* **1998**, *36*, 1723–1728. [CrossRef]

133. Kwon-Chung, K.J.; Fraser, J.A.; Doering, T.L.; Wang, Z.; Janbon, G.; Idnurm, A.; Bahn, Y.-S. Cryptococcus Neoformans and Cryptococcus Gattii, the Etiologic Agents of Cryptococcosis. *Cold Spring Harb. Perspect. Med.* **2014**, *4*, a019760. [CrossRef] [PubMed]
134. Gushiken, A.C.; Saharia, K.K.; Baddley, J.W. Cryptococcosis. *Infect. Dis. Clin. N. Am.* **2021**, *35*, 493–514. [CrossRef] [PubMed]
135. Kwon-Chung, K.J.; Bennett, J.E. Epidemiologic Differences between the Two Varieties of Cryptococcus Neoformans. *Am. J. Epidemiol.* **1984**, *120*, 123–130. [CrossRef] [PubMed]
136. Rajasingham, R.; Wake, R.M.; Beyene, T.; Katende, A.; Letang, E.; Boulware, D.R. Cryptococcal Meningitis Diagnostics and Screening in the Era of Point-of-Care Laboratory Testing. *J. Clin. Microbiol.* **2019**, *57*, e01238-18. [CrossRef] [PubMed]
137. Alspaugh, J.A. Virulence Mechanisms and Cryptococcus Neoformans Pathogenesis. *Fungal Genet. Biol. FG B* **2015**, *78*, 55–58. [CrossRef] [PubMed]
138. Vanselow, M.; Brandt, M.E.; Park, B.J. Diagnosis and Management of Cryptococcal Disease in Resource-Limited Settings. *Curr. Fungal Infect. Rep.* **2012**, *6*, 35–40. [CrossRef]
139. Kozel, T.R.; Bauman, S.K. CrAg Lateral Flow Assay for Cryptococcosis. *Expert Opin. Med. Diagn.* **2012**, *6*, 245–251. [CrossRef]
140. Speed, B.R.; Kaldor, J.; Cairns, B.; Pegorer, M. Serum Antibody Response to Active Infection with *Cryptococcus Neoformans* and Its Varieties in Immunocompetent Subjects. *Med. Mycol.* **1996**, *34*, 187–193. [CrossRef]
141. de Andrade, I.B.; de Almeida, M.A.; Figueiredo-Carvalho, M.H.G.; Coelho, R.A.; da Chaves, A.L.S.; Frases, S.; Zancopé-Oliveira, R.M.; Brito-Santos, F.; Almeida-Paes, R. Production of Secreted Carbohydrates That Present Immunologic Similarities with the *Cryptococcus* Glucuronoxylomannan by Members of the Trichosporonaceae Family: A Comparative Study Among Species of Clinical Interest. *Mycopathologia* **2021**, *186*, 377–385. [CrossRef]
142. Gade, W.; Hinnefeld, S.W.; Babcock, L.S.; Gilligan, P.; Kelly, W.; Wait, K.; Greer, D.; Pinilla, M.; Kaplan, R.L. Comparison of the PREMIER Cryptococcal Antigen Enzyme Immunoassay and the Latex Agglutination Assay for Detection of Cryptococcal Antigens. *J. Clin. Microbiol.* **1991**, *29*, 1616–1619. [CrossRef]
143. Tintelnot, K.; Hagen, F.; Han, C.O.; Seibold, M.; Rickerts, V.; Boekhout, T. Pitfalls in Serological Diagnosis of Cryptococcus Gattii Infections. *Med. Mycol.* **2015**, *53*, 874–879. [CrossRef] [PubMed]
144. Mitchell, T.G.; Perfect, J.R. Cryptococcosis in the Era of AIDS–100 Years after the Discovery of Cryptococcus Neoformans. *Clin. Microbiol. Rev.* **1995**, *8*, 515–548. [CrossRef] [PubMed]
145. Rajasingham, R.; Meya, D.B.; Boulware, D.R. Integrating Cryptococcal Antigen Screening and Pre-Emptive Treatment into Routine HIV Care. *J. Acquir. Immune Defic. Syndr. 1999* **2012**, *59*, e85–e91. [CrossRef] [PubMed]
146. Jarvis, J.N.; Percival, A.; Bauman, S.; Pelfrey, J.; Meintjes, G.; Williams, G.N.; Longley, N.; Harrison, T.S.; Kozel, T.R. Evaluation of a Novel Point-of-Care Cryptococcal Antigen Test on Serum, Plasma, and Urine from Patients with HIV-Associated Cryptococcal Meningitis. *Clin. Infect. Dis. Off. Publ. Infect. Dis. Soc. Am.* **2011**, *53*, 1019–1023. [CrossRef] [PubMed]
147. Percival, A.; Thorkildson, P.; Kozel, T.R. Monoclonal Antibodies Specific for Immunorecessive Epitopes of Glucuronoxylomannan, the Major Capsular Polysaccharide of Cryptococcus Neoformans, Reduce Serotype Bias in an Immunoassay for Cryptococcal Antigen. *Clin. Vaccine Immunol. CVI* **2011**, *18*, 1292–1296. [CrossRef] [PubMed]
148. Blasich, N.P.; Wake, R.M.; Rukasha, I.; Prince, Y.; Govender, N.P. Association of Semi-Quantitative Cryptococcal Antigen Results in Plasma with Subclinical Cryptococcal Meningitis and Mortality among Patients with Advanced HIV Disease. *Med. Mycol.* **2021**, *59*, 1041–1047. [CrossRef]
149. Vidal, J.E.; Boulware, D.R. Lateral Flow Assay for Cryptococcal Antigen: An Important Advance to Improve the Continuum of HIV Care and Reduce Cryptococcal Meningitis-Related Mortality. *Rev. Inst. Med. Trop. Sao Paulo* **2015**, *57* (Suppl. 19), 38–45. [CrossRef]
150. World Health Organization. *Guidelines for Diagnosing, Preventing and Managing Cryptococcal Disease among Adults, Adolescents and Children Living with HIV*; World Health Organization: Geneva, Switzerland, 2022; ISBN 978-92-4-005217-8.
151. French, N.; Gray, K.; Watera, C.; Nakiyingi, J.; Lugada, E.; Moore, M.; Lalloo, D.; Whitworth, J.A.G.; Gilks, C.F. Cryptococcal Infection in a Cohort of HIV-1-Infected Ugandan Adults. *AIDS Lond. Engl.* **2002**, *16*, 1031–1038. [CrossRef]
152. Micol, R.; Lortholary, O.; Sar, B.; Laureillard, D.; Ngeth, C.; Dousset, J.-P.; Chanroeun, H.; Ferradini, L.; Guerin, P.J.; Dromer, F.; et al. Prevalence, Determinants of Positivity, and Clinical Utility of Cryptococcal Antigenemia in Cambodian HIV-Infected Patients. *J. Acquir. Immune Defic. Syndr.* **2007**, *45*, 555–559. [CrossRef]
153. Sathongdejwisit, P.; Pruksaphon, K.; Intaramat, A.; Aiumurai, P.; Sookrung, N.; Ratanabanangkoon, K.; Nosanchuk, J.D.; Youngchim, S. A Novel, Inexpensive In-House Immunochromatographic Strip Test for Cryptococcosis Based on the Cryptococcal Glucuronoxylomannan Specific Monoclonal Antibody 18B7. *Diagnostics* **2021**, *11*, 758. [CrossRef]
154. Kakeya, H.; Udono, H.; Maesaki, S.; Sasaki, E.; Kawamura, S.; Hossain, M.A.; Yamamoto, Y.; Sawai, T.; Fukuda, M.; Mitsutake, K.; et al. Heat Shock Protein 70 (Hsp70) as a Major Target of the Antibody Response in Patients with Pulmonary Cryptococcosis. *Clin. Exp. Immunol.* **1999**, *115*, 485–490. [CrossRef] [PubMed]
155. de Serpa Brandão, R.M.S.; Faria, A.R.; de Andrade, H.M.; Soares Martins, L.M.; da Silva, A.S.; do Monte, S.J.H. Novel Recombinant Multiepitope Proteins for the Detection of Anti-Cryptococcus Antibodies. *Future Microbiol.* **2018**, *13*, 429–436. [CrossRef] [PubMed]
156. Orofino-Costa, R.; de Macedo, P.M.; Rodrigues, A.M.; Bernardes-Engemann, A.R. Sporotrichosis: An Update on Epidemiology, Etiopathogenesis, Laboratory and Clinical Therapeutics. *An. Bras. Dermatol.* **2017**, *92*, 606–620. [CrossRef] [PubMed]
157. de Albornoz, M.B.; Villanueva, E.; de Torres, E.D. Application of Immunoprecipitation Techniques to the Diagnosis of Cutaneous and Extracutaneous Forms of Sporotrichosis. *Mycopathologia* **1984**, *85*, 177–183. [CrossRef] [PubMed]

158. Blumer, S.O.; Kaufman, L.; Kaplan, W.; McLaughlin, D.W.; Kraft, D.E. Comparative Evaluation of Five Serological Methods for the Diagnosis of Sporotrichosis. *Appl. Microbiol.* **1973**, *26*, 4–8. [CrossRef]
159. Casserone, S.; Conti-Díaz, I.A.; Zanetta, E.; Peña de Pereira, M.E. Serologia de la esporotrichosis cutanea. *Sabouraudia* **1983**, *21*, 317–321. [CrossRef]
160. Bernardes-Engemann, A.R.; Costa, R.C.O.; Miguens, B.R.; Penha, C.V.L.; Neves, E.; Pereira, B.a.S.; Dias, C.M.P.; Mattos, M.; Gutierrez, M.C.; Schubach, A.; et al. Development of an Enzyme-Linked Immunosorbent Assay for the Serodiagnosis of Several Clinical Forms of Sporotrichosis. *Med. Mycol.* **2005**, *43*, 487–493. [CrossRef]
161. Almeida-Paes, R.; Pimenta, M.A.; Pizzini, C.V.; Monteiro, P.C.F.; Peralta, J.M.; Nosanchuk, J.D.; Zancopé-Oliveira, R.M. Use of Mycelial-Phase Sporothrix Schenckii Exoantigens in an Enzyme-Linked Immunosorbent Assay for Diagnosis of Sporotrichosis by Antibody Detection. *Clin. Vaccine Immunol. CVI* **2007**, *14*, 244–249. [CrossRef]
162. Bernardes-Engemann, A.R.; de Lima Barros, M.; Zeitune, T.; Russi, D.C.; Orofino-Costa, R.; Lopes-Bezerra, L.M. Validation of a Serodiagnostic Test for Sporotrichosis: A Follow-up Study of Patients Related to the Rio de Janeiro Zoonotic Outbreak. *Med. Mycol.* **2015**, *53*, 28–33. [CrossRef]
163. Scott, E.N.; Muchmore, H.G. Immunoblot Analysis of Antibody Responses to Sporothrix Schenckii. *J. Clin. Microbiol.* **1989**, *27*, 300–304. [CrossRef]
164. Lopes-Bezerra, L.M.; Mora-Montes, H.M.; Zhang, Y.; Nino-Vega, G.; Rodrigues, A.M.; de Camargo, Z.P.; de Hoog, S. Sporotrichosis between 1898 and 2017: The Evolution of Knowledge on a Changeable Disease and on Emerging Etiological Agents. *Med. Mycol.* **2018**, *56*, 126–143. [CrossRef] [PubMed]
165. Almeida-Paes, R.; Pimenta, M.A.; Monteiro, P.C.F.; Nosanchuk, J.D.; Zancopé-Oliveira, R.M. Immunoglobulins G, M, and A against Sporothrix Schenckii Exoantigens in Patients with Sporotrichosis before and during Treatment with Itraconazole. *Clin. Vaccine Immunol. CVI* **2007**, *14*, 1149–1157. [CrossRef] [PubMed]
166. Fernandes, G.F.; Lopes-Bezerra, L.M.; Bernardes-Engemann, A.R.; Schubach, T.M.P.; Dias, M.A.G.; Pereira, S.A.; de Camargo, Z.P. Serodiagnosis of Sporotrichosis Infection in Cats by Enzyme-Linked Immunosorbent Assay Using a Specific Antigen, SsCBF, and Crude Exoantigens. *Vet. Microbiol.* **2011**, *147*, 445–449. [CrossRef] [PubMed]
167. Bernardes-Engemann, A.R.; de Almeida, M.A.; Bison, I.; de Rabello, V.B.S.; Ramos, M.L.M.; Pereira, S.A.; Almeida-Paes, R.; de Lima Brasil, A.W.; Zancopé-Oliveira, R.M. Anti-Sporothrix Antibody Detection in Domestic Cats as an Indicator of a Possible New Occurrence Area for Sporotrichosis in North Brazil. *Mycopathologia* **2022**, *187*, 375–384. [CrossRef]
168. Alvarado, P.; Ostos, A.; Franquiz, N.; Roschman-González, A.; Zambrano, E.A.; Mendoza, M. Diagnóstico serológico de la esporotricosismediante el empleo del antígeno de miceliode Sporothrix schenckii sensu stricto. *Investig. Clin.* **2015**, *56*, 111–122.
169. Grisolia, J.C.; Santos, L.A.; Coelho, L.M.L.; Silva, R.R.; de Camargo, Z.P.; Velloso, T.R.G.; Coelho, L.F.L.; Chavasco, J.K.; Malaquias, L.C.C. Seroepidemiological Survey on Sporotrichosis-Infection in Rural Areas of the South of Minas Gerais State, Brazil. *Braz. J. Microbiol. Publ. Braz. Soc. Microbiol.* **2021**, *52*, 41–47. [CrossRef]
170. Almeida-Paes, R.; Bailão, A.M.; Pizzini, C.V.; Reis, R.S.; de Soares, C.M.A.; Peralta, J.M.; Gutierrez-Galhardo, M.C.; Zancopé-Oliveira, R.M. Cell-Free Antigens of Sporothrix Brasiliensis: Antigenic Diversity and Application in an Immunoblot Assay. *Mycoses* **2012**, *55*, 467–475. [CrossRef]
171. Cognialli, R.; Bloss, K.; Weiss, I.; Caceres, D.H.; Davis, R.; Queiroz-Telles, F. A Lateral Flow Assay for the Immunodiagnosis of Human Cat-Transmitted Sporotrichosis. *Mycoses* **2022**. [CrossRef]
172. de Macedo, P.M.; Sztajnbok, D.C.N.; Camargo, Z.P.; Rodrigues, A.M.; Lopes-Bezerra, L.M.; Bernardes-Engemann, A.R.; Orofino-Costa, R. Dacryocystitis Due to Sporothrix Brasiliensis: A Case Report of a Successful Clinical and Serological Outcome with Low-Dose Potassium Iodide Treatment and Oculoplastic Surgery. *Br. J. Dermatol.* **2015**, *172*, 1116–1119. [CrossRef]
173. Costa, R.O.; de Mesquita, K.C.; Damasco, P.S.; Bernardes-Engemann, A.R.; Dias, C.M.P.; Silva, I.C.; Lopes-Bezerra, L.M. Infectious Arthritis as the Single Manifestation of Sporotrichosis: Serology from Serum and Synovial Fluid Samples as an Aid to Diagnosis. *Rev. Iberoam. Micol.* **2008**, *25*, 54–56. [CrossRef]
174. Orofino-Costa, R.; Unterstell, N.; Carlos Gripp, A.; de Macedo, P.M.; Brota, A.; Dias, E.; de Melo Teixeira, M.; Felipe, M.S.; Bernardes-Engemann, A.R.; Lopes-Bezerra, L.M. Pulmonary Cavitation and Skin Lesions Mimicking Tuberculosis in a HIV Negative Patient Caused by Sporothrix Brasiliensis. *Med. Mycol. Case Rep.* **2013**, *2*, 65–71. [CrossRef] [PubMed]
175. Narayanasamy, S.; Dat, V.Q.; Thanh, N.T.; Ly, V.T.; Chan, J.F.-W.; Yuen, K.-Y.; Ning, C.; Liang, H.; Li, L.; Chowdhary, A.; et al. A Global Call for Talaromycosis to Be Recognised as a Neglected Tropical Disease. *Lancet Glob. Health* **2021**, *9*, e1618–e1622. [CrossRef]
176. Pruksaphon, K.; Intaramat, A.; Ratanabanangkoon, K.; Nosanchuk, J.D.; Vanittanakom, N.; Youngchim, S. Diagnostic Laboratory Immunology for Talaromycosis (Penicilliosis): Review from the Bench-Top Techniques to the Point-of-Care Testing. *Diagn. Microbiol. Infect. Dis.* **2020**, *96*, 114959. [CrossRef] [PubMed]
177. Sekhon, A.S.; Li, J.S.; Garg, A.K. *Penicillosis marneffei*: Serological and Exoantigen Studies. *Mycopathologia* **1982**, *77*, 51–57. [CrossRef]
178. Imwidthaya, P.; Sekhon, A.S.; Mastro, T.D.; Garg, A.K.; Ambrosie, E. Usefulness of a Microimmunodiffusion Test for the Detection of *Penicillium marneffei* Antigenemia, Antibodies, and Exoantigens. *Mycopathologia* **1997**, *138*, 51–55. [CrossRef]
179. Cao, L.; Chen, D.L.; Lee, C.; Chan, C.M.; Chan, K.M.; Vanittanakom, N.; Tsang, D.N.; Yuen, K.Y. Detection of Specific Antibodies to an Antigenic Mannoprotein for Diagnosis of *Penicillium marneffei* Penicilliosis. *J. Clin. Microbiol.* **1998**, *36*, 3028–3031. [CrossRef]

180. Wang, Y.; Zeng, L.; Chen, X.; Hao, W.; Yang, M.; Cai, J.; Wang, Y.; Yuan, G.; Che, X. A double-antigen sandwich ELISA for detecting Penicillium marneffei Mp1p-specific antibody. *Nan Fang Yi Ke Da Xue Xue Bao* **2013**, *33*, 439–443.
181. Richardson, M.; Page, I. Role of Serological Tests in the Diagnosis of Mold Infections. *Curr. Fungal Infect. Rep.* **2018**, *12*, 127–136. [CrossRef]
182. Tobar Vega, P.; Erramilli, S.; Lee, E. *Talaromyces marneffei* Laboratory Cross Reactivity with *Histoplasma* and *Blastomyces* Urinary Antigen. *Int. J. Infect. Dis. IJID Off. Publ. Int. Soc. Infect. Dis.* **2019**, *86*, 15–17. [CrossRef]
183. Ning, C.; Lai, J.; Wei, W.; Zhou, B.; Huang, J.; Jiang, J.; Liang, B.; Liao, Y.; Zang, N.; Cao, C.; et al. Accuracy of Rapid Diagnosis of *Talaromyces marneffei*: A Systematic Review and Meta-Analysis. *PLoS ONE* **2018**, *13*, e0195569. [CrossRef]
184. Thu, N.T.M.; Chan, J.F.W.; Ly, V.T.; Ngo, H.T.; Hien, H.T.A.; Lan, N.P.H.; Chau, N.V.V.; Cai, J.-P.; Woo, P.C.Y.; Day, J.N.; et al. Superiority of a Novel Mp1p Antigen Detection Enzyme Immunoassay Compared to Standard BACTEC Blood Culture in the Diagnosis of Talaromycosis. *Clin. Infect. Dis. Off. Publ. Infect. Dis. Soc. Am.* **2021**, *73*, e330–e336. [CrossRef] [PubMed]
185. Desakorn, V.; Simpson, A.J.H.; Wuthiekanun, V.; Sahassananda, D.; Rajanuwong, A.; Pitisuttithum, P.; Howe, P.A.; Smith, M.D.; White, N.J. Development and Evaluation of Rapid Urinary Antigen Detection Tests for Diagnosis of Penicilliosis Marneffei. *J. Clin. Microbiol.* **2002**, *40*, 3179–3183. [CrossRef] [PubMed]
186. Wang, Y.-F.; Cai, J.-P.; Wang, Y.-D.; Dong, H.; Hao, W.; Jiang, L.-X.; Long, J.; Chan, C.; Woo, P.C.Y.; Lau, S.K.P.; et al. Immunoassays Based on *Penicillium marneffei* Mp1p Derived from Pichia Pastoris Expression System for Diagnosis of Penicilliosis. *PLoS ONE* **2011**, *6*, e28796. [CrossRef] [PubMed]
187. Mendoza, L.; Belone, A.F.F.; Vilela, R.; Rehtanz, M.; Bossart, G.D.; Reif, J.S.; Fair, P.A.; Durden, W.N.; St Leger, J.; Travassos, L.R.; et al. Use of Sera from Humans and Dolphins with Lacaziosis and Sera from Experimentally Infected Mice for Western Blot Analyses of Lacazia Loboi Antigens. *Clin. Vaccine Immunol. CVI* **2008**, *15*, 164–167. [CrossRef]
188. Camargo, Z.P.; Baruzzi, R.G.; Maeda, S.M.; Floriano, M.C. Antigenic Relationship between Loboa Loboi and *Paracoccidioides Brasiliensis* as Shown by Serological Methods. *Med. Mycol.* **1998**, *36*, 413–417. [CrossRef]
189. Beltrame, A.; Danesi, P.; Farina, C.; Orza, P.; Perandin, F.; Zanardello, C.; Rodari, P.; Staffolani, S.; Bisoffi, Z. Case Report: Molecular Confirmation of Lobomycosis in an Italian Traveler Acquired in the Amazon Region of Venezuela. *Am. J. Trop. Med. Hyg.* **2017**, *97*, 1757–1760. [CrossRef]
190. Anstead, G.M.; Sutton, D.A.; Graybill, J.R. Adiaspiromycosis Causing Respiratory Failure and a Review of Human Infections Due to Emmonsia and *Chrysosporium* spp. *J. Clin. Microbiol.* **2012**, *50*, 1346–1354. [CrossRef]
191. Stěrba, J.; Tomsíková, A.; Prokopic, J.; Novácková, D. A Serological Proof of Adiaspiromycosis in Wild Rodents. *Folia Parasitol. (Praha)* **1980**, *27*, 377–381.
192. Schwartz, I.S.; Govender, N.P.; Sigler, L.; Jiang, Y.; Maphanga, T.G.; Toplis, B.; Botha, A.; Dukik, K.; Hoving, J.C.; Muñoz, J.F.; et al. Emergomyces: The Global Rise of New Dimorphic Fungal Pathogens. *PLoS Pathog.* **2019**, *15*, e1007977. [CrossRef]
193. Schwartz, I.S.; Sanche, S.; Wiederhold, N.P.; Patterson, T.F.; Sigler, L. Emergomyces Canadensis, a Dimorphic Fungus Causing Fatal Systemic Human Disease in North America. *Emerg. Infect. Dis.* **2018**, *24*, 758–761. [CrossRef]
194. Samaddar, A.; Sharma, A. Emergomycosis, an Emerging Systemic Mycosis in Immunocompromised Patients: Current Trends and Future Prospects. *Front. Med.* **2021**, *8*, 670731. [CrossRef] [PubMed]
195. Wang, P.; Kenyon, C.; de Hoog, S.; Guo, L.; Fan, H.; Liu, H.; Li, Z.; Sheng, R.; Yang, Y.; Jiang, Y.; et al. A Novel Dimorphic Pathogen, Emergomyces Orientalis (Onygenales), Agent of Disseminated Infection. *Mycoses* **2017**, *60*, 310–319. [CrossRef] [PubMed]
196. Schwartz, I.S.; Govender, N.P.; Corcoran, C.; Dlamini, S.; Prozesky, H.; Burton, R.; Mendelson, M.; Taljaard, J.; Lehloenya, R.; Calligaro, G.; et al. Clinical Characteristics, Diagnosis, Management, and Outcomes of Disseminated Emmonsiosis: A Retrospective Case Series. *Clin. Infect. Dis. Off. Publ. Infect. Dis. Soc. Am.* **2015**, *61*, 1004–1012. [CrossRef]
197. Maphanga, T.G.; Naicker, S.D.; Gómez, B.L.; Mhlanga, M.; Mpembe, R.S.; Schwartz, I.S.; Bamford, C.; Nel, J.; Govender, N.P. Cross-Reactivity of a *Histoplasma capsulatum* Antigen Enzyme Immunoassay in Urine Specimens from Persons with Emergomycosis in South Africa. *Med. Mycol.* **2021**, *59*, 672–682. [CrossRef]
198. Schwartz, I.S.; Kenyon, C.; Lehloenya, R.; Claasens, S.; Spengane, Z.; Prozesky, H.; Burton, R.; Parker, A.; Wasserman, S.; Meintjes, G.; et al. AIDS-Related Endemic Mycoses in Western Cape, South Africa, and Clinical Mimics: A Cross-Sectional Study of Adults with Advanced HIV and Recent-Onset, Widespread Skin Lesions. *Open Forum Infect. Dis.* **2017**, *4*, ofx186. [CrossRef]

Journal of
Fungi

MDPI

Review

Sporothrix brasiliensis Causing Atypical Sporotrichosis in Brazil: A Systematic Review

Vanice Rodrigues Poester [1,2,†], Melissa Orzechowski Xavier [1,2,†], Lívia Silveira Munhoz [1,2], Rossana Patricia Basso [1,2,3], Rosely Maria Zancopé-Oliveira [4], Dayvison Francis Saraiva Freitas [4] and Alessandro Comarú Pasqualotto [5,6,*]

1 Programa de Pós-Graduação em Ciências da Saúde, Faculdade de Medicina (FAMED), Universidade Federal do Rio Grande (FURG), Rio Grande 96200-190, Rio Grande do Sul (RS), Brazil; vanicerp@gmail.com (V.R.P.); melissaxavierfurg@gmail.com (M.O.X.); liviasmunhoz@gmail.com (L.S.M.); rossanabasso3@gmail.com (R.P.B.)
2 Mycology Laboratory of FAMED-FURG, Rio Grande 96200-190, RS, Brazil
3 Hospital Universitário Dr. Miguel Riet Correa Jr., FURG/Empresa Brasileira de Serviços Hospitalares (EBSERH), Rio Grande 96200-190, RS, Brazil
4 Instituto Nacional de Infectologia Evandro Chagas, Fundação Oswaldo Cruz, Rio de Janeiro 21040-360, Rio de Janeiro (RJ), Brazil; rosely.zancope@ini.fiocruz.br (R.M.Z.-O.); dayvison.freitas@ini.fiocruz.br (D.F.S.F.)
5 Molecular Biology Laboratory, Hospital Dom Vicente Scherer, Santa Casa de Misericórdia de Porto Alegre, Porto Alegre 90020-090, Brazil
6 Medicine Department, Universidade Federal de Ciências da Saúde de Porto Alegre, Porto Alegre 90035-075, RS, Brazil
* Correspondence: pasqualotto@ufcspa.edu.br
† These authors contributed equally to this work.

Citation: Poester, V.R.; Xavier, M.O.; Munhoz, L.S.; Basso, R.P.; Zancopé-Oliveira, R.M.; Freitas, D.F.S.; Pasqualotto, A.C. *Sporothrix brasiliensis* Causing Atypical Sporotrichosis in Brazil: A Systematic Review. *J. Fungi* **2024**, *10*, 287. https://doi.org/10.3390/jof10040287

Academic Editor: Fernando José Santos Rodrigues

Received: 9 March 2024
Revised: 3 April 2024
Accepted: 11 April 2024
Published: 13 April 2024

Abstract: Zoonotic sporotrichosis, a subcutaneous mycosis caused by *Sporothrix brasiliensis*, has become hyperendemic and a serious public health issue in Brazil and an emerging disease throughout the world. Typical sporotrichosis is defined as fixed or lymphocutaneous lesion development, however, reports of atypical presentations have been described in hyperendemic areas, which may result in a worse prognosis. Thus, considering an increase in atypical cases and in more severe extracutaneous cases and hospitalizations reported in Brazil, we aimed to perform a systematic review to search for hypersensitivity reactions (HRs) and extracutaneous presentations associated with zoonotic sporotrichosis. A systematic review was performed, following the PRISMA guidelines to search for atypical/extracutaneous cases (mucosal, osteoarthritis, HRs, pulmonary, meningeal) of zoonotic sporotrichosis. A total of 791 published cases over 26 years (1998–2023) in eleven Brazilian states were reviewed. Most cases corresponded to a HR (47%; n = 370), followed by mucosal (32%; n = 256), multifocal (8%; n = 60), osteoarthritis (7%; n = 59), meningeal (4%; n = 32), and pulmonary (2%; n = 14) infections. When available (n = 607), the outcome was death in 7% (n = 43) of cases. Here, we show a frequent and worrisome scenario of zoonotic sporotrichosis in Brazil, with a high and dispersed incidence of atypical/extracutaneous cases throughout the Brazilian territory. Therefore, educational measures are necessary to make health professionals and the overall population aware of this fungal pathogen in Brazil as well as in other countries in the Americas.

Keywords: zoonosis; ocular; nasal; hypersensitivity; osteoarthritis; pulmonary; meningeal disease

1. Introduction

Sporothrix brasiliensis is one of the major causes of sporotrichosis, a neglected implantation mycosis that has become a significant public health problem in South America, especially in Brazil. Thousands of cases of zoonotic sporotrichosis have been described, and, in the last decade, this disease has spread all over the Brazilian territory [1–3]. Not restricted to Brazil anymore, nowadays *S. brasiliensis* is also an emergent pathogen in other South American countries, with autochthonous cases being described in Chile, Argentina, Paraguay, and Uruguay. Furthermore, cases of zoonotic sporotrichosis have been reported in recent years in other continents (North America and Europe) where patients were infected by cats imported from Brazil [4–9].

Domestic cats are the main victims of sporotrichosis caused by *S. brasiliensis* and the main source of the dissemination of this fungal species. Affected animals usually develop a severe manifestation of the disease, often with disseminated cutaneous and nasal lesions, which may potentially lead to death [10,11]. The high fungal burden found in feline cutaneous lesions explains how easily *S. brasiliensis* can be transmitted by bites or scratches to other cats or humans [12]. Since feline sporotrichosis outbreaks are increasing in number and spreading geographically, zoonotic sporotrichosis nowadays involves several epidemiological scenarios, including (i) areas of hyperendemicity (in many regions of Brazil and some Argentinian provinces); (ii) large outbreaks (in distinct regions of South America); and (iii) occasional cases occurring in other continents, resulting in an increased awareness.

Although most zoonotic sporotrichosis is clinically evidenced by the typical known fixed or lymphocutaneous lesions, reports of atypical presentations have been described in hyperendemic areas, which may result in a worse prognosis [13–21]. Mucosal involvement, hypersensitivity reactions (HRs), osteoarthritis, and pulmonary and meningeal involvement are highlighted as atypical presentations of sporotrichosis [17–22]. *S. brasiliensis* is more likely to cause atypical clinical manifestations of sporotrichosis than the sapronotic species, *S. schenckii* and *S. globosa* [14].

HRs in zoonotic sporotrichosis can occur as (i) Sweet's syndrome, characterized as painful erythematous papules, plaques, or nodules with neutrophil infiltrate, associated with fever; (ii) erythema nodosum, defined as the appearance of painful erythematous subcutaneous nodules in the lower limbs, with fever, arthralgia, and myalgia; (iii) erythema multiforme, an acute inflammatory reaction, showing erythematous plaques on the skin or mucosa, typically target-shaped; and (iv) arthritis, characterized as an aseptic inflammatory articular process resulting in pain and edema, frequently affecting the knees, wrists, elbows, or ankles [17,23–28]. Although atypical, the HRs fortunately have a good prognosis, but they require a correct diagnosis by healthcare professionals [17,23].

In mucosal involvement, ocular sporotrichosis is described in a diversity of ophthalmologic patterns, which occur as ocular adnexal or intraocular infections. Sporotrichosis in ocular adnexa is subclassified as eyelid involvement, granulomatous conjunctivitis, Parinaud syndrome, or disorders of the lacrimal system (dacryocystitis). Intraocular sporotrichosis can be presented as endophthalmitis, granulomatous uveitis, scleritis, retinitis, choroiditis, or iridocyclitis [15,29–32]. Nasal involvement is another manifestation of mucosal zoonotic sporotrichosis, which varies from mild to severe, progressing to a perforation of the septum [21].

Sporotrichosis in bones or causing septic articular involvement occurs mostly as a consequence of the extension of a cutaneous lesion, through the classical traumatic route, or due to *Sporothrix* sp. hematogenous dissemination [33]. Similarly, pulmonary and meningeal infection can follow *Sporothrix* conidia inhalation, after fungal dissemination [18,19,34–36]. Severe atypical presentations of sporotrichosis are often associated with comorbidities, such as AIDS, alcoholism, chronic obstructive pulmonary disease, or diabetes [13,18,37].

Considering the increase in the reports of these atypical manifestations in patients with zoonotic sporotrichosis, which can lead to a higher rate of hospitalization and a worse prognosis [3,9,23,38], we compiled data from all cases reported and/or series of cases published in the literature in a unique and combined approach, to better understand their impact. Therefore, here we perform a systematic review of the literature on zoonotic sporotrichosis involving HR and extracutaneous presentations.

2. Materials and Methods

A systematic review was performed, following the PRISMA guidelines [39], using the databases Pubmed, SciELO, Web of Science, and LILACS, which were last consulted in February 2024. Aiming to expand the search, references of the selected articles were reviewed to find additional articles. Descriptors were (extracutaneous and sporotrichosis) OR (epidemiological and *Sporothrix* and *brasiliensis*) OR (disseminated and *Sporothrix* and *brasiliensis*) OR (pulmonary and *Sporothrix* and *brasiliensis*) OR (ocular and *Sporothrix* and *brasiliensis*) OR (nasal and *Sporothrix* and *brasiliensis*) OR (hypersensitivity and *Sporothrix* and *brasiliensis*) OR (osteoarthritis and *Sporothrix* and *brasiliensis*) OR (osteomyelitis and *Sporothrix* and *brasiliensis*) OR (meningeal and *Sporothrix* and *brasiliensis*). No automation tools were used to search articles, and two researchers performed the search independently, reducing the bias.

We included cases of extracutaneous sporotrichosis (mucosal, osteoarthritis, pulmonary, meningeal) and HRs. Both cases of proven (with *Sporothrix* spp. isolation in culture from the primary site of infection) and probable (patients with clinical–epidemiological characteristics of zoonotic sporotrichosis) sporotrichosis were included [40]. Each case was categorized as (i) mucosal: patients with ocular (adnexal and/or intraocular), nasal and/or oral involvement; (ii) osteoarthritis: patients with osteomyelitis and/or septic arthritis; (iii) a HR: patients showing Sweet's syndrome, erythema nodosum, erythema multiforme, or aseptic arthritis in consequence of *Sporothrix* spp. infection; (iv) pulmonary: patients with lung involvement; (v) or meningeal: patients with central nervous system involvement. Cases of sporotrichosis were also classified as unifocal (involving a single site or two contiguous sites of infection/type of atypical presentation) or multifocal (two or more non-contiguous sites).

Articles were initially selected by title, then by abstract, without language restriction. Eligible articles were read in full and included when they met the inclusion criteria. Criteria for inclusion were articles (i) from 1990 (beginning of zoonotic outbreaks) to December 2023 and (ii) describing atypical sporotrichosis manifestations (mucosal, osteoarthritis, HR, pulmonary, or meningeal). Review articles and case series with no data regarding the site of infection were excluded.

Cases were grouped and analyzed in each category (HRs, mucosal, osteoarthritis, pulmonary, meningeal) regarding sex, age, dispersion in time and location, primary site of infection and sites of dissemination, presence of comorbidities, and outcome. Data compilation and analyses were performed with the Excel software (2013 version, Microsoft Corporation®, Redmond, WA, USA). Geographical analyses were performed using Google Maps (https://maps.google.com/, accessed on 28 February 2024) (Google®, Mountain View, CA, USA) superimposed on a world map, generating a Geographic Information System (GIS). QGIS software version 3.30.0 (Open Source Geospatial Foundation—OSGeo, Anchorage, AK, USA) was used to analyze the geographical distribution of cases.

3. Results

3.1. Article Searching

A total of 442 articles were selected for analysis (404 from databases and 38 from references of articles), and 206 were excluded due to duplicate records, with no other reason for excluding articles. After reviewing titles and abstracts, no additional records were retrieved, and 89 articles were selected for full-text reading. Of these, 64 articles were included in our systematic review (Figure 1).

The first case of atypical zoonotic sporotrichosis was published in 2002. Nearly half of the articles (55%; 35/64) were published in the last five years (2019–2023), which correspond to an increase of 700% in comparison with the number of publications from the first 5 years of our review (Figure 2) [13–15,17–22,24–29,31,32,34–37,41–83].

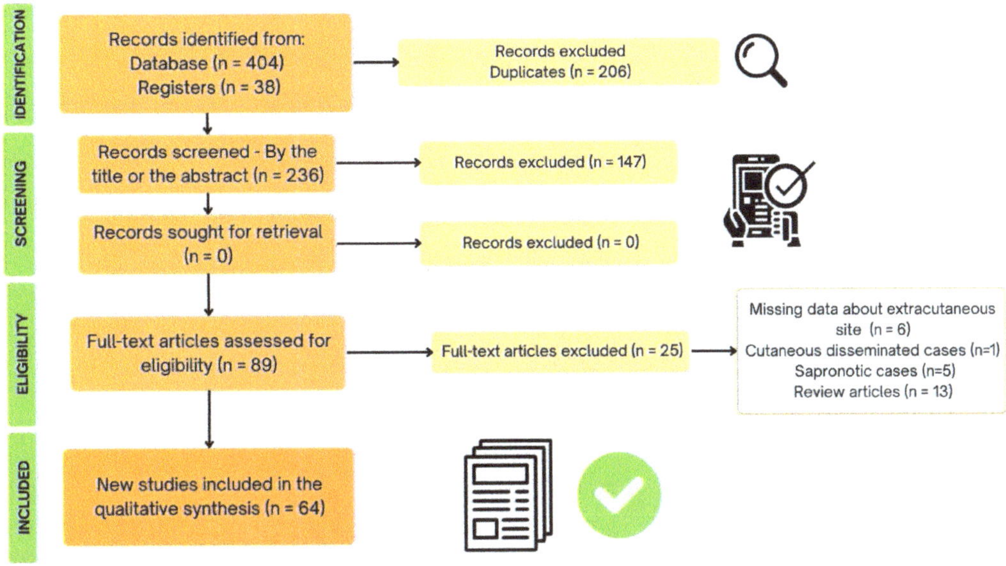

Figure 1. Flowchart describing the total of scientific articles obtained by the database searches for atypical cases of sporotrichosis by *Sporothrix brasiliensis* that occurred in Brazil and included in this study.

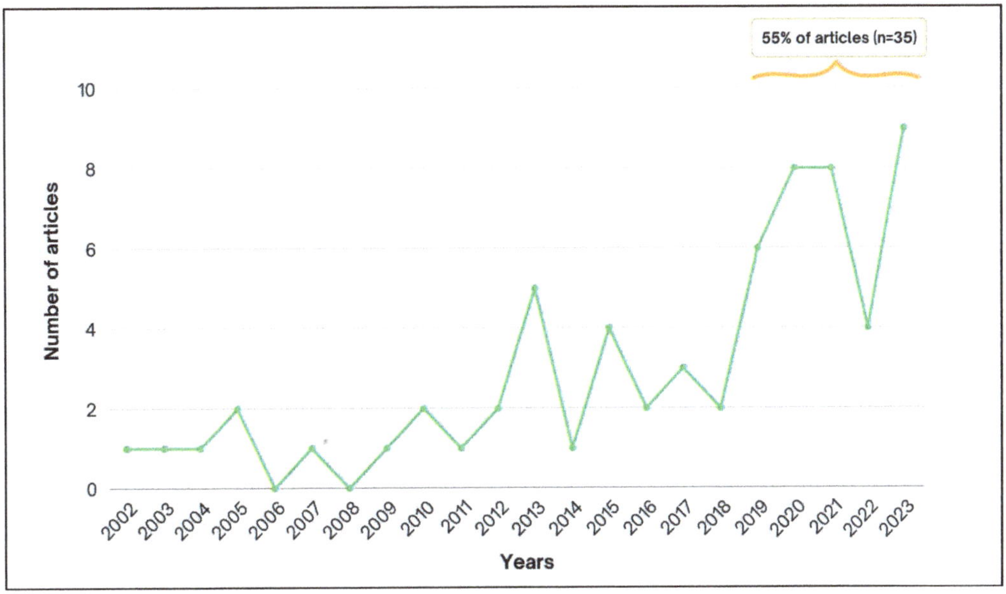

Figure 2. Temporal distribution of articles describing atypical zoonotic sporotrichosis caused by *Sporothrix brasiliensis* between the years of 2002 and 2023, highlighting an increase of 700% in the number of publications comparing two periods of five years (beginning and most recent period).

3.2. Extracutaneous and HR Cases

A total of 791 cases of patients showing atypical zoonotic sporotrichosis were described in the 64 articles included in this review, all of them in patients from Brazil. Geographic distribution showed the occurrence throughout eleven Brazilian states, with 88% (n = 695) of cases occurring in Rio de Janeiro. Other states with the description of patients with atypical zoonotic sporotrichosis were Minas Gerais (5%; n = 36), Paraná (3%; n = 23), Pernambuco (1%; n = 11), São Paulo (1%; n = 8), Rio Grande do Sul (<1%; n = 7), Paraíba (<1%; n = 7), Espírito Santo, Bahia, Distrito Federal, and Rio Grande do Norte (<1%; n = 4; one case from each of the last four states described) (Figure 3).

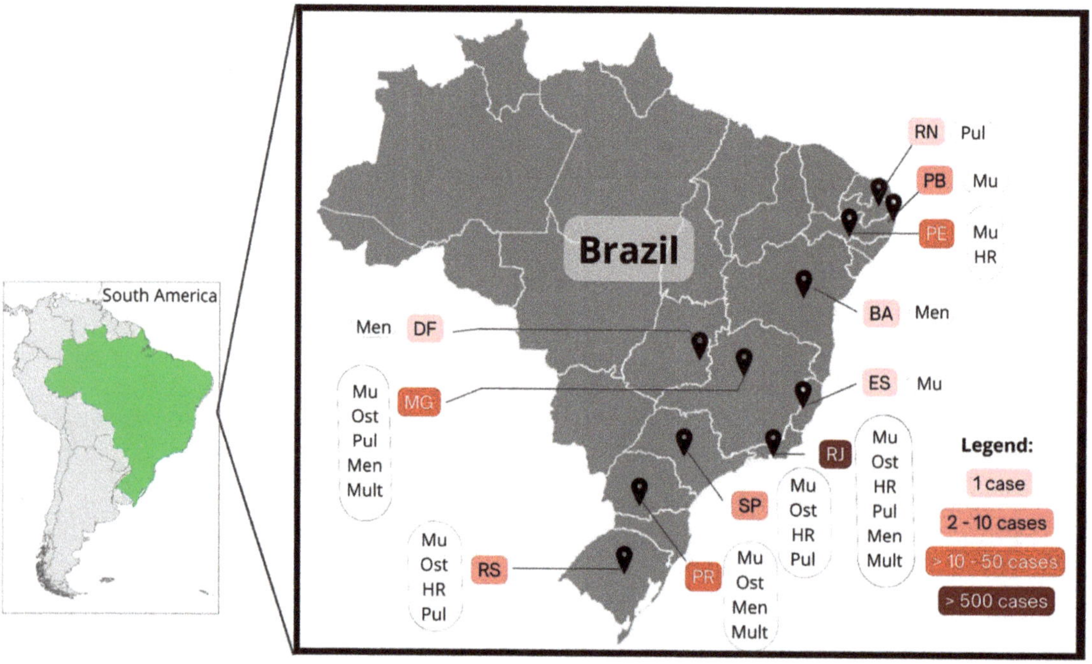

Figure 3. South America map highlighting (green color) the only country (Brazil) where atypical zoonotic sporotrichosis has been reported and a Brazilian map showing the distribution of cases reported through states. DF: Distrito Federal; MG: Minas Gerais; RS: Rio Grande do Sul; PR: Paraná; SP: São Paulo; RJ: Rio de Janeiro; ES: Espírito Santo; BA: Bahia; PE: Pernambuco; PB: Paraíba; RN: Rio Grande do Norte. Mu: mucosal; Ost: osteoarthritis; HR: hypersensitivity reaction; Pul: pulmonary; Men: meningeal; Mult: multifocal.

Most patients, 66% (453/689), were female, and the mean age was 38 years old (ranging from 2 to 80, standard deviation: 18; n = 682). More than 90% of cases involved close contact with cats, and in 98% a proven diagnosis of sporotrichosis was achieved by the isolation of the fungus in mycological culture.

Into the group of patients with unifocal atypical zoonotic sporotrichosis (n = 731), a HR was the main presentation occurring in 51% of the patients (n = 370), with mucosal involvement being the second mainly atypical condition, occurring in 35% (n = 256) of the patients. Patients with more severe patterns of atypical zoonotic sporotrichosis totalized 105, including 8% (n = 59) with osteoarthritis, 4% (n = 32) with meningeal involvement, and 2% (n = 14) with pulmonary involvement. The other 60 patients had multifocal disease, often (53 patients) involving two sites/types of presentation: osteoarthritis + mucosal (n = 32); osteoarthritis + pulmonary (n = 14); osteoarthritis + meningeal (n = 3); mucosal + meningeal (n = 2); and meningeal + pulmonary (n = 2). There were six patients with three sites: osteoarthritis + meningeal + pulmonary (n = 3); osteoarthritis + meningeal + mucosal (n = 2); and osteoarthritis + mucosal + pulmonary (n = 1). One patient showed osteoarthritis + mucosal + meningeal + pulmonary involvement (Tables 1 and 2).

Table 1. Extracutaneous zoonotic sporotrichosis data.

Extracutaneous Type	Clinical Manifestation/Site of Infection	Number	Immunosuppression/ Comorbidities (%)	Primary Site of Infection (%)	Outcome (%)	Refs.
Hypersensitivity		370	Yes 11/29 (38%) No 18/29 (62%)	Not applicable	Cure 318/318 (100%)	[14,17,24–28,42, 47,54,69,76,81]
	Erythema nodosum	184				
	Erythema multiforme	143				
	Sweet syndrome	35				
	Arthritis	8				
Mucosal		256	Yes 29/65 (45%) No 36/65 (55%)	Yes 229/250 (92%) No 21/250 (8%)	Cure 193/193 (100%); # Sequelae 23/193 (12%)	[15,20,21,29,31,32, 41,42,47,48,52,53, 57,60,61,63–65,67,70,71,73–75,78,79,81,82]
	Ocular	221				
	Nasal	28				
	Oral	2				
	Ocular and nasal	1				
	Oral and nasal	4				
Osteoarthritis	Hand Upper and lower limb Foot and/or ankle Knee Elbow Clavicle	59 26 9 6 3 1 1	Yes 26/28 (93%) No 2/28 (7%)	Yes 22/49 (45%) No 27/49 (55%)	Cure 22/26 (85%); ## Sequelae 3/22 (14%) Death 4/26 (15%)	[22,37,46,50,56,76, 79–81,83]
Meningeal		32	Yes 29/32 (91%) No 3/32 (9%)	Yes 3/32 (9%) No 29/32 (91%)	Cure 9/30 (30%) ### Sequelae 1/9 (11%); Death 21/30 (70%)	[19,34–36,43–45,47,55,62]
Pulmonary	Cavitary nodules Bronchiectasis Infiltrate Pleural effusion Fibrosis	14 4 2 1 1 1	Yes 7/9 (78%) No 2/9 (22%)	Yes 10/14 (71%) No 4/14 (29%)	Cure 3/5 (60%) Death 2/5 (40%)	[18,37,49,51,59,66, 68,72,77,81]
Multifocal		60	Yes 46/46 (100%) No 0/46 (0%)	Not applicable	Cure 19/35 (54%) #### Sequelae 2/19 (11%); Death 16/35 (46%)	[18,19,21,22,37,47, 58,71,79]
	* Two sites	53				
	** Three sites	6				
	*** Four sites	1				

Mucosal sequelae: ocular (chronic dacryocystitis, corneal changes, cutaneous fistula, lagophthalmos, ectropion, entropion, pannus 180°, symblepharon, conjunctival fibrosis, paracentral leucoma, or eyelid retraction) or nasal (hyperrhynolalia and retraction of the right ala nasi); ## sequelae osteoarthritis: amputation, total or partial; ### meningeal sequelae: neurological sequelae (ataxia and extrinsic ocular motor paresis related to basal meningitis); #### multifocal sequelae: amputation. * Two sites: osteoarthritis + mucosal (n = 30); osteoarthritis + pulmonary (n = 13); hypersensitivity + meningeal (n = 6); osteoarthritis + meningeal (n = 3); meningeal + pulmonary (n = 2); mucosal + meningeal (n = 2); hypersensitivity + mucosal (n = 1). ** Three sites: osteoarthritis + meningeal + pulmonary (n = 3); osteoarthritis + meningeal + mucosal (n = 2); osteoarthritis + mucosal + pulmonary (n = 1). *** Four sites: osteoarthritis + mucosal + meningeal + pulmonary. REF: reference.

Comorbidities were described in 53% (110/209) of the total of patients for whom these data were available. The predominant condition was HIV infection (n = 74), followed by systemic arterial hypertension (n = 20), alcohol abuse (n = 19), and diabetes (n = 12). Other comorbidities described were tuberculosis (n = 7), renal transplantation or chronic renal failure (n = 4), chronic obstructive pulmonary disease (n = 3), asthma (n = 2), bacteremia (n = 2), tobacco abuse (n = 2), corticosteroid use (n = 2), hepatitis C infection (n = 2), malnutrition (n = 2), cytomegalovirus infection (n = 2), and epilepsy (n = 2). Additional conditions included (n = 1 each) drug abuse, chronic anemia, bronchiectasis, post-COVID-19 with asthma, deep vein thrombosis, pneumocystosis, neurotoxoplasmosis, nocardiosis, sarcoidosis, dyslipidemia, Takayasu arthritis, lepra, fibromyalgia, hypothyroidism, hepatic steatosis, benign prostatic hyperplasia, and/or pulmonary arterial hypertension. Most

(62%; 18/29) patients with a HR had no underlying condition, while in the mucosal form of sporotrichosis 45% (29/65) of the patients had comorbidities. In the more severe forms of sporotrichosis (osteoarthritis, meningeal, pulmonary, and multifocal), 94% (108/115) of patients had comorbidities.

Table 2. Outcome of multifocal cases of extracutaneous zoonotic sporotrichosis.

Number of Sites	Types	N	Outcome (%)	Refs.
Two	Osteoarthritis + mucosal	32	Cure (68%; 13/19) Death (32%; 6/19)	[18,19,21,22,37,47,58,71,79]
	Osteoarthritis + pulmonary	14	Cure (67%; 6/9) Death (33%; 3/9)	
	Osteoarthritis + meningeal	3	Death (100%; 2/2)	
	Meningeal + pulmonary	2	Death (100%; 1/1)	
	Mucosal + meningeal	2	---	
Three	Osteoarthritis + meningeal + pulmonary	3	Death (100%; 2/2)	
	Osteoarthritis + meningeal + mucosal	2	Death (100%; 1/1)	
	Osteoarthritis + mucosal + pulmonary	1	Death (100%; 1/1)	
Four	Osteoarthritis + mucosal + meningeal + pulmonary	1	---	

N: number; Refs.: reference.

3.2.1. HR Cases

Erythema nodosum or multiforme represented 88% of the HRs (327/370), followed by Sweet's syndrome in 10% (35/370) and arthritis in 2% (8/370) (Table 1). The patients being cured was the outcome in all zoonotic sporotrichosis cases that developed HRs (n = 318) [14,17,24–28,42,47,54,69,76,81].

3.2.2. Mucosal Cases

Mucosal cases were predominantly ocular (86%; 221/256), followed by nasal (11%; 28/256) and oral (<1%; 2/256). Four patients (2%) had oral and nasal sporotrichosis associated, and one patient (<1%) showed ocular and nasal lesions. Data of the pattern of ocular disease were available from 199/221 patients, 158 (83%) with only one pattern (conjunctivitis, n = 123; eyelids, n = 27; dacryocystitis, n = 7; retinal granuloma, n = 1), 39 presenting two associated patterns (conjunctivitis + Parinaud syndrome, n = 23; conjunctivitis + dacryocystitis, n = 5; conjunctivitis + eyelids, n = 3; Parinaud syndrome + eyelids, n = 3; Parinaud syndrome + dacryocystitis, n = 3; or dacryocystitis + eyelids, n = 2), and two patients showing three associated patterns (conjunctivitis + eyelids + dacryocystitis, n = 2). One patient presented ocular involvement (Parinaud syndrome + dacryocystitis) together with a HR.

Mucosal involvement occurred as a primary disease in the majority of the patients (92%, n = 230/251), while the remaining (8%, n = 21) developed mucosal sporotrichosis due to a hematogenous dissemination of *Sporothrix* spp. Even though all such patients were cured, 12% (23/193) had sequelae of the mycosis, such as chronic dacryocystitis, corneal changes, cutaneous fistula, lagophthalmos, ectropion, entropion, pannus 180°, symblepharon, conjunctival fibrosis, paracentral leucoma or eyelid retraction (ocular), and hyperrhynolalia and retraction of the right ala nasi (nasal) (Table 1) [15,20,21,29,31,32,41,42,47,48,52,53,57,60,61,63–65,67,70,71,73–75,78,79,81,82].

3.2.3. Osteoarthritis Cases

Osteoarthritis as an extracutaneous manifestation of zoonotic sporotrichosis was reported in 59 patients, 42 as osteomyelitis and 17 as septic arthritis. Information on the site of osteoarthritis was available for 46 patients. This most frequently involved the hands (56%; n = 26), followed by upper and/or lower limbs (20%; n = 9), feet and/or ankles (13%; n = 6), knees (7%; n = 3), elbows (2%; n = 1), and clavicles (2%; n = 1). Fifty-five percent (27/49) occurred after fungal hematogenous dissemination, with the bone involvement a consequence (contiguity infection) of a cutaneous lesion in the other 22 patients (45%); in 10 patients these data were unavailable. The outcomes were described in case reports of 26 patients, with 85% (22/26) being cured and 15% (4/26) dying. From the 22 cured patients, sequelae (total or partial amputation) occurred in 14% (3/22) (Table 1) [22,37,46,50,56,76,79–81,83].

3.2.4. Pulmonary Cases

Pulmonary zoonotic sporotrichosis occurred as a primary disease in 71% of patients (10/14), and in 29% (4/14) it occurred because of a hematogenous fungal dissemination. The infection resulted in cavitary nodules (n = 4), bronchiectasis (n = 2), lung infiltrate (n = 1), pleural effusion (n = 1), and/or fibrosis (n = 1) (data available for eight patients only). In 93% (13/14) of cases, *Sporothrix* spp. was isolated from respiratory samples in mycological cultures (Table 1). In total, 60% of these patients were cured (3/5), including two patients with disseminated disease and one patient with primary lung involvement [18,37,49,51,59,66,68,72,77,81].

3.2.5. Meningeal Cases

Meningeal involvement was reported in 32 cases, the majority (91%, 29/32) occurring in immunosuppressed patients, with sporotrichosis disseminating to the central nervous system. Underlying conditions for these patients included HIV infection (n = 25), drug/alcohol abuse (n = 3), liver transplantation (n = 1), leprosy (n = 1), or chronic steroid use (n = 1), with an overlap of diseases in two cases (drug/alcohol abuse and liver transplantation; leprosy and chronic steroid use). The three (9%) other patients with neurological sporotrichosis as a primary disease were immunocompetent, and the diagnosis was confirmed by the isolation of *Sporothrix* spp. from the cerebrospinal fluid. Death was the predominant outcome (70%, 21/30), with one immunocompetent individual, and 30% (9/30) of patients were cured, including two immunocompetent patients, with sequelae (ataxia and extrinsic ocular motor paresis) reported in one (11%; 1/9) [19,34–36,43–45,47,55,62].

3.2.6. Multifocal Cases

Among 60 patients with multifocal atypical sporotrichosis, 88% (53/60) presented two sites of involvement, while others had three (10%; 6/60) or four (2%; 1/60) noncontiguous sites of infection (Table 1). The predominant outcome of multifocal cases was the patient being cured (54%; 19/35); sequelae occurred at a rate of 11% (2/19), associated with bone involvement; and death occurred in 46% (16/35) of patients. It was reported that 68% and 67% of patients with osteoarthritis plus mucosal or pulmonary involvement were cured, respectively. However, all patients with the following presentations died: osteoarthritis + meningeal; meningeal + pulmonary; osteoarthritis + meningeal + pulmonary; osteoarthritis + meningeal + mucosal; and osteoarthritis + mucosal + pulmonary [18,19,21,22,37,47,58,71,79] (Table 2).

4. Discussion

In this systematic review, we summarized ~800 atypical zoonotic sporotrichosis cases described in the literature, with an exponential increase in the number of papers reporting these atypical cases of zoonotic sporotrichosis in the last five years of study, which coincides with the significant expansion of the disease throughout the Brazilian territory [84].

All the atypical cases of zoonotic sporotrichosis occurred in Brazil, which is considered the epicenter of the disease in the world [9], and the majority occurred in the state of Rio de Janeiro, where this mycosis has a hyperendemic status, in a scenario that began as an outbreak at the end of the 1990s [3,42]. Besides Rio de Janeiro, atypical zoonotic sporotrichosis cases were also reported in patients from ten additional Brazilian states. Two of these states (Rio Grande do Sul and São Paulo) are nowadays hyperendemic areas for sporotrichosis, starting with outbreak reports since the 2000s [76,85–87], and in eight other states (Paraná, Distrito Federal, Minas Gerais, Pernambuco, Bahia, Rio Grande do Norte,

Paraíba, and Espírito Santo) outbreaks of *S. brasiliensis* infection have been reported in the last five/ten years [79,88–93].

Thus, we can speculate that, in general, atypical cases of zoonotic sporotrichosis are not directly related to areas with a high endemicity already established for long periods. Instead, atypical cases can occur from the beginning of *S. brasiliensis* reports, which can be attributed to fungi strain characteristics, considering that they exhibit distinct virulence profiles and genotypes [94,95]. In fact, it has been demonstrated that at least seven genotypes of *S. brasiliensis* are currently circulating in the Brazilian territory [95,96] and that these different genotypes can vary in terms of virulence profiles and susceptibility to antifungal drugs. Considering the incidence of different clinical patterns in distinct Brazilian states, which can be associated with the genotypes and the evolution of fungi from different environmental conditions, in addition to the well-known importance of cats on the sporotrichosis epidemiological chain, a One Health approach is necessary together with the investigation of clinical patterns in humans and distinct genotypes of *S. brasiliensis* to clarify these points.

A potential limitation of our systematic review was the bias caused by descriptor delimitation and by the non-registration of the systematic review before the beginning of the methods. In addition, the impossibility of recovering other nonscientific material using databases, like epidemiological bulletins or informal notices, which could report more cases from the same or other areas, is highlighted as an additional limitation.

HRs, with a predominance of erythema (nodosum or multiforme), described in patients from four Brazilian states (Rio de Janeiro, São Paulo, Rio Grande do Sul, and Pernambuco), have been associated with continuous exposure to the *S. brasiliensis* antigen, resulting in immune sensitization by the patients. The exposure to high levels of *S. brasiliensis* antigens commonly occurs in hyperendemic areas of zoonotic sporotrichosis. Close and prolonged contact of patients with lesions that are rich in fungal propagules from infected cats contribute to the overexpression of an immunological reaction during the disease development, which probably explains the high number of HRs in patients with zoonotic sporotrichosis in Brazil [12,23]. However, the exact immunological mechanism that promotes this inflammatory exacerbation still needs to be unveiled, contributing to research in the immune field of sporotrichosis [97]. In particular, the role of the host defense in determining disease predisposition is currently unknown, along with the protagonism of variables such as genetic predisposition, inoculum size, genotype, and *S. brasiliensis* virulence factors. Although HRs have a good prognosis and are not potentially severe, they need to be promptly and correctly diagnosed for the adequate management and treatment of the cases.

Mucosal sporotrichosis, a prevalent condition that usually affects immunocompetent patients, occurs mainly as conjunctivitis, dacryocystitis, and Parinaud syndrome. These unspecific manifestations argue for the necessity of ophthalmologists' awareness to include the infection by *S. brasiliensis* in the roll of the diagnostic hypotheses in patients from endemic areas. The clinical features are also distinct from ocular sporotrichosis caused by *S. schenckii* and *S. globosa*, which often manifest as eyelid lesions after an accidental traumatic fungal inoculation [98,99]. An explanation for this difference can be attributed to the source of infection, since in zoonotic sporotrichosis the mucosal surface is directly infected through contact with fungal propagules carried by sneezes of infected cats, which can carry a high quantity of fungal cells of *S. brasiliensis* [20,30,66,100].

As demonstrated in this systematic review, osteoarthritis caused by *S. brasiliensis* was an important atypical manifestation of zoonotic sporotrichosis, resulting in relevant impacts on the quality of life and prognosis of the patients [22,37,50,56,76,79,81,83]. Its consequences, like partial or total amputation, are unfortunately also common in patients with other neglected fungal diseases, such as mycetoma and chromoblastomycosis, and these are often associated with a late diagnosis [101]. Bone sporotrichosis was rarely reported until the emergence of the species *S. brasiliensis* and is frequently associated with various degrees of immunosuppression. Considering that 59 cases of osteoarthritis were published in Brazil over 24 years (1998 to 2021), while a previous review of the literature (1980–2015) found 20 cases over 36 years [33], we can hypothesize that *S. brasiliensis* is associated with more severe manifestations of sporotrichosis, in comparison to other species of *Sporothrix* [14,33]. This hypothesis is supported by in vivo studies. Arrillaga-Moncrieff et al. [102] showed a more expressive virulence of *S. brasiliensis* in a murine model in comparison with *S. globosa* and *S. schenckii*, stimulating a more severe form of sporotrichosis with the highest rate of dissemination and more severely cutaneous lesions. In addition, an in vivo study with different isolates recovered from one unique human patient in different periods showed that *S. brasiliensis* can increase its virulence during pathogenesis [103].

In the same line, systemic sporotrichosis with the involvement of internal organs (brain and lungs) has severe patterns of sporotrichosis and results in a poor prognosis, warning of a worrisome situation in Brazil [34,36,66]. Pulmonary sporotrichosis is, fortunately, still considered a rare condition, but it is associated with unspecific symptoms that overlap with other lung infections (including tuberculosis and histoplasmosis) [104]. This factor contributes to delays in its diagnosis, potentially leading to severe lung parenchyma destruction, which is associated with high rates of death, bringing the necessity to add sporotrichosis to the differential diagnosis of pulmonary fungal infections, particularly in hyperendemic areas in Brazil [72].

Meningitis caused by *S. brasiliensis* also showed unspecific symptoms (i.e., headache, mental confusion, weight loss, lethargy, and vomiting), similar to other meningeal infections, including tuberculosis, syphilis, and cryptococcosis [105]. Meningitis usually follows hematogenous dissemination, mostly in immunosuppressed patients. However, it may also occur in immunocompetent patients as a primary disease with brain involvement, which brings the necessity of in vivo studies to evaluate if some *S. brasiliensis* strains could have a primary neurological tropism, which has already been described in other fungi such as *Cryptococcus neoformans* [106,107] and evidenced in in vivo experimental studies [108,109].

The high rate of deaths and sequelae evidenced after the compilation and analyses of ~800 cases in our study can be attributed to poor host conditions, such as immunosuppression and other co-infections and comorbidities, and may have been aggravated by the unfamiliarity of health professionals with the unusual clinical presentations of sporotrichosis, with delayed diagnosis and treatment. Thus, it is urgently necessary to invest in educational programs for health professionals and the general population regarding zoonotic sporotrichosis, to increase the knowledge about individual control and preventive measures and to reduce underdiagnosis of patients. In addition, it is necessary to make efforts to continue educational activities to highlight the necessity of including sporotrichosis as a differential diagnosis of these atypical clinical manifestations in hyperendemic areas [110–112]. In addition, in the Brazilian territory, sporotrichosis is a disease of non-compulsory notification, which certainly makes cases underestimated; thus, our compiled cases are only a part of the real cases. It is necessary to implement public politics at a national level to know the real scenario of sporotrichosis in Brazil.

Another gap to work with is the development of novel diagnostic tests for sporotrichosis, since cultures (current gold standard) require 7–21 days for a conclusive positive result [12,113]. Even though such a delay in non-severe cases may not be a problem, in systemic cases this can directly influence the prognosis, sequelae, and outcome of patients. Therefore, the development of rapid tests is urged to diagnose extracutaneous and atypical cases of zoonotic sporotrichosis, being a promising field for research and innovation.

5. Conclusions

In conclusion, here we have compiled and summarized data on atypical zoonotic sporotrichosis, showing that severe and unusual manifestations of this mycosis are not uncommon and are geographically dispersed in the Brazilian territory, supporting previous in vivo and clinical studies that hypothesize that *S. brasiliensis* is a species that is more virulent than sapronotic *Sporothrix* species. Therefore, educational measures are needed to make health professionals and the general population aware of different zoonotic sporotrichosis manifestations that can occur in endemic areas of Brazil. In addition, research should evolve to improve diagnosis, aiming for a quicker diagnostic method in severe cases, and the role of new antifungal agents should also be evaluated. A better understanding of the relevance of *S. brasiliensis* genotypes in relation to these diverse clinical presentations is ultimately needed.

Author Contributions: Visualization, A.C.P. and M.O.X.; conceptualization, V.R.P. and R.P.B.; methodology, V.R.P. and L.S.M.; formal analysis, V.R.P., L.S.M., R.P.B. and D.F.S.F.; writing—original draft preparation, V.R.P. and L.S.M.; writing—review and editing, A.C.P., R.M.Z.-O., D.F.S.F. and M.O.X. All authors have read and agreed to the published version of the manuscript.

Funding: This research received no external funding.

Institutional Review Board Statement: Not applicable.

Informed Consent Statement: Not applicable.

Data Availability Statement: Data sharing is not applicable to this article.

Acknowledgments: The authors are grateful to Coordenação de Aperfeiçoamento de Pessoal de Nível Superior (CAPES), to Conselho Nacional de Desenvolvimento Científico e Tecnológico (CNPq), and to Fundação de Amparo à pesquisa do Estado do Rio Grande do Sul (FAPERGS).

Conflicts of Interest: A.C. Pasqualotto has received research grants and/or consulted on behalf of Gilead, Pfizer, MSD, Knight Therapeutics, IMMY, and PAHO.

References

1. Poester, V.R.; Mattei, A.S.; Madrid, I.M.; Pereira, J.T.B.; Klafke, G.B.; Sanchotene, K.O.; Brandolt, T.M.; Xavier, M.O. Sporotrichosis in Southern Brazil, towards an Epidemic? *Zoonoses Public Health* **2018**, *65*, 815–821. [CrossRef] [PubMed]
2. Falcão, E.M.M.; Romão, A.R.; Magalhães, M.d.A.F.M.; Filho, J.B.d.L.; Valle, A.C.F.d.; Bastos, F.I.; Gutierrez-Galhardo, M.C.; Freitas, D.F.S. A Spatial Analysis of the Spread of Hyperendemic Sporotrichosis in the State of Rio de Janeiro, Brazil. *J. Fungi* **2022**, *8*, 434. [CrossRef]
3. Rabello, V.B.S.; Almeida, M.A.; Bernardes-Engemann, A.R.; Almeida-Paes, R.; de Macedo, P.M.; Zancopé-Oliveira, R.M. The Historical Burden of Sporotrichosis in Brazil: A Systematic Review of Cases Reported from 1907 to 2020. *Braz. J. Microbiol.* **2022**, *53*, 231–244. [CrossRef] [PubMed]
4. Kaadan, M.I.; Dennis, M.; Desai, N.; Yadavalli, G.; Lederer, P. One Health Education for Future Physicians: A Case Report of Cat-Transmitted Sporotrichosis. *Open Forum Infect. Dis.* **2020**, *7*, ofaa049. [CrossRef] [PubMed]
5. Etchecopaz, A.; Toscanini, M.A.; Gisbert, A.; Mas, J.; Scarpa, M.; Iovannitti, C.A.; Bendezú, K.; Nusblat, A.D.; Iachini, R.; Cuestas, M.L. *Sporothrix brasiliensis*: A Review of an Emerging South American Fungal Pathogen, Its Related Disease, Presentation and Spread in Argentina. *J. Fungi* **2021**, *7*, 170. [CrossRef] [PubMed]
6. Rachman, R.; Ligaj, M.; Chinthapalli, S.; Wani, R.S. Zoonotic Acquisition of Cutaneous *Sporothrix Braziliensis* Infection in the UK. *BMJ Case Rep.* **2022**, *15*, e248418. [CrossRef]
7. do Prado, C.M.; Razzolini, E.; Santacruz, G.; Ojeda, L.; Geraldo, M.R.; Segovia, N.; Brunelli, J.P.; Vicente, V.A.; Svoboda, W.K.; Queiroz-Telles, F. First Cases of Feline Sporotrichosis Caused by *Sporothrix brasiliensis* in Paraguay. *J. Fungi* **2023**, *9*, 972. [CrossRef]
8. Thomson, P.; González, C.; Blank, O.; Ramírez, V.; del Río, C.; Santibáñez, S.; Pena, P. Sporotrichosis Outbreak Due to *Sporothrix brasiliensis* in Domestic Cats in Magallanes, Chile: A One-Health-Approach Study. *J. Fungi* **2023**, *9*, 226. [CrossRef] [PubMed]
9. Xavier, M.O.; Poester, V.R.; Trápaga, M.R.; Stevens, D.A. *Sporothrix brasiliensis*: Epidemiology, Therapy, and Recent Developments. *J. Fungi* **2023**, *9*, 921. [CrossRef]
10. Schubach, T.M.; Schubach, A.; Okamoto, T.; Barros, M.B.; Figueiredo, F.B.; Cuzzi, T.; Fialho-Monteiro, P.C.; Reis, R.S.; Perez, M.A.; Wanke, B. Evaluation of an Epidemic of Sporothricosis in Cats: 347 Cases (1998–2001). *J. Am. Vet. Med. Assoc.* **2004**, *224*, 1623–1629. [CrossRef]
11. Sanchotene, K.O.; Madrid, I.M.; Klafke, G.B.; Bergamashi, M.; Terra, P.P.D.; Rodrigues, A.M.; de Camargo, Z.P.; Xavier, M.O. *Sporothrix brasiliensis* Outbreaks and the Rapid Emergence of Feline Sporotrichosis. *Mycoses* **2015**, *58*, 652–658. [CrossRef] [PubMed]
12. Gremião, I.D.F.; Martins da Silva da Rocha, E.; Montenegro, H.; Carneiro, A.J.B.; Xavier, M.O.; de Farias, M.R.; Monti, F.; Mansho, W.; de Macedo Assunção Pereira, R.H.; Pereira, S.A.; et al. Guideline for the Management of Feline Sporotrichosis Caused by *Sporothrix brasiliensis* and Literature Revision. *Braz. J. Microbiol.* **2021**, *52*, 107–124. [CrossRef] [PubMed]
13. Xavier, M.O.; Bittencourt, L.R.; Silva, C.M.; Vieira, R.S.; Pereira, H.C. Atypical Presentation of Sporotrichosis: Report of Three Cases. *Rev. Soc. Bras. Med. Trop.* **2013**, *46*, 116–118. [CrossRef] [PubMed]

14. Almeida-Paes, R.; de Oliveira, M.M.E.; Freitas, D.F.S.; do Valle, A.C.F.; Zancopé-Oliveira, R.M.; Gutierrez-Galhardo, M.C. Sporotrichosis in Rio de Janeiro, Brazil: *Sporothrix brasiliensis* Is Associated with Atypical Clinical Presentations. *PLoS Neglected Trop. Dis.* **2014**, *8*, e3094. [CrossRef] [PubMed]
15. Brandolt, T.M.; Madrid, I.M.; Poester, V.R.; Sanchotene, K.O.; Basso, R.P.; Klafke, G.B.; de Lima Rodrigues, M.; Xavier, M.O. Human Sporotrichosis: A Zoonotic Outbreak in Southern Brazil, 2012–2017. *Med. Mycol.* **2018**, *57*, 527–533. [CrossRef] [PubMed]
16. Poester, V.R.; Munhoz, L.S.; Basso, R.P.; Roca, B.M.; Vieira, M.U.; Melo, A.M.; Klafke, G.B.; Sanchotene, K.O.; Silveira, J.M.; Stevens, D.; et al. Disseminated Sporotrichosis with Immune Reconstitution Inflammatory Syndrome in an HIV Patient: Case Report and Review of the Literature. *Rev. Iberoam. Micol.* **2020**, *37*, 97–99. [CrossRef] [PubMed]
17. Procópio-Azevedo, A.C.; Rabello, V.B.S.; Muniz, M.M.; Figueiredo-Carvalho, M.H.G.; Almeida-Paes, R.; Zancopé-Oliveira, R.M.; Silva, J.C.A.L.; Macedo, P.M.; Valle, A.C.F.; Gutierrez-Galhardo, M.C.; et al. Hypersensitivity Reactions in Sporotrichosis: A Retrospective Cohort of 325 Patients from a Reference Hospital in Rio de Janeiro, Brazil (2005–2018). *Br. J. Dermatol.* **2021**, *185*, 1272–1274. [CrossRef]
18. Fichman, V.; Mota-Damasceno, C.G.; Procópio-Azevedo, A.C.; Almeida-Silva, F.; de Macedo, P.M.; Medeiros, D.M.; Astacio, G.S.-M.; Zancopé-Oliveira, R.M.; Almeida-Paes, R.; Freitas, D.F.S.; et al. Pulmonary Sporotrichosis Caused by *Sporothrix brasiliensis*: A 22-Year, Single-Center, Retrospective Cohort Study. *J. Fungi* **2022**, *8*, 536. [CrossRef] [PubMed]
19. Lima, M.A.; Freitas, D.F.S.; Oliveira, R.V.C.; Fichman, V.; Varon, A.G.; Freitas, A.D.; Lamas, C.C.; Andrade, H.B.; Veloso, V.G.; Almeida-Paes, R.; et al. Meningeal Sporotrichosis Due to *Sporothrix brasiliensis*: A 21-Year Cohort Study from a Brazilian Reference Center. *J. Fungi* **2023**, *9*, 17. [CrossRef]
20. Arinelli, A.; Aleixo, A.L.Q.C.; Freitas, D.F.S.; do Valle, A.C.F.; Almeida-Paes, R.; Guimarães, A.L.N.; Oliveira, R.V.C.; Gutierrez-Galhardo, M.C.; Curi, A.L.L. Ocular Manifestations of Sporotrichosis in a Hyperendemic Region in Brazil: Description of a Series of 120 Cases. *Ocul. Immunol. Inflamm.* **2023**, *31*, 329–337. [CrossRef]
21. Izoton, C.F.G.; de Brito Sousa, A.X.; Valete, C.M.; de Oliveira Schubach, A.; Procópio-Azevedo, A.C.; Zancopé-Oliveira, R.M.; de Macedo, P.M.; Gutierrez-Galhardo, M.C.; Castro-Alves, J.; Almeida-Paes, R.; et al. Sporotrichosis in the Nasal Mucosa: A Single-Center Retrospective Study of 37 Cases from 1998 to 2020. *PLOS Neglected Trop. Dis.* **2023**, *17*, e0011212. [CrossRef] [PubMed]
22. Ramos, V.; Astacio, G.S.-M.; do Valle, A.C.F.; de Macedo, P.M.; Lyra, M.R.; Almeida-Paes, R.; Oliveira, M.M.E.; Zancopé-Oliveira, R.M.; Brandão, L.G.P.; Quintana, M.S.B.; et al. Bone Sporotrichosis: 41 Cases from a Reference Hospital in Rio de Janeiro, Brazil. *PLOS Neglected Trop. Dis.* **2021**, *15*, e0009250. [CrossRef] [PubMed]
23. Schechtman, R.C.; Falcão, E.M.M.; Carard, M.; García, M.S.C.; Mercado, D.S.; Hay, R.J. Sporotrichosis: Hyperendemic by Zoonotic Transmission, with Atypical Presentations, Hypersensitivity Reactions and Greater Severity. *An. Bras. Dermatol.* **2022**, *97*, 1–13. [CrossRef] [PubMed]
24. Gutierrez-Galhardo, M.C.; Barros, M.B.L.; Schubach, A.O.; Cuzzi, T.; Schubach, T.M.P.; Lazéra, M.S.; Valle, A.C.F. Erythema Multiforme Associated with Sporotrichosis. *J. Eur. Acad. Dermatol. Venereol.* **2005**, *19*, 507–509. [CrossRef]
25. Gutierrez Galhardo, M.C.; de Oliveira Schubach, A.; de Lima Barros, M.B.; Blanco, T.C.M.; Cuzzi-Maya, T.; Schubach, T.M.P.; dos Santos Lazéra, M.; Francesconi do Valle, A.C. Erythema nodosum associated with sporotrichosis. *Int. J. Dermatol.* **2002**, *41*, 114–116. [CrossRef] [PubMed]
26. Orofino-Costa, R.; Bóia, M.N.; Magalhães, G.A.P.; Damasco, P.S.; Bernardes-Engemann, A.R.; Benvenuto, F.; Silva, I.C.; Lopes-Bezerra, L.M. Arthritis as a Hypersensitivity Reaction in a Case of Sporotrichosis Transmitted by a Sick Cat: Clinical and Serological Follow up of 131 Months. *Mycoses* **2010**, *53*, 81–83. [CrossRef] [PubMed]
27. de Lima, Í.M.F.; Ferraz, C.E.; de Lima-Neto, R.G.; Takano, D.M. Case Report: Sweet Syndrome in Patients with Sporotrichosis: A 10-Case Series. *Am. J. Trop. Med. Hyg.* **2020**, *103*, 2533–2538. [CrossRef] [PubMed]
28. Freitas, D.F.S.; Valle, A.C.F.; Cuzzi, T.; Brandão, L.G.P.; Zancope-Oliveira, R.M.; Galhardo, M.C.G. Sweet Syndrome Associated with Sporotrichosis. *Br. J. Dermatol.* **2012**, *166*, 212–213. [CrossRef]
29. de Macedo, P.M.; Sztajnbok, D.C.N.; Camargo, Z.P.; Rodrigues, A.M.; Lopes-Bezerra, L.M.; Bernardes-Engemann, A.R.; Orofino-Costa, R. Dacryocystitis Due to *Sporothrix brasiliensis*: A Case Report of a Successful Clinical and Serological Outcome with Low-Dose Potassium Iodide Treatment and Oculoplastic Surgery. *Br. J. Dermatol.* **2015**, *172*, 1116–1119. [CrossRef]
30. Zhang, Y.; Wang, Y.; Cong, L.; Yang, H.; Cong, X. Eyelid Sporotrichosis: Unique Clinical Findings in 72 Patients. *Australas. J. Dermatol.* **2016**, *57*, 44–47. [CrossRef]
31. Paiva, A.d.C.M.; Biancardi, A.L.; Curi, A.L.L. Different Clinical Manifestations of Ocular Sporotrichosis in the Same Patient: An Alert to Ophthalmologists in Nonendemic Areas. *Arq. Bras. Oftalmol.* **2020**, *83*, 457–458. [CrossRef] [PubMed]
32. Ramírez-Oliveros, J.F.; Schechtman, R.C.; de Vries, H.J.; Lora, L.; Arinelli, A.C.; da Costa Nery, J.A.; Freitas, D.F.S. Ocular Adnexal Sporotrichosis: A Case Series. *JAAD Case Rep.* **2021**, *13*, 52–56. [CrossRef] [PubMed]
33. Lederer, H.T.; Sullivan, E.; Crum-Cianflone, N.F. Sporotrichosis as an Unusual Case of Osteomyelitis: A Case Report and Review of the Literature. *Med. Mycol. Case Rep.* **2016**, *11*, 31–35. [CrossRef] [PubMed]
34. Mialski, R.; de Oliveira, J.N.; da Silva, L.H.; Kono, A.; Pinheiro, R.L.; Teixeira, M.J.; Gomes, R.R.; de Queiroz-Telles, F.; Pinto, F.G.; Benard, G. Chronic Meningitis and Hydrocephalus Due to *Sporothrix brasiliensis* in Immunocompetent Adults: A Challenging Entity. *Open Forum Infect. Dis.* **2018**, *5*, ofy081. [CrossRef] [PubMed]
35. Duani, H.; Palmerston, M.F.; Júnior, J.F.R.; Ribeiro, V.T.; Neves, P.L.A. Meningeal and Multiorgan Disseminated Sporotrichosis: A Case Report and Autopsy Study. *Med. Mycol. Case Rep.* **2019**, *26*, 47–52. [CrossRef] [PubMed]

36. Lima, M.A.; Vallier, R.; Silva, M.M. *Sporothrix brasiliensis* Meningitis in an Immunocompetent Patient. *Pract. Neurol.* **2021**, *21*, 241–242. [CrossRef] [PubMed]
37. Magalhães, V.C.R.; Colombo, S.A.; Peres, N.T.A.; Moura, A.S.; Lyon, A.C.; Lyon, S.; Dutra, M.R.T.; Fereguetti, T.O.; Andrade, V.A.; Azevedo, M.I.; et al. Clinical Factors Associated with Systemic Sporotrichosis in Brazil. *Mycoses* **2023**, *67*, e13656. [CrossRef]
38. Falcão, E.M.M.; de Lima Filho, J.B.; Campos, D.P.; do Valle, A.C.F.; Bastos, F.I.; Gutierrez-Galhardo, M.C.; Freitas, D.F.S. Hospitalizações e Óbitos Relacionados à Esporotricose No Brasil (1992–2015). *Cad. Saúde Pública* **2019**, *35*, e00109218. [CrossRef] [PubMed]
39. Page, M.J.; McKenzie, J.E.; Bossuyt, P.M.; Boutron, I.; Hoffmann, T.C.; Mulrow, C.D.; Shamseer, L.; Tetzlaff, J.M.; Akl, E.A.; Brennan, S.E.; et al. The PRISMA 2020 statement: An updated guideline for reporting systematic reviews. *BMJ* **2021**, *29*, 372. [CrossRef]
40. Queiroz-Telles, F.; Bonifaz, A.; Cognialli, R.; Lustosa, B.P.R.; Vicente, V.A.; Ramírez-Marín, H.A. Sporotrichosis in Children: Case series and Narrative Review. *Curr. Fungal Infect. Rep.* **2022**, *16*, 33–46. [CrossRef]
41. Curi, A.L.L.; Félix, S.; Azevedo, K.M.L.; Estrela, R.; Villar, E.G.; Saraça, G. Retinal Granuloma Caused by *Sporothrix schenckii*. *Am. J. Ophthalmol.* **2003**, *136*, 205–207. [CrossRef] [PubMed]
42. Barros, M.B.; de Oliveira Schubach, A.; do Valle, A.C.; Galhardo, M.C.G.; Conceição-Silva, F.; Schubach, T.M.; Reis, R.S.; Wanke, B.; Marzochi, K.B.; Conceição, M.J. Cat-Transmitted Sporotrichosis Epidemic in Rio de Janeiro, Brazil: Description of a Series of Cases. *Clin. Infect. Dis.* **2004**, *38*, 529–535. [CrossRef] [PubMed]
43. Silva-Vergara, M.L.; Maneira, F.R.Z.; Oliveira, R.M.D.; Santos, C.T.B.; Etchebehere, R.M.; Adad, S.J. Multifocal Sporotrichosis with Meningeal Involvement in a Patient with AIDS. *Med. Mycol.* **2005**, *43*, 187–190. [CrossRef] [PubMed]
44. Vilela, R.; Souza, G.F.; Cota, G.F.; Mendoza, L. Esporotricosis Cutánea y Meningea En Un Paciente Con VIH. *Rev. Iberoam. Micol.* **2007**, *24*, 161–163. [CrossRef] [PubMed]
45. Galhardo, M.C.G.; Silva, M.T.T.; Lima, M.A.; Nunes, E.P.; Schettini, L.E.C.; de Freitas, R.F.; de Almeida Paes, R.; Neves, E.d.S.; do Valle, A.C.F. *Sporothrix schenckii* Meningitis in AIDS during Immune Reconstitution Syndrome. *J. Neurol. Neurosurg. Psychiatry* **2010**, *81*, 696–699. [CrossRef] [PubMed]
46. Gutierrez-Galhardo, M.C.; Valle, A.C.F.D.; Fraga, B.L.B.; Schubach, A.O.; Hoagland, B.R.D.S.; Monteiro, P.C.F.; Barros, M.B.D.L. Disseminated Sporotrichosis as a Manifestation of Immune Reconstitution Inflammatory Syndrome. *Mycoses* **2010**, *53*, 78–80. [CrossRef] [PubMed]
47. Freitas, D.F.S.; de Siqueira Hoagland, B.; do Valle, A.C.F.; Fraga, B.B.; de Barros, M.B.; de Oliveira Schubach, A.; de Almeida-Paes, R.; Cuzzi, T.; Rosalino, C.M.V.; Zancopé-Oliveira, R.M.; et al. Sporotrichosis in HIV-Infected Patients: Report of 21 Cases of Endemic Sporotrichosis in Rio de Janeiro, Brazil. *Med. Mycol.* **2012**, *50*, 170–178. [CrossRef] [PubMed]
48. Silva-Vergara, M.L.; de Camargo, Z.P.; Silva, P.F.; Ferreira-Paim, K.; Abdalla, M.R.; Sgarbieri, R.N.; Rodrigues, A.M.; dos Santos, K.C.; Barata, C.H.; Ferreira-Paim, K. Disseminated *Sporothrix brasiliensis* Infection with Endocardial and Ocular Involvement in an HIV-Infected Patient. *Am. J. Trop. Med. Hyg.* **2012**, *86*, 477–480. [CrossRef] [PubMed]
49. da Silva, L.M.C.; Oliveira, F.D.M.; Hochhegger, B.; Severo, L.C. Pulmonary sporotrichosis in Brazil: A case report and review. *Rev. Patol. Trop.* **2013**, *42*, 121–125. [CrossRef]
50. Eustace, K.; Sampaio, F.; Lyra, M.; Quintella, L.; Valle, A. Cutaneous Disseminated Sporotrichosis Complicated by Osteomyelitis. *Acta Derm. Venereol.* **2013**, *93*, 192–193. [CrossRef]
51. Orofino-Costa, R.; Unterstell, N.; Gripp, A.C.; de Macedo, P.M.; Brota, A.; Dias, E.; de Melo Teixeira, M.; Felipe, M.S.; Bernardes-Engemann, A.R.; Lopes-Bezerra, L.M. Pulmonary Cavitation and Skin Lesions Mimicking Tuberculosis in a HIV Negative Patient Caused by *Sporothrix brasiliensis*. *Med. Mycol. Case Rep.* **2013**, *2*, 65–71. [CrossRef] [PubMed]
52. Freitas, D.F.S.; Lima, I.A.R.; Curi, C.L.; Jordão, L.; Zancopé-Oliveira, R.M.; do Valle, A.C.F.; Galhardo, M.C.G.; Curi, A.L.L. Acute Dacryocystitis: Another Clinical Manifestation of Sporotrichosis. *Memórias Inst. Oswaldo Cruz* **2014**, *109*, 262–264. [CrossRef] [PubMed]
53. de Araujo, M.L.; Rodrigues, A.M.; Fernandes, G.F.; de Camargo, Z.P.; Hoog, G.S.D. Human Sporotrichosis beyond the Epidemic Front Reveals Classical Transmission Types in Espírito Santo, Brazil. *Mycoses* **2015**, *58*, 485–490. [CrossRef] [PubMed]
54. Papaiordanou, F.; da Silveira, B.R.L.; Abulafia, L.A. Hypersensitivity Reaction to *Sporothrix schenckii*: Erythema Nodosum Associated with Sporotrichosis. *Rev. Soc. Bras. Med. Trop.* **2015**, *48*, 504. [CrossRef]
55. Freitas, D.F.S.; Lima, M.A.; de Almeida-Paes, R.; Lamas, C.C.; do Valle, A.C.F.; Oliveira, M.M.E.; Zancopé-Oliveira, R.M.; Gutierrez-Galhardo, M.C. Sporotrichosis in the Central Nervous System Caused by *Sporothrix brasiliensis*. *Clin. Infect. Dis.* **2015**, *61*, 663–664. [CrossRef] [PubMed]
56. Ferreira, L.C.; Barroso, P.F.; Tonomura, E.; Akiti, T.; de Pinho Rodrigues, K.M. Osteomyelitis Caused by *Sporothrix schenckii* in an Immunocompetent Patient. *Rev. Soc. Bras. Med. Trop.* **2016**, *49*, 527–529. [CrossRef] [PubMed]
57. Medeiros, K.B.; Landeiro, L.G.; Diniz, L.M.; Falqueto, A. Disseminated Cutaneous Sporotrichosis Associated with Ocular Lesion in an Immunocompetent Patient. *An. Bras. Dermatol.* **2016**, *91*, 537–539. [CrossRef] [PubMed]
58. Biancardi, A.L.; Freitas, D.F.S.; Valviesse, V.R.G.d.A.; Andrade, H.B.; de Oliveira, M.M.E.; do Valle, A.C.F.; Zancope-Oliveira, R.M.; Galhardo, M.C.G.; Curi, A.L.L. Multifocal choroiditis in disseminated sporotrichosis in patients with hiv/aids. *Retin. Cases Brief Rep.* **2017**, *11*, 67–70. [CrossRef] [PubMed]
59. da Silva, R.F.; Bonfitto, M.; da Silva Junior, F.I.M.; de Ameida, M.T.G.; de CMA da Silva, R. Sporotrichosis in a Liver Transplant Patient: A Case Report and Literature Review. *Med. Mycol. Case Rep.* **2017**, *17*, 25–27. [CrossRef]

60. Yamagata, J.P.M.; Rudolph, F.B.; Nobre, M.C.L.; Nascimento, L.V.; Sampaio, F.M.S.; Arinelli, A.; Freitas, D.F. Ocular Sporotrichosis: A Frequently Misdiagnosed Cause of Granulomatous Conjunctivitis in Epidemic Areas. *Am. J. Ophthalmol. Case Rep.* **2017**, *8*, 35–38. [CrossRef]
61. Eyer-Silva, W.A.; de Azevedo, M.C.V.M.; da Silva, G.A.R.; Basílio-de-Oliveira, R.P.; de Araujo, L.F.; Lago, I.V.d.; Pereira, F.C.F.; Fernandes, M.B.T.; Figueiredo-Carvalho, M.H.G.; de Souza Rabello, V.B.; et al. Palate Ulcer, Uvular Destruction and Nasal Septal Perforation Caused by *Sporothrix brasiliensis* in an HIV-Infected Patient. *Med. Mycol. Case Rep.* **2019**, *23*, 16–19. [CrossRef] [PubMed]
62. Ferreira, G.S.A.; Watanabe, A.L.C.; Trevizoli, N.C.; Jorge, F.M.F.; Cajá, G.O.N.; Diaz, L.G.G.; Meireles, L.P.; Araújo, M.C.C.L. Disseminated Sporotrichosis in a Liver Transplant Patient: A Case Report. *Transplant. Proc.* **2019**, *51*, 1621–1624. [CrossRef] [PubMed]
63. de Oliveira Furtado, L.; Biancardi, A.L.; de Souza Cravo, L.M.; Anjo, R.P.P.; Junior, H.V.d.M. Ocular Sporotrichosis: Atypical Manifestations. *Rev. Bras. Oftalmol.* **2019**, *78*, 59–61. [CrossRef]
64. Lacerda Filho, A.M.; Cavalcante, C.M.; Silva, A.B.D.; Inácio, C.P.; de Lima-Neto, R.G.; de Andrade, M.C.L.; Magalhães, O.M.C.; Santos, F.d.A.G.d.; Neves, R.P. High-Virulence Cat-Transmitted Ocular Sporotrichosis. *Mycopathologia* **2019**, *184*, 547–549. [CrossRef] [PubMed]
65. Matos, A.; Moreira, L.; Barczewski, B.; de Matos, L.; de Oliveira, J.; Pimentel, M.I.; Almeida-Paes, R.; Oliveira, M.; Pinto, T.; Lima, N.; et al. Identification by MALDI-TOF MS of *Sporothrix brasiliensis* Isolated from a Subconjunctival Infiltrative Lesion in an Immunocompetent Patient. *Microorganisms* **2019**, *8*, 22. [CrossRef] [PubMed]
66. do Monte Alves, M.; Milan, E.P.; da Silva-Rocha, W.P.; de Sena da Costa, A.S.; Maciel, B.A.; Vale, P.H.C.; de Albuquerque, P.R.; Lima, S.L.; de Azevedo Melo, A.S.; Rodrigues, A.M.; et al. Fatal Pulmonary Sporotrichosis Caused by *Sporothrix brasiliensis* in Northeast Brazil. *PLOS Neglected Trop. Dis.* **2020**, *14*, e0008141. [CrossRef] [PubMed]
67. Arinelli, A.; do Couto Aleixo, A.L.Q.; Freitas, D.F.S.; do Valle, A.C.F.; Almeida-Paes, R.; Gutierrez-Galhardo, M.C.; Curi, A.L.L. Ocular Sporotrichosis: 26 Cases with Bulbar Involvement in a Hyperendemic Area of Zoonotic Transmission. *Ocul. Immunol. Inflamm.* **2020**, *28*, 764–771. [CrossRef] [PubMed]
68. de Moura Barros, N.; de Souza Pessoa, A.; Brotas, A.M. Systemic Sporotrichosis in an Alcoholic Patient. *An. Bras. Dermatol.* **2020**, *95*, 376–378. [CrossRef]
69. de Sá Menezes Carvalho, G.; Veasey, J.V. Immunoreactive Cutaneous Sporotrichosis. *An. Bras. Dermatol.* **2020**, *95*, 737–739. [CrossRef]
70. Ribeiro, C.R.; Silva, B.P.; de Alcantara Almeida Costa, A.; Neto, A.B.; Vieira, L.A.; de Lima, M.A.; de Lima, M.H.C. Ocular Sporotrichosis. *Am. J. Ophthalmol. Case Rep.* **2020**, *19*, 100865. [CrossRef]
71. Fichman, V.; de Macedo, P.M.; Freitas, D.F.S.; do Valle, A.C.F.; Almeida-Silva, F.; Bernardes-Engemann, A.R.; Zancopé-Oliveira, R.M.; Almeida-Paes, R.; Gutierrez-Galhardo, M.C. Zoonotic Sporotrichosis in Renal Transplant Recipients from Rio de Janeiro, Brazil. *Transpl. Infect. Dis.* **2021**, *23*, e13485. [CrossRef] [PubMed]
72. Lauar, I.D.; Nascimento, V.A.M.; Soares, J.E.P.; de Matos, M.A.; Faria, K.E.P.; Nogueira, L.P.A.; Parreiras, G.C. Esporotricose Pulmonar Em Paciente Imunocompetente: Relato de Caso. *Rev. Médica Minas Gerais* **2021**, *31*, e31415. [CrossRef]
73. Lemes, L.R.; Veasey, J.V.; Mayor, S.S.; Proença, C.C. Ocular Involvement in Sporotrichosis: Report of Two Cases in Children. *An. Bras. Dermatol.* **2021**, *96*, 349–351. [CrossRef] [PubMed]
74. Liborio Neto, A.O.; Caetano, T.R.; Gervasio, N.H.P.; Carneiro, R.C. Conjunctival and Bulbar Sporotrichosis as Parinaud's Oculoglandular Syndrome Acquired by Blood Inoculation. *GMS Ophthalmol. Cases* **2021**, *11*, Doc02. [CrossRef] [PubMed]
75. Xavier, J.R.B.; Waller, S.B.; da Gama Osório, L.; Vives, P.S.; Albano, A.P.N.; de Aguiar, E.S.V.; Ferreira, M.R.A.; da Conceição, F.R.; de Faria, R.O.; Meireles, M.C.A.; et al. Human Sporotrichosis Outbreak Caused by *Sporothrix brasiliensis* in a Veterinary Hospital in Southern Brazil. *J. Med. Mycol.* **2021**, *31*, 101163. [CrossRef] [PubMed]
76. Bittencourt, A.A.; Oyafuso, L.K.M.; Cavalin, R.F.; Palhares, R.B.; Benard, G.; Gimenes, V.M.F.; Negro, G.M.B.D.; Siqueira, L.P.M.; de Freitas Xavier, R.S.; Lopes-Bezerra, L.M.; et al. A Neglected Disease. Human Sporotrichosis in a Densely Populated Urban Area in São Paulo, Brazil: Clinical–Epidemiological and Therapeutic Aspects. *Braz. J. Microbiol.* **2022**, *53*, 739–748. [CrossRef]
77. de Castro Sant'Anna, L.; Sousa, M.J.A.D.; Reis, M.F.D.; Silva, M.L.; Neto, J.A.D.C.; Gontijo, C.M.; Neto, J.E.J.; Costa, M.S. Esporotricose Pulmonar: Relato de Caso. *Braz. J. Health Rev.* **2022**, *5*, 21440–21443. [CrossRef]
78. Abrahão, A.; Agostini, M.; de Oliveira, T.; Noce, C.; Júnior, A.; Cabral, M.; Romañach, M.; Torres, S. Oral Manifestations of Sporotrichosis: A Neglected Disease. *J. Clin. Exp. Dent.* **2023**, *15*, e82–e87. [CrossRef]
79. Cognialli, R.C.R.; Cáceres, D.H.; Bastos, F.d.A.G.D.; Cavassin, F.B.; Lustosa, B.P.R.; Vicente, V.A.; Breda, G.L.; Santos-Weiss, I.; Queiroz-Telles, F. Rising Incidence of *Sporothrix brasiliensis* Infections, Curitiba, Brazil, 2011–2022. *Emerg. Infect. Dis.* **2023**, *29*, 1330–1339. [CrossRef]
80. de Sousa, B.I.C.; Ferreira, L.C.F.; de Lima, M.M.R.; Netto, J.d.S.B.; Astacio, G.S.-M.; Bernardes-Engemann, A.R.; de Macedo, P.M.; Gutierrez-Galhardo, M.C.; Freitas, D.F.S. Osteoarticular Sporotrichosis of the Knee Caused by *Sporothrix brasiliensis*: Two Similar Cases with Different Outcomes. *J. Fungi* **2023**, *9*, 956. [CrossRef]
81. Gomes, R.d.S.R.; do Valle, A.C.F.; Freitas, D.F.S.; de Macedo, P.M.; Oliveira, R.d.V.C.; Almeida-Paes, R.; Zancopé-Oliveira, R.M.; Gutierrez-Galhardo, M.C. Sporotrichosis in Older Adults: A Cohort Study of 911 Patients from a Hyperendemic Area of Zoonotic Transmission in Rio de Janeiro, Brazil. *J. Fungi* **2023**, *9*, 804. [CrossRef] [PubMed]

82. Silva Neto, F.; Gadelha, C.; Silva, L.; Pazos, N.; Silva, A.; Silva, M.; Sousa, B.; Dulgheroff, A.C.B.; Guerra, F.; da Silveira, V.M.; et al. Case Report: Ocular Sporotrichosis—A Seven-Case Series. *Am. J. Trop. Med. Hyg.* **2023**, *108*, 1109–1114. [CrossRef] [PubMed]
83. Siqueira, A.M.; D'Angioli, W.M.; Lapera, B.; de Souza, I.G.C.; Löwenthal, N.; Rossit, J.; Salles, S.d.A.N.; Machado, R.L.D.; Rocha, E.M.d.S.d.; Baptista, A.R.d.S. Case Report: Osteomyelitis Due to *Sporothrix brasiliensis* in Two Immunocompetent Patients Requiring Surgical Amputation. *Am. J. Trop. Med. Hyg.* **2023**, *109*, 1351–1355. [CrossRef] [PubMed]
84. Gremião, I.D.F.; Oliveira, M.M.E.; de Miranda, L.H.M.; Freitas, D.F.S.; Pereira, S.A. Geographic Expansion of Sporotrichosis, Brazil. *Emerg. Infect. Dis.* **2020**, *26*, 621–624. [CrossRef] [PubMed]
85. Fleury, R.N.; Taborda, P.R.; Gupta, A.K.; Fujita, M.S.; Rosa, P.S.; Weckwerth, A.C.; Negrão, M.S.; Bastazini, I. Zoonotic Sporotrichosis. Transmission to Humans by Infected Domestic Cat Scratching: Report of Four Cases in São Paulo, Brazil. *Int. J. Dermatol.* **2001**, *40*, 318–322.
86. Nobre, M.O.; Potter-de-Castro, A.; Caetano, D.L.; Souza, L.L.; Meireles, M.C.; Ferreiro, L. Araujo-Meireles. Recurrence of Sporotrichosis in Cats with Zoonotic Involvement. *Rev. Iberoam. Micol.* **2001**, *18*, 137–140.
87. Munhoz, L.S.; Poester, V.R.; Severo, C.B.; Trápaga, M.R.; Madrid, I.M.; Benelli, J.L.; Stevens, D.A.; Xavier, M.O. Update of the Epidemiology of the Sporotrichosis Epidemic in the State of Rio Grande Do Sul, Brazil. *Mycoses* **2022**, *65*, 1112–1118. [CrossRef] [PubMed]
88. State of Bahia, Brazil. *Institutes the Inclusion of Sporotrichosis in the List of Diseases and Conditions of Compulsory Notification in the City of Salvador*; Ordinance No. 191; Municipal Health Department: Salvador, Bahia, Brazil, 2018.
89. Eudes Filho, J.; dos Santos, I.B.; Reis, C.M.S.; Patané, J.S.L.; Paredes, V.; Bernardes, J.P.R.A.; dos Santos Costa Poggiani, S.; de Cássia Borges Castro, T.; Gomez, O.M.; Pereira, S.A.; et al. A novel *Sporothrix brasiliensis* genomic variant in Midwestern Brazil: Evidence for an older and wider sporotrichosis epidemic. *Emerg. Microbes Infect.* **2020**, *9*, 2515–2525. [CrossRef] [PubMed]
90. Paiva, M.T.; de Oliveira, C.S.F.; Nicolino, R.R.; Bastos, C.V.; Lecca, L.O.; de Azevedo, M.I.; Keller, K.M.; Salvato, L.A.; Brandão, S.T.; de Oliveira, H.M.R.; et al. Spatial Association between Sporotrichosis in Cats and in Human during a Brazilian Epidemics. *Prev. Vet. Med.* **2020**, *183*, 105125. [CrossRef]
91. de Oliveira Bento, A.; de Sena Costa, A.S.; Lima, S.L.; do Monte Alves, M.; de Azevedo Melo, A.S.; Rodrigues, A.M.; da Silva-Rocha, W.P.; Milan, E.P.; Chaves, G.M. The Spread of Cat-Transmitted Sporotrichosis Due to *Sporothrix brasiliensis* in Brazil towards the Northeast Region. *PLOS Neglected Trop. Dis.* **2021**, *15*, e0009693. [CrossRef]
92. Silva, C.E.; Valeriano, C.A.; Ferraz, C.E.; Neves, R.P.; Oliveira, M.M.; Silva, J.C.; Magalhães, V.; Lima-Neto, R.G. Epidemiological Features and Geographical Expansion of Sporotrichosis in the State of Pernambuco, Northeastern Brazil. *Future Microbiol.* **2021**, *16*, 1371–1379. [CrossRef]
93. Rediguieri, B.C.; da Cruz Bahiense, I.; de Carvalho, J.A.; Leite, G.R.; Falqueto, A.; Rodrigues, A.M.; Gonçalves, S.S. Clinical, Epidemiological, and Epizootic Features of *Sporothrix brasiliensis* in Espírito Santo, Brazil. *EcoHealth* **2022**, *19*, 124–134. [CrossRef]
94. Della Terra, P.P.; Rodrigues, A.M.; Fernandes, G.F.; Nishikaku, A.S.; Burger, E.; de Camargo, Z.P. Exploring virulence and immunogenicity in the emerging pathogen *Sporothrix brasiliensis*. *PLOS Neglected Trop. Dis.* **2017**, *11*, e0005903. [CrossRef] [PubMed]
95. de Souza Rabello, V.B.; de Melo Teixeira, M.; Meyer, W.; Irinyi, L.; Xavier, M.O.; Poester, V.R.; Pereira Brunelli, J.G.; Almeida-Silva, F.; Bernardes-Engemann, A.R.; Ferreira Gremião, I.D.; et al. Multi-locus sequencing typing reveals geographically related intraspecies variability of *Sporothrix brasiliensis*. *Fungal Genet. Biol.* **2023**, *170*, 103845. [CrossRef]
96. Spruijtenburg, B.; Bombassaro, A.; Meijer, E.F.J.; Rodrigues, A.M.; Grisolia, M.E.; Vicente, V.A.; de Queiroz-Telles, F.; Meis, J.F.; de Groot, T. *Sporothrix brasiliensis* Genotyping Reveals Numerous Independent Zoonotic Introductions in Brazil. *J. Infect.* **2023**, *86*, 610–613. [CrossRef] [PubMed]
97. Martínez-Álvarez, J.A.; Pérez-García, L.A.; Mellado-Mojica, E.; López, M.G.; Martínez-Duncker, I.; Lópes-Bezerra, L.M.; Mora-Montes, H.M. *Sporothrix schenckii* sensu stricto and *Sporothrix brasiliensis* Are Differentially Recognized by Human Peripheral Blood Mononuclear Cells. *Front Microbiol.* **2017**, *10*, 843. [CrossRef]
98. Ramírez-Soto, M.C.; Tirado-Sánchez, A.; Bonifaz, A. Ocular Sporotrichosis. *J. Fungi* **2021**, *7*, 951. [CrossRef]
99. Ramírez Soto, M.C. Sporotrichosis in the Ocular Adnexa: 21 Cases in an Endemic Area in Peru and Review of the Literature. *Am. J. Ophthalmol.* **2016**, *162*, 173–179.e3. [CrossRef] [PubMed]
100. de Andrade Galliano Daros Bastos, F.; Cognialli, R.C.R.; de Farias, M.R.; dos Santos Monti, F.; Wu, K.; Queiroz-Telles, F. Spread of *Sporothrix* spp. through Respiratory Droplets from Infected Cats: A Potential Route of Transmission. *Med. Mycol.* **2022**, *60*, myac079. [CrossRef]
101. World Health Organization, WHO. *Neglected Tropical Diseases*; WHO: Genève, Switzerland, 2024.
102. Arrillaga-Moncrieff, I.; Capilla, J.; Mayayo, E.; Marimon, R.; Mariné, M.; Gené, J.; Cano, J.; Guarro, J. Different virulence levels of the species of *Sporothrix* in a murine model. *Clin. Microbiol. Infect.* **2009**, *15*, 651–655. [CrossRef]
103. Freitas, D.F.; Santos, S.S.; Almeida-Paes, R.; de Oliveira, M.M.; do Valle, A.C.; Gutierrez-Galhardo, M.C.; Zancopé-Oliveira, R.M.; Nosanchuk, J.D. Increase in virulence of *Sporothrix brasiliensis* over five years in a patient with chronic disseminated sporotrichosis. *Virulence* **2015**, *6*, 112–120. [CrossRef]
104. Bonifaz, A.; Tirado-Sánchez, A. Cutaneous Disseminated and Extracutaneous Sporotrichosis: Current Status of a Complex Disease. *J. Fungi* **2017**, *3*, 6. [CrossRef]
105. Kauffman, C.A. Central Nervous System Infection with Other Endemic Mycoses: Rare Manifestation of Blastomycosis, Paracoccidioidomycosis, Talaromycosis, and Sporotrichosis. *J. Fungi* **2019**, *5*, 64. [CrossRef]

106. Tseng, H.-K.; Huang, T.-Y.; Wu, A.Y.-J.; Chen, H.-H.; Liu, C.-P.; Jong, A. How Cryptococcus Interacts with the Blood–Brain Barrier. *Future Microbiol.* **2015**, *10*, 1669–1682. [CrossRef]
107. Góralska, K.; Blaszkowska, J.; Dzikowiec, M. Neuroinfections Caused by Fungi. *Infection* **2018**, *46*, 443–459. [CrossRef]
108. Munhoz, L.S.; Poester, V.R.; Benelli, J.L.; Melo, A.M.; Trápaga, M.R.; Nogueira, C.W.; Zeni, G.; Flores, M.M.; Stevens, D.A.; Xavier, M.O. Effectiveness of Diphenyl Diselenide against Experimental Sporotrichosis Caused by *Sporothrix brasiliensis*. *Med. Mycol.* **2023**, *61*, myad035. [CrossRef]
109. Poester, V.R.; Munhoz, L.S.; Stevens, D.A.; Melo, A.M.; Trápaga, M.R.; Flores, M.M.; Larwood, D.J.; Xavier, M.O. Nikkomycin Z for the Treatment of Experimental Sporotrichosis Caused by *Sporothrix brasiliensis*. *Mycoses*, 2023; *in press*.
110. Poester, V.R.; Brandolt, T.M.; Klafke, G.B.; Xavier, M.O. Population Knowledge on Sporotrichosis in an Endemic Area in Southern Brazil. *Rev. Bras. Pesqui. Saúde* **2018**, *20*, 25–30. [CrossRef]
111. Poester, V.R.; Stevens, D.A.; Basso, R.P.; Munhoz, L.S.; Zanchi, M.; Benelli, J.L.; Klafke, G.B.; Cardone, S.; Xavier, M.O. CATastrophe: Response to the Challenges of Zoonotic Sporotrichosis in Southern Brazil. *Mycoses* **2022**, *65*, 30–34. [CrossRef]
112. Poester, V.R.; Saraiva, L.A.; Pretto, A.C.; Klafke, G.B.; Sanchotene, K.O.; Melo, A.M.; Cardone, S.; Xavier, M.O. Desconhecimento de Profissionais e Ações de Extensão Quanto à Esporotricose No Extremo Sul Do Brasil. *Vittalle-Rev. Ciências Saúde* **2019**, *31*, 8–14. [CrossRef]
113. Poester, V.R.; Basso, R.P.; Stevens, D.A.; Munhoz, L.S.; Rabello, V.B.d.S.; Almeida-Paes, R.; Zancopé-Oliveira, R.M.; Zanchi, M.; Benelli, J.L.; Xavier, M.O. Treatment of Human Sporotrichosis Caused by *Sporothrix brasiliensis*. *J. Fungi* **2022**, *8*, 70. [CrossRef]

Journal of *Fungi*

MDPI

Review

What's New in *Cryptococcus gattii*: From Bench to Bedside and Beyond

Justin Beardsley [1,2,3,*], Aiken Dao [1,2,3], Caitlin Keighley [1], Katherine Garnham [1,4], Catriona Halliday [1,2,5], Sharon C.-A. Chen [1,2,5] and Tania C. Sorrell [1,2,3]

1 Sydney Infectious Disease Institute, University of Sydney, Sydney, NSW 2145, Australia
2 Westmead Hospital, New South Wales Health, Sydney, NSW 2145, Australia
3 Westmead Institute for Medical Research, Sydney, NSW 2145, Australia
4 Sunshine Coast University Hospital, Sunshine Coast University, Birtinya, QLD 4575, Australia
5 Institute of Clinical Pathology and Medical Research (ICPMR), NSW Health Pathology, Sydney, NSW 2145, Australia
* Correspondence: justin.beardsley@sydney.edu.au

Abstract: *Cryptococcus* species are a major cause of life-threatening infections in immunocompromised and immunocompetent hosts. While most disease is caused by *Cryptococcus neoformans*, *Cryptococcus gattii*, a genotypically and phenotypically distinct species, is responsible for 11–33% of global cases of cryptococcosis. Despite best treatment, *C. gattii* infections are associated with early mortality rates of 10–25%. The World Health Organization's recently released Fungal Priority Pathogen List classified *C. gattii* as a medium-priority pathogen due to the lack of effective therapies and robust clinical and epidemiological data. This narrative review summarizes the latest research on the taxonomy, epidemiology, pathogenesis, laboratory testing, and management of *C. gattii* infections.

Keywords: *Cryptococcus gattii*; cryptococcosis; fungal infection; medical mycology; diagnostic tools; epidemiology; antifungal; antimicrobial resistance

Citation: Beardsley, J.; Dao, A.; Keighley, C.; Garnham, K.; Halliday, C.; Chen, S.C.-A.; Sorrell, T.C. What's New in *Cryptococcus gattii*: From Bench to Bedside and Beyond. *J. Fungi* **2023**, *9*, 41. https://doi.org/10.3390/jof9010041

Academic Editor: Alessandro C. Pasqualotto

Received: 9 December 2022
Revised: 19 December 2022
Accepted: 20 December 2022
Published: 27 December 2022

1. Introduction

Cryptococcus can cause life threatening infections in both immunocompromised and immunocompetent hosts. Although most disease is caused by *Cryptococcus neoformans*, *C. gattii*—a genotypically and phenotypically distinct species—causes 11–33% of invasive cryptococcosis globally [1–3]. Even with best available therapy, *C. gattii* is associated with 28 day case fatality rates in range 10–25% [4–7]. The World Health Organisation's recently released Fungal Priority Pathogen List categorised *C. gattii* as a medium priority pathogen in light of the poor availability of safe and effective therapies and a lack of robust clinical and epidemiological data [8].

We conducted a narrative review of the literature, summarising what is new in epidemiology, pathogenesis, diagnosis, and management of *Cryptococcus gattii* infections—contrasting this, where relevant, with what is known about the more common pathogen, *Cryptococcus neoformans*. Data, as far as possible, are from the last 10 years. Only English-language publications were included.

2. Latest on Taxonomy and Molecular Epidemiology of Cryptococcus

Cryptococcus neoformans and *Cryptococcus gattii* are haploid encapsulated yeast-like fungi of the class *Trellomycetes* [9]. First described in 1895, there have been numerous revisions in nomenclature since [9]. *Cryptococcus neoformans* includes two varieties, *Cryptococcus neoformans* var. *grubii* and *Cryptococcus neoformans* var. *neoformans*, though further sub-divisions have been proposed; *Cryptococcus gattii* is a separate species [9–11]. The term species complex is used for both in recognition of the substantial genetic variation within the *Cryptococcus neoformans* and *Cryptococcus gattii* complexes [12].

Historically, *Cryptococcus* spp. were distinguished by serotype with serotypes A, B and C recognized two decades before serotype D [9]. Serotyping initially used rabbit sera to react with capsular polysaccharides before moving on to PCR targeting the laccase gene (*LAC1*) and capsule gene (*CAP64*) [13–15]. Serotypes A and D are both *Cryptococcus neoformans*: A is *C. neoformans* var *grubii* and D is var *neoformans*. Serotypes B and C are *Cryptococcus gattii*. Hybrids also exist. A minority of *Cryptococcus* spp. (ranging from 2–21%), are acapsular or pauci-capsular, and so untypeable [16–18].

The advent of molecular taxonomy has revolutionized fungal nomenclature. Various techniques have been used including PCR-fingerprinting, Restriction Fragment Length Polymorphism (RFLP), Amplified Fragment Length Polymorphism (AFLP) and Multi-Locus Sequence Typing (MLST) to identify distinct lineages within the two species complexes: *C. gattii* (VGI-VGIV) and *C. neoformans* (VNI, VNII, VNB) (now separated into VNBI and II–all serotype A), VNIII (hybrid of serotypes A and D), VNIV–serotype D) [19] (see Table 1).

Adding a further genotype based on a single clinical isolate from Mexico, Hagen et al. in 2015 proposed that *C. neoformans* and *C. gattii* be divided into 7 sibling species based on the phylogenetic (evolutionary) species concept. Namely, for *neoformans*: *Cryptococcus neoformans* (genotype VNI), *Cryptococcus deneoformans* (genotype VNIV), plus *C. neoformans* × *C. deoneoformans* hybrid (genotype VNIII). For *gattii*: *Cryptococcus gattii* (genotype VGI), *Cryptococcus deuterogattii* (VGII), *Cryptococcus bacillispora* (VGIII), *Cryptococcus tetragattii* (VGIV), and *Cryptococcus decagattii* [20]. Subsequently, a genetically distinct lineage of *C. gattii* discovered in Zambian woodland environments in association with nitrogen-rich middens of a small mammal known as Hyrax, has been designated VGV, and *C. decagattii* as VGVI [21,22].

Debate continues within the scientific community as to whether *C. neoformans* and *C. gattii* should be considered species complexes, supported by clinical, epidemiological and ecological data, or divided into separate clades (species) with corresponding changes in nomenclature, based on genotyping alone, without clearcut biological differences [12,20,23]. It was argued by Kwon-Chung and colleagues in 2017 [12] and later by Farrer et al. [21] that species designations as proposed are not yet stable. The recent characterization of multiple environmental isolates of a new *C. gattii* lineage in Zambian woodlands introduced above suggests that further lineages will be discovered [21].

MLST-based DNA Barcoding, which uses short diverse genetic sequences that are stable and unique to a given species, is currently the gold standard for cryptococcal typing [21,24,25]. A multitude of genetic loci were initially proposed for MLST, complicating its utility and comparability [9,22]. A single schema is now supported by the International Society for Human and Animal Mycology (ISHAM) and includes seven loci (*URA5*, *CAP59*, *GPD1*, *LAC1*, *SOD1*, *PLB1*, and *IGS*). A database of quality-controlled reference sequences is available at https://mlst.mycologylab.org/ (accessed on 30 June 2022). and in the CBS-KNAW database (Westerdijk Fungal Bioversity Institute, https://www.knaw.nl (accessed on 30 June 2022)). As of 30 June 2022, this data base contained 566 entries for *C. gattii* and 662 for *C. neoformans*.

Accurate species identification from fungal colonies can also be achieved reliably and rapidly by matrix-assisted laser desorption/ionization-time of flight mass spectrometry (MALDI-TOF MS) [25] with minimal sample preparation or it can be combined with nucleic acid-based testing. Using MALDI-TOF MS methods, *C. neoformans* and *C. gattii* can be distinguished readily [24,26,27]. However, because of lack of standardization of sample preparative procedures and limitations of commercial databases, the use of MALDI-TOF MS for detailed fungal identification is generally limited to reference or research laboratories.

MLST has revealed geographical associations with clonal populations with increased heterogeneity possibly related to the origins of the species; for example, *C. gattii*, may have spread from Australia and Northern Brazil to North America and Canada [28,29] whereas for *C. neoformans*, ST5 is the predominant serotype in China and Taiwan, ST93 in Brazil, ST41 in Japan and the US, with greater diversity in African countries [16,28–32].

Table 1. Classification of *Cryptococcus neoformans* and *Cryptococcus gattii* species complexes based on serotype, molecular sub-type, and genomic sequence.

Species Complex	Proposed Species Name	Variety	Serotype	PCR Fingerprinting	Reference Strain—Accession ID of Reference Assembly
Cryptococcus neoformans species complex	*C. neoformans*	var. *grubii*	A	VNI	H99-ASM301198v1
				VNB	Bt88-BROAD_CneoA_Bt88_1
				VNII	PMHc1023.ENR BROAD_CneoA_PMHc1023.Enr_1
	C. deneoformans	var. *neoformans*	D	VNIV	JEC21-ASM9104v1
	C. neoformans × *C. deneoformans hybrid*	AD hybrid	AD	VNIII	
Cryptocccus gattii species complex	*C. gattii*		B/C	VGI	WM276-ASM18594v1 Ru294-Cryp_gatt_Ru294_V1
	C. deuterogattii		B/C	VGII, VGIIa, b, c, d	R265-R265.1
	C. bacillisporus		B/C	VGIII	CA1280-Cryp_gatt_CA1280_V1
			B		CA1873-Cryp_gatt_CA1873_V1
	C. tetragattii		B/C	VGIV	IND107-Cryp_gatt_IND107_V2
			B		MF34-Cryp_gatt_MF34
	-		B/C	VGV	
	C. decagattii		B/C	VGVI	

Adapted from Hong 2021 Frontiers in Cellular and Infection Microbiology [28].

Despite the value and comparatively low cost of MLST, the highest resolution is achieved with whole genome (next-generation) sequencing (WGS or NGS). This allows greater discrimination between strains and offers a method for species identification, detection of drug resistance mutations and epidemiological information. WGS has broadly confirmed the relationships between the different genotypes identified by MLST. The first whole genome sequence for *C. neoformans* was published in 2005 [33]. There are now 115 genomes incorporated in NCBI belonging to what was previously known as *C. neoformans* (NCBI accessed 30 June 2022). This has enabled further elucidation of origins, evolution, and genetic variation amongst the species complexes [13,34]. Associations between genomic differences and virulence or habitat have been postulated [35,36]. Because of the tendency of these yeasts towards aneuploidy and to become diploid, as well as more minor structural rearrangements, accuracy requires a higher genome coverage than that accepted for simpler organisms, and complete assessment may require long read sequencing to supplement short read sequencing [28,37]. Nonetheless, WGS enabled the delineation of VGV *C. gattii* isolates in Zambia, and sub-lineages VNBI and VNBII within VNB of sub-Saharan African isolates associated with phenotypic differences, as well as the new designation of VGVI (putative *C. decagattii*) [19–21].

In summary, debate persists regarding the delineation of species within both *C. gattii* and *C. neoformans* species complexes. For clinical purposes, the use of species complexes is the more useful terminology and will be used from this point. Meanwhile, important research continues into the implications of different genetic groupings.

3. Ecology, Epidemiology and Clinical Features

C. gattii and *C. neoformans* are adapted to different ecological niches and exhibit differences in epidemiology, clinical features, and virulence mechanisms [38,39].

3.1. Ecology

Several arboreal species have been identified as the predominant environmental reservoir of *C. gattii*, often eucalypts but including a huge range from oaks to baobabs [40]. *C. neoformans*, on the other hand, is primarily associated with weathered pigeon guano.

Despite potential sampling bias across the globe, evidence to date indicates that *C. gattii* infections occur predominantly in Australia, British Columbia, Canada, the Pacific Northwest of the USA, parts of South America and Africa. Molecular type VGI, has a global distribution but is most prevalent in Africa, Australia, and Europe. Molecular type VGII occurs globally but is mostly reported from Australia and the Americas. Molecular type VGIII is concentrated in, but not limited to, the Americas. The relatively rare molecular type VGIV has been reported in southern Africa and India, while the very rare molecular type VGVI has been reported from Mexico [9,20,21,40,41].

There is some evidence that subgroups (subclades) within the VG lineages are also geographically defined, for example VGIa is found in Australia and the Pacific Northwest of the US, whereas VGI b has been found in Zambia; VGII a and c predominate in the Vancouver area and the Pacific Northwest of the US whereas VGIIb has been reported in far Northern Australia. These associations may reflect microevolution within these lineages over time.

Interest has recently turned to environmental factors affecting the distribution and virulence of fungal species, including *C. gattii*. Modelling performed by Cogliati et al. determined that in Europe, *C. gattii* was concentrated along the Mediterranean coast, with its distribution limited by low temperatures during the coldest season, and by heavy precipitations in the driest season; *C. neoformans* var. *grubii* colonized the same areas but it tolerated cold winter temperatures and summer precipitations better [42]. More extensive modelling using Maxent analysis showed a gradual expansion of *C gattii* from 1980 to 2009 in the same region followed by doubling of the potential environmental niche in 2010–2019 and a predicted further extension inland from the coastal Mediterranean basin as a result of climatic change and global warming [43].

In relation to evolution of *C. gattii*, the discovery of VGV coexisting with VGI and VGII in Zambia in an entirely new ecological niche, namely, middens of the Southern tree hyrax (*Dendrohyrax arboreus*) [21] is of great interest. Hyrax middens concentrate faecal material and are extremely stable and long-lasting structures that can exist in the same place for thousands of years [44]. They have a high nitrogen content which is known to aid cryptococcal growth. Hence due to their extreme environmental stability, these middens may provide stable, long-term, evolutionary niches that facilitate the evolution of genetic diversity within *Cryptococcus*.

3.2. Epidemiology

Although it was traditionally believed that *C. gattii* infection occurred predominantly in immunocompetent people, evidence increasingly points to potential immune and genetic factors predisposing to infection. In Australia the incidence of cryptococcosis in aboriginal populations is 10.4/million/year compared with 0.7/million/year in non-indigenous populations, and the difference cannot be wholly explained by differences in places of domicile or work [45]. In the outbreak centred on the Pacific Northwest of the USA, some form of immune deficiency or underlying health problem has been identified in approximately half of patients [46,47]. Subtle immune defects can play an important role. For example, clear links have been established between high titres of autoantibodies to granulocyte-macrophage colony-stimulating factor and both acquisition and severity of *C. gattii* infection [48,49]. Over the last five years, a plethora of new immunological and genetic factors predisposing to cryptococcosis have been described—more research is needed to understand their role in *C. gattii*.

When dealing with *C. gattii* infections, clinicians must consider the possibility of an underlying immune defect. Although many major causes of immunocompromise may be obvious on history taking, such as transplant or haematological malignancy, others require specific questioning about history of prior infections, family history of unusual infections, early deaths and/or malignancies. Positive findings may indicate further testing is desirable, such as blood smear/film, CD4/CD8/natural killer (NK) cell subsets by flow cytometry, immunoglobulin levels, complement testing and others. Where inborn errors of immunity are suspected, genetic testing will be required, and a clinical immunologist should be involved in discussions.

3.3. Clinical Features and Diagnosis

Although *C. gattii* is less likely to present with CNS involvement (varying by genotype), when it does, there is a higher incidence of CNS imaging abnormalities, mass lesions and hydrocephalus requiring neurosurgical intervention. There appear to be significant differences in cerebrospinal fluid (CSF) examination between the two species, with *C. gattii* CNS infections demonstrating lower CSF glucose and protein, higher median white blood cell counts and more frequent India ink positivity. Immune reconstitution inflammatory syndrome (IRIS) is more frequent in *C. gattii* infections than *C. neoformans*. Despite all this, CNS *C. gattii* infection has a lower mortality rate than *C. neoformans*. It must be noted, however, that mortality in patients with cryptococcal infections is often related to the predisposing disease rather than direct cryptococcal-induced mortality and this may partially explain this observation [6,39]. See Table 2 for a summary.

Table 2. Comparison of morbidity, mortality, and antifungal susceptibilities between *C. gattii* and *C. neoformans* [4–7,39,50].

Pathogen	Hospitalization and Complications	Mortality by 12 Months	Antifungal Geometric Mean MIC
C. neoformans	Length of hospital stay: 2–210 days Renal impairment: 28% Elevated ICP: 18% Blindness: 12%	PLWHIV: 20–61% Non-HIV: 8–50%	Fluconazole: 0.5–9.7 µg/mL Voriconazole: 0.021–0.5 µg/mL Posaconazole: 0.027–0.10 µg/mL Amphotericin B: 0.098–0.69 µg/mL
C. gattii	Length of ICU stay: 1–29 days. Neurological sequelae: 17–27% at 12 months IRIS: 44%	10–43%	Fluconazole: 1.46–8.6 µg/mL Voriconazole: 0.02–0.10 µg/mL Posaconazole: 0.04–0.36 µg/mL Amphotericin B: 0.2726–0.39 µg/mL

Abbreviations: ICP, Intracranial pressure; ICU, Intensive Care Unit; IRIS, immune reconstitution inflammatory syndrome; MIC, mean inhibitory concentration; PLWHIV, person (s) living with Human Immunodeficiency Virus.

4. Advances in Understanding of Virulence and Pathogenesis

After inhaled Cryptococcus cells are phagocytosed by alveolar macrophages, they alter the macrophage transcriptome, preventing significant acidification, calcium efflux and protease activity, and thereby allowing intracellular pathogen proliferation [51].

Classically activated M1 macrophages that result from the usual Th1 cytokine immune profiles are essential for clearing Cryptococcus, while alternatively activated M2 macrophages resulting from Th2 cytokine stimulation are associated with a higher burden of disease [52,53]. *C. gattii* can modulate macrophage polarisation to M2, thereby evading immune recognition and clearance [51–53]. There are observed species-dependent in vitro differences in M1-polarised macrophage transcriptomic gene regulation and subsequent bioprocess modulation. Both Cryptococcus species affect the Akt/mTOR pathway and TNF-alpha gene expression, reducing protective M1 macrophage fungicidal activity. Subsequently, there is dysregulation of normal immunologic dendritic cell development, NK cell activation and proliferation, other cellular cytokine profiles and nitric oxide production, leading to the alternative Th2-activated M2 macrophage polarisation [51]. *C. gattii* impacts signalling, cell differentiation and regulation of immune system processes, while *C. neoformans* modulates the cellular response to stimulus, response to DNA damage and cell death [51].

Deficiencies in alveolar macrophage functions, including phagocytosis and pattern recognition receptors, result in an intracellular proliferation of Cryptococcus. Subsequent abnormalities in cytokine and protein signalling pathways then initiate a suboptimal or potentially pathogenic immune response, depending on where the deficiency lies [53]. Advances in defining these pathways have led to diagnoses of immunodeficiencies in otherwise seemingly healthy hosts and stimulated significant interest in therapeutic approaches.

Endogenous interferon (IFN)-gamma production is associated with protection against Cryptococcus disease, leading to increased phagocytosis and fungicidal activity of phagocytes. Although exogneous IFN-gamma is sometimes useful as a therapy for refractory IRIS [54], as discussed in the treatment section, it has not been proven effective as a treatment adjunct for cryptococcal meningitis. Anti-granulocyte-macrophage colony-stimulating factor autoantibodies, introduced as a predisposing factor above, result in defective surfactant clearance by pulmonary macrophages [55] offering a mechanistic explanation for increased risk of infection, although as yet this has not been translated into a therapeutic option.

C. gattii VGI and VGII genotypes cause most infections in immunocompetent hosts, while VGIII and VGIV are mainly isolated from immunocompromised hosts. Genotype VGII is the most heat-resistant of the known strains, especially at 37 °C. Different subtypes have different gene expressions identified in vitro, which correlate with their observed virulence clinically. Notably, the outbreak strain VGIIa, in a whole genome sequencing study, demonstrated genetic transformation and mitotic microevolution from 12 missense mutations and one shift mutation. Hypervirulent strains of *C. gattii*, but not *C. neoformans*, demonstrate enhanced mitochondrial tubularisation, which positively correlates with intracellular proliferative capacity within macrophages, a possible species-specific virulence mechanism [56].

Both species elicit a robust host immune response. However, murine studies note a considerably different species-dependent host transcriptomic profile [52]. There is variation in the molecular and chemical properties of extracellular enzymes secreted by *C. gattii* compared with *C. neoformans*, perhaps reflecting their different environmental niches and contributing to different clinical manifestations [57].

Another difference is related to the cryptococcal capsule. The cryptococcal capsular polysaccharides, glucuronoxylomannan (GXM) and galactoxylomannan, simultaneously evoke an immune response, enhancing pathogen virulence and allowing evasion of immune recognition by coating antigens on *C. gattii* [58]. The GXM of *C. gattii* has an additional xylose residue compared to *C. neoformans* [59].

Two structural factors associated with increased virulence, the larger size of the dynamic capsule and increased melanin production protecting against reactive oxygen species, are impacted by environmental conditions, with the former shown to be altered by some common herbicides [60]. The different environmental niches of *C. neoformans* and *C. gattii*, hypothetically, are subjected to different herbicidal and environmental pressures, which may differentially alter capsule size and melanin production.

Capsule size correlates with virulence, although not case-fatalit [61]. Notably, *C. gattii* seems to have a greater propensity to form titan cells [62]. Titan cells are abnormally sized (>10 μm) yeast cells that have an immunomodulatory role and potential roles in the pathogenesis of cryptococcosis [63].

C. gattii melanise more slowly than *C. neoformans* [61]. Melanin modulates susceptibility to amphotericin and fluconazole [61], and the speed of melanisation may correlate with virulence and subsequent morbidity and mortality, making this another important avenue of investigation into differences in disease phenotype between the species. It is important to note that there is no evidence showing a relationship between susceptibility and treatment outcome.

Some strains of *C. gattii* have a greater ability to form biofilms in vitro, which may also contribute to differences in disease phenotype between the species [64]. Along with higher levels of the enzymes phospholipase and haemolysin activity, the ability to form biofilms has been associated with human pathogenicity, and correlates with histopathologic pulmonary tissue damage in animal models [64–66].

A recent in-depth review summarising *C. gattii* genotypes, phenotypes, virulence and regulatory mechanisms is recommended for those interested in further reading on this topic [57].

5. Novel Therapeutics for *C. gattii*

Six international consensus guidelines have been released in the past decade covering cryptococcal infections [67–72]. None are targeted explicitly at *C. gattii* infection, and recommendations are based on expert opinion as, to date, no prospective randomized controlled treatment trials have been conducted for *C. gattii* infection. The rarity of *C. gattii* infection presents a significant challenge to conducting such trials. For severe infection, many clinicians favour a prolonged induction therapy with liposomal amphotericin B and 5-flucytosine (4–6 weeks) followed by consolidation therapy for 12–18 months, and this is supported by an Australian case series [5]. There have been no significant new data to guide best use of existing antifungal therapies in *C. gattii* infections in the last decade—more research is needed to optimise this aspect of care.

A comprehensive review of the antifungal drug pipeline in 2021 [73] reported the spectra of activity for fosmanogepix, ibrexafungerp, olorofim, opelconazole, and rezafungin (all agents in late-stage clinical development). Fosmanogepix and opelconazole have potent in vitro activity against *C. gattii* and *C. neoformans*. Olorofim has no activity against either. Rezafungin, as consistent with data for the echinocandin class of antifungal, is not active against *C. neoformans* and is unlikely to be effective for *C. gattii* (although data are awaited). There were no data for ibrexafungerp. Not included in that review was oteseconazole, a novel tetrazole, which shows activity against *C. neoformans*, though no data are available for *C. gattii* [74,75]. A less developed agent, T-2307, also shows evidence of activity against both species in vitro and in animal models [76,77]. Several other agents are undergoing pre-clinical investigation for activity against *C. neoformans* but not *C. gattii*.

Given the relative dearth of agents in the antifungal pipeline and the ongoing high mortality and morbidity rates with existing antifungal therapy, researchers have enthusiastically pursued adjunctive therapies. Despite their theoretical promise, trials have returned disappointing results. However, it must be highlighted that the following studies for "cryptococcal meningitis" were not species-specific and mainly recruited *C. neoformans* patients.

Both pathogen and host play roles in tissue damage, so immunomodulation is a natural target for adjunctive therapies. Under this category, both IFN-gamma and dexamethasone have been subject to clinical trials in HIV co-infected patients. In 2004, a double-blind placebo-controlled trial (n = 70) of adjunctive IFN-gamma Ib (100 or 200 μg thrice weekly for ten weeks) showed a non-significant trend towards improved CSF clearance of cryptococci [78]. However, adverse events increased significantly. In 2012, a randomised, open-label study in Malawi (n = 88) (again in HIV co-infection) compared two/six adjunctive doses of 100 μg IFN-gamma Ib. Although patients receiving IFN-gamma cleared cryptococci from their CSF more quickly, there was no mortality benefit [79]. There are currently no further trials registered for IFN-gamma at ClinicalTrials.gov as of 26 November 2022.

Dexamethasone was considered a promising therapy for HIV-associated CM based on mortality benefits in tuberculous meningitis and confirmed bacterial meningitis [80], and supportive evidence from cryptococcal meningitis observational studies (particularly in *C. gattii*) and animal models. In 2016, results of a double-blind, randomised placebo-controlled trial ("Crypto-Dex", n = 451) which recruited patients in Vietnam, Thailand, Indonesia, Uganda, and Malawi, were released [81]. The trial was stopped early by the data safety and monitoring board as clear evidence emerged that the same dose of dexamethasone was used successfully in tuberculous meningitis (starting at 0.3 mg/kg/day and weaning weekly over six weeks to 1 mg/day) was harmful in this group [81]. Dexamethasone caused higher rates of disability, slower fungal clearance, increased frequency of adverse events, and a statistically non-significant increase in 10-week mortality (47% vs. 41%) [81]. Notably, the hazards were non-proportional, with early signs of benefit outweighing later signs of harm. It remains unclear whether steroids could be helpful in specific subgroups (e.g., space-occupying lesions), whether a shorter course at induction would have led to a different outcome, or indeed whether outcomes would differ for patients with *C. gattii* infection.

Unfortunately, there is insufficient evidence to inform clinicians about the harms or benefits of steroids in HIV-uninfected cryptococcal meningitis patients. These patients have a very different baseline immune response to HIV-co-infected patients, so responses are likely to differ. Case series have described benefits, especially in *C. gattii* infections and patients with mass CNS lesions [82,83], but there is no robust trial evidence. Recent guidelines recommend against general use but support short courses in specific indications such as space-occupying lesions with surrounding mass effect [84].

Other agents have been identified as promising non-immunomodulating adjuncts, although again, results have been disappointing. Between 2018–2020 three trials were published reporting on the efficacy of the selective serotonin reuptake inhibitor sertraline in cryptococcosis [85–87]. A significant phase III double-blind, randomised placebo-controlled trial (n = 460) in Uganda [85] and in a smaller study in Mexico (n = 12) [86] both targeted HIV-associated CM and showed neither microbiological nor clinical benefit. Another trial looked at the efficacy of sertraline in HIV-associated asymptomatic antigenaemia but was ceased after just 21 patients due to an excess of adverse events [87]. Based on in vitro evidence for efficacy, adjunctive tamoxifen 300 mg/day was trialled in a randomised open-label trial (n = 50), but results published in 2021 show no benefit [88].

Work on cryptococcal vaccinations has been undertaken since the 1950s, with many different antigens and delivery systems protecting against infection or attenuation of disease in animal studies. More recently, acapsular strains of *C. gattii* and dendritic cell-based vaccines are under investigation as vaccine antigens [58,89,90]. Adjuvants are likely to play an important role in stimulating an immune response that is effective at multiple sites without causing undue toxicities. They are the subject of intense investigation, as summarised in a 2021 paper by Oliveira et al. [91]. Although vaccines for some fungal pathogens have reach clinical trials, none directed against Cryptococcus have yet reached this stage. Recently, chimeric antigen receptor (CAR) cytotoxic T cells targeting glucuronoxylomannan have been demonstrated in vitro to bind to *C. gattii* and *C. neoformans* regardless of whether the cells are titan or not, but with a more pronounced affinity for the yeast form and reduced the size and number of pulmonary titan cells in cryptococcus infected mice [63,92].

6. Updates in Diagnostics

Although there are multiple strategies to diagnose cryptococcosis, with and without fungal culture, diagnosis can be challenging. Delays in treatment because of missed early diagnoses or incorrect diagnoses, lead to worse clinical outcomes [93].

Traditionally, fungal culture has been used to diagnose cryptococcosis (see Figure 1). Whilst reasonably sensitive, this requires an equipped laboratory and is time consuming [94]. The species complexes can be distinguished by their growth on canavanine-glycine-bromthymol blue (CGB) agar; *C. gattii* turns the culture medium blue, whereas *C. neoformans* does not change the colour of the medium [95].

In terms of microscopy, Cryptococcus can be seen in body fluid with India ink (see Figure 1) examination, histopathology of infected tissue with specific stains to identify capsule (mucicarmine and alcian blue) or presence of Fontana Masson's melanin detection [93] in addition to standard histopathological stains. Although the India ink detection method of cerebrospinal fluid (CSF) is simple and fast, the sensitivity is only 50% and it is unable to distinguish between the two species complexes.

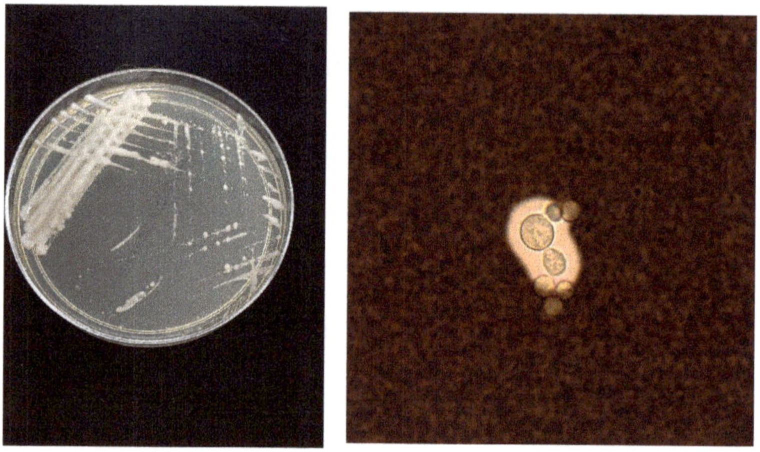

(**a**) (**b**)

Figure 1. (**a**) Cryptococcus gattii growing on Sabouraud dextrose agar (**b**) *C. gattii* strained with India Ink.

X-ray, computerised tomography (CT) scans and magnetic resonance images (MRI) of the lungs (pulmonary cryptococcosis), brain (cryptococcal meningoencephalitis), and other parts of the body are valuable adjunctive tools for diagnosing cryptococcosis and especially the extent of disease, but radiography results alone are unlikely to be diagnostic [96–98].

Serological diagnostic tests are used to detect cryptococcal capsular polysaccharide in serum and CSF by Latex agglutination test (LATs), enzyme-linked immunoassays (ELISA). These tests have an overall sensitivity and specificity of 93–100% and 93–98%, respectively [93]. The false positive rate is less than 1% mostly due to technical issues or other infections (e.g., a cross reaction with antigens from Trichosporon species). A possible limitation is that LATs based on monoclonal antibodies can be less sensitive to *C. gattii* infections, including infections caused by *C. gattii* and *C. neoformans* hybrids [99].

The introduction of cryptococcal antigen lateral flow assays (CrAg-LFA) marked a major revolution in the diagnosis of cryptococcal infections [100]. CrAG-LFA is a dipstick immunochromatographic assay that is simple to use and does not require any pathology laboratory infrastructure (see Figure 2) [101]. Additionally, the LFA shows excellent concordance with Latex Agglutination, ELISA and cultures. It works well in resource-limited settings and effectively diagnoses *C. gattii* infections which may be missed by other serological tests.

Although molecular assays, such as polymerase chain reactions (PCRs), are potentially a more rapid diagnostic method than culture and antigen testing, such assays are not routienly required. When in use, commercial PCR-based assays that target Cryptococcus cannot distinguish between *C. neoformans* and *C. gattii*. Recently, a real-time PCR assay targeting the multicopy mitochondrial cytochrome b (cyt b) gene to detect *C. neoformans* and *C. gattii* has been developed [102]. The assay was tested in clinical specimens, showing that the cyt b-directed assay accurately detected and identified all eight molecular genotypes of *C. neoformans* and *C. gattii*. The overall reported assay sensitivity was 96.4%, and the specificity was 100%, and it can diagnose cryptococcosis in patients within four hours [102]. The targeted assay is cost-saving (USD 40 per sample) and applicable to a diverse range of clinical specimens, including respiratory tract specimens and formalin-fixed paraffin-embedded tissue that is not feasible for culture.

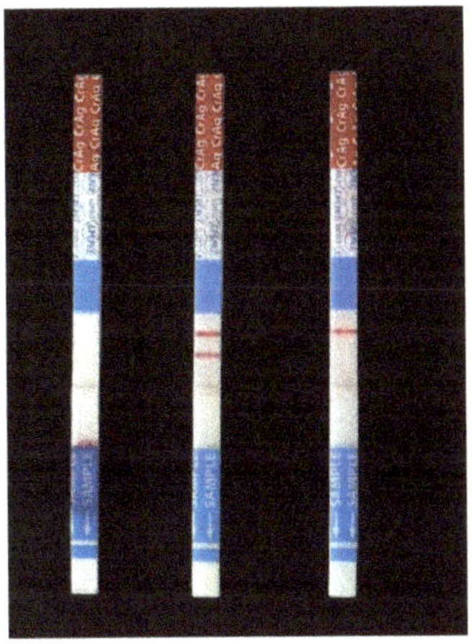

Figure 2. Cryptococcal antigen lateral flow assays (CrAg-LFA).

The novel lateral flow strips combined with recombinase polymerase amplification (LF-RPA) assays were developed to detect the specific DNA sequences of *C. neoformans* and *C. gattii* in clinical CSF specimens [103,104]. The LF-RPA assay could detect 0.64 pg of genomic DNA of *C. neoformans* per reaction within 10 min and be highly specific to Cryptococcus species. The assay sensitivity was 95.2%, and the specificity was 95.8% [103]. Meanwhile, a separate LF-RPA assay was developed to amplify the capsule-associated gene, CAP64, of *C. neoformans* or *C. gattii* to detect cryptococcosis in CSF [104]. Nonetheless, while the LF-RPA assays provide a more rapid approach for screening cryptococcal meningoencephalitis, they cannot distinguish *C. neoformans* and *C. gattii*, and may not be able to detect cases where CSF is clear of infection.

Other recent advances include identification systems such as MALDI-TOF MS. This is a rapid identification tool that can identify both *C. gattii* and *C. neoformans* [105]. It offers a simple method for the separation of the eight major molecular types and the detection of hybrid strains within this species complex in the clinical laboratory. Nonetheless, a limitation of this technology is that it requires a positive culture and can only identify new isolates if the spectral database contains peptide mass fingerprints of the type strains of specific genera, species, subspecies or strains [106].

Surface-enhanced Raman scattering (SERS) and spectral analysis offer an additional potential diagnostic tool for *C. neoformans* and *C. gattii* [94]. This novel technology uses positively charged silver nanoparticles (AgNPs) as a substrate to distinguish *C. neoformans* and *C. gattii* in clinical samples directly. SERS is rapid and nondestructive and has relatively low equipment cost. Briefly, it was shown that AgNPs-self-assembled on the fungal cell wall surface via electrostatic aggregation, leading to enhanced SERS signals that were better than the standard substarte negatively charged AgNPs. The SERS spectra could then be used as a sample database in the multivariate analysis via orthogonal partial least-squares discriminant analysis. The SERS detection method can accurately (was shown to be 100%) distinguish between *C. neoformans* and *C. gattii* using principal component analysis. SERS seems to be a breakthrough, though further evaluation will be required before it can be introduced into routine clinical practice.

7. Conclusions

In conclusion, it is valuable to distinguish between *C. gattii* and *C. neoformans* at the individual patient level given their differences in patient predisposition, disease phenotype, and treatment approaches. Treatment trials should include patients infected with both strains and, where possible, run subgroup analyses to identify any differences.

At the population/public health level, molecular characterization of the strains circulating in specific geographies can provide valuable information on virulence and impacts on at-risk groups. Although there is currently no evidence of differences in MICs impacting treatment outcome it is important that in vitro and in vivo studies of new antifungals include both *C. gattii* and *C. neoformans*. Ongoing basic science research is vital to identify potential new targets for both organisms.

Ensuring that diagnostic tools are effective against both strains is also crucial. Further research comparing IRIS manifestations and therapies between the two species would be valuable in reducing morbidity and mortality.

The vast majority of high-grade evidence on the management of cryptococcosis comes from HIV-associated cryptococcal meningitis trials, which are increasingly conducted in low- and middle-income countries with the highest disease burden and a predominance of *C. neoformans* as the causative pathogen. Consequently, our understanding of cryptococcosis in other settings remains poor, and we struggle to know how much can be directly translated. These gaps can only be filled by establishing large international collaborative networks designed to describe the protean manifestations of cryptococcosis in well-established and emerging niche host groups. These large networks would be fertile ground for investigating novel antifungal or adjunctive therapies, including proper assessment of pharmacokinetic-pharmacodynamic and drug-drug interactions in disparate groups.

Author Contributions: All authors (J.B., A.D., C.K., K.G., C.H., S.C.-A.C. & T.C.S.) shared the literature review and contributed to the original draft of the manuscript. Editing for consistency by J.B. and A.D. All Authors have read and approved this final manuscript. All authors have read and agreed to the published version of the manuscript.

Funding: Justin Beardsley is supported by an NHMRC Emerging Leader Fellowship. Otherwise, this research received no external funding.

Data Availability Statement: Not applicable.

Conflicts of Interest: The authors declare no conflict of interest.

References

1. Hurtado, J.C.; Castillo, P.; Fernandes, F.; Navarro, M.; Lovane, L.; Casas, I.; Quintó, L.; Marco, F.; Jordao, D.; Ismail, M.R.; et al. Mortality due to Cryptococcus neoformans and Cryptococcus gattii in low-income settings: An autopsy study. *Sci. Rep.* **2019**, *9*, 7493. [CrossRef] [PubMed]
2. Lee, G.A.; Arthur, I.; Merritt, A.; Leung, M. Molecular types of Cryptococcus neoformans and Cryptococcus gattii in Western Australia and correlation with antifungal susceptibility. *Med. Mycol.* **2019**, *57*, 1004–1010. [CrossRef] [PubMed]
3. Lahiri, S.; Manjunath, N.; Bhat, M.; Hagen, F.; Bahubali, V.H.; Palaniappan, M.; Maji, S.; Chandrashekar, N. Clinical insights and epidemiology of central nervous system infection due to Cryptococcus neoformans/gattii species complexes: A prospective study from South India. *Med. Mycol.* **2020**, *58*, 600–608. [CrossRef] [PubMed]
4. Smith, R.M.; Mba-Jonas, A.; Tourdjman, M.; Schimek, T.; DeBess, E.; Marsden-Haug, N.; Harris, J.R. Treatment and outcomes among patients with Cryptococcus gattii infections in the United States Pacific Northwest. *PLoS ONE* **2014**, *9*, e88875. [CrossRef]
5. Chen, S.C.; Korman, T.M.; Slavin, M.A.; Marriott, D.; Byth, K.; Bak, N.; Currie, B.J.; Hajkowicz, K.; Heath, C.H.; Kidd, S.; et al. Antifungal therapy and management of complications of cryptococcosis due to Cryptococcus gattii. *Clin. Infect. Dis.* **2013**, *57*, 543–551. [CrossRef]
6. Chen, S.C.; Slavin, M.A.; Heath, C.H.; Playford, E.G.; Byth, K.; Marriott, D.; Kidd, S.E.; Bak, N.; Currie, B.; Hajkowicz, K.; et al. Clinical manifestations of Cryptococcus gattii infection: Determinants of neurological sequelae and death. *Clin. Infect. Dis.* **2012**, *55*, 789–798. [CrossRef]
7. Phillips, P.; Galanis, E.; MacDougall, L.; Chong, M.Y.; Balshaw, R.; Cook, V.J.; Bowie, W.; Steiner, T.; Hoang, L.; Morshed, M.; et al. Longitudinal clinical findings and outcome among patients with Cryptococcus gattii infection in British Columbia. *Clin. Infect. Dis.* **2015**, *60*, 1368–1376. [CrossRef]
8. WHO. *Fungal Priority Pathogens List to Guide Research, Development and Public Health Action*; World Health Organization: Geneva, Switzerland, 2022; pp. 1–48.
9. Chen, S.C.; Meyer, W.; Sorrell, T.C. Cryptococcus gattii infections. *Clin. Microbiol. Rev.* **2014**, *27*, 980–1024. [CrossRef]
10. Maziarz, E.K.; Perfect, J.R. Cryptococcosis. *Infect. Dis. Clin. N. Am.* **2016**, *30*, 179–206. [CrossRef]
11. Firacative, C.; Roe, C.C.; Malik, R.; Ferreira-Paim, K.; Escandon, P.; Sykes, J.E.; Castanon-Olivares, L.R.; Contreras-Peres, C.; Samayoa, B.; Sorrell, T.C.; et al. MLST and Whole-Genome-Based Population Analysis of Cryptococcus gattii VGIII Links Clinical,

Veterinary and Environmental Strains, and Reveals Divergent Serotype Specific Sub-populations and Distant Ancestors. *PLoS Negl. Trop. Dis.* **2016**, *10*, e0004861. [CrossRef]
12. Kwon-Chung, K.J.; Bennett, J.E.; Wickes, B.L.; Meyer, W.; Cuomo, C.A.; Wollenburg, K.R.; Bicanic, T.A.; Castañeda, E.; Chang, Y.C.; Chen, J.; et al. The Case for Adopting the "Species Complex" Nomenclature for the Etiologic Agents of Cryptococcosis. *mSphere* **2017**, *2*, e00357-16. [CrossRef] [PubMed]
13. Firacative, C.; Trilles, L.; Meyer, W. Recent Advances in Cryptococcus and Cryptococcosis. *Microorganisms* **2021**, *10*, 13. [CrossRef] [PubMed]
14. Ito-Kuwa, S.; Nakamura, K.; Aoki, S.; Vidotto, V. Serotype identification of Cryptococcus neoformans by multiplex PCR. *Mycoses* **2007**, *50*, 277–281. [CrossRef] [PubMed]
15. Rathore, S.S.; Sathiyamoorthy, J.; Lalitha, C.; Ramakrishnan, J. A holistic review on Cryptococcus neoformans. *Microb. Pathog.* **2022**, *166*, 105521. [CrossRef] [PubMed]
16. Bive, B.Z.; Sacheli, R.; Situakibanza Nani-Tuma, H.; Kabututu Zakayi, P.; Ka, A.; Mbula Mambimbi, M.; Muendele, G.; Boreux, R.; Landu, N.; Nzanzu Mudogo, C.; et al. Clinical epidemiology and high genetic diversity amongst Cryptococcus spp. isolates infecting people living with HIV in Kinshasa, Democratic Republic of Congo. *PLoS ONE* **2022**, *17*, e0267842. [CrossRef] [PubMed]
17. Nishikawa, M.M.; Lazera, M.S.; Barbosa, G.G.; Trilles, L.; Balassiano, B.R.; Macedo, R.C.; Bezerra, C.C.; Pérez, M.A.; Cardarelli, P.; Wanke, B. Serotyping of 467 Cryptococcus neoformans isolates from clinical and environmental sources in Brazil: Analysis of host and regional patterns. *J. Clin. Microbiol.* **2003**, *41*, 73–77. [CrossRef]
18. Pfeiffer, T.J.; Ellis, D.H. Serotypes of Australian environmental and clinical isolates of Cryptococcus neoformans. *J. Med. Vet. Mycol.* **1993**, *31*, 401–404. [CrossRef]
19. Desjardins, C.A.; Giamberardino, C.; Sykes, S.M.; Yu, C.H.; Tenor, J.L.; Chen, Y.; Yang, T.; Jones, A.M.; Sun, S.; Haverkamp, M.R.; et al. Population genomics and the evolution of virulence in the fungal pathogen Cryptococcus neoformans. *Genome Res.* **2017**, *27*, 1207–1219. [CrossRef]
20. Hagen, F.; Khayhan, K.; Theelen, B.; Kolecka, A.; Polacheck, I.; Sionov, E.; Falk, R.; Parnmen, S.; Lumbsch, H.T.; Boekhout, T. Recognition of seven species in the Cryptococcus gattii/Cryptococcus neoformans species complex. *Fungal Genet. Biol.* **2015**, *78*, 16–48. [CrossRef]
21. Farrer, R.A.; Chang, M.; Davis, M.J.; van Dorp, L.; Yang, D.H.; Shea, T.; Sewell, T.R.; Meyer, W.; Balloux, F.; Edwards, H.M.; et al. A New Lineage of Cryptococcus gattii (VGV) Discovered in the Central Zambezian Miombo Woodlands. *mBio* **2019**, *10*, e02306-19. [CrossRef]
22. Meyer, W.; Aanensen, D.M.; Boekhout, T.; Cogliati, M.; Diaz, M.R.; Esposto, M.C.; Fisher, M.; Gilgado, F.; Hagen, F.; Kaocharoen, S.; et al. Consensus multi-locus sequence typing scheme for Cryptococcus neoformans and Cryptococcus gattii. *Med. Mycol.* **2009**, *47*, 561–570. [CrossRef] [PubMed]
23. Hagen, F.; Lumbsch, H.T.; Arsic Arsenijevic, V.; Badali, H.; Bertout, S.; Billmyre, R.B.; Bragulat, M.R.; Cabañes, F.J.; Carbia, M.; Chakrabarti, A.; et al. Importance of Resolving Fungal Nomenclature: The Case of Multiple Pathogenic Species in the Cryptococcus Genus. *mSphere* **2017**, *2*, e00238-17. [CrossRef] [PubMed]
24. McTaggart, L.R.; Lei, E.; Richardson, S.E.; Hoang, L.; Fothergill, A.; Zhang, S.X. Rapid identification of Cryptococcus neoformans and Cryptococcus gattii by matrix-assisted laser desorption ionization-time of flight mass spectrometry. *J. Clin. Microbiol.* **2011**, *49*, 3050–3053. [CrossRef] [PubMed]
25. Fraser, M.; Brown, Z.; Houldsworth, M.; Borman, A.M.; Johnson, E.M. Rapid identification of 6328 isolates of pathogenic yeasts using MALDI-ToF MS and a simplified, rapid extraction procedure that is compatible with the Bruker Biotyper platform and database. *Med. Mycol.* **2016**, *54*, 80–88. [CrossRef]
26. Posteraro, B.; Vella, A.; Cogliati, M.; De Carolis, E.; Florio, A.R.; Posteraro, P.; Sanguinetti, M.; Tortorano, A.M. Matrix-assisted laser desorption ionization-time of flight mass spectrometry-based method for discrimination between molecular types of Cryptococcus neoformans and Cryptococcus gattii. *J. Clin. Microbiol.* **2012**, *50*, 2472–2476. [CrossRef]
27. Siqueira, L.P.M.; Gimenes, V.M.F.; de Freitas, R.S.; Melhem, M.S.C.; Bonfietti, L.X.; da Silva, A.R., Jr.; Souza Santos, L.B.; Motta, A.L.; Rossi, F.; Benard, G.; et al. Evaluation of Vitek MS for Differentiation of Cryptococcus neoformans and Cryptococcus gattii Genotypes. *J. Clin. Microbiol.* **2019**, *57*, e01282-18. [CrossRef]
28. Hong, N.; Chen, M.; Xu, J. Molecular Markers Reveal Epidemiological Patterns and Evolutionary Histories of the Human Pathogenic Cryptococcus. *Front. Cell. Infect. Microbiol.* **2021**, *11*, 683670. [CrossRef]
29. Kan, A.; Schmertmann, L.J.; McArthur, C.; Mella, V.S.A.; Crowther, M.S.; Miranda, L.; Malik, R.; Meyer, W.; Krockenberger, M.B. A Possible Link between the Environment and Cryptococcus gattii Nasal Colonisation in Koalas (*Phascolarctos cinereus*) in the Liverpool Plains, New South Wales. *Int. J. Environ. Res. Public Health* **2022**, *19*, 4603. [CrossRef]
30. Chen, Y.C.; Kuo, S.F.; Lin, S.Y.; Lin, Y.S.; Lee, C.H. Epidemiological and Clinical Characteristics, Antifungal Susceptibility, and MLST-Based Genetic Analysis of Cryptococcus Isolates in Southern Taiwan in 2013–2020. *J. Fungi* **2022**, *8*, 287. [CrossRef]
31. Fan, X.; Xiao, M.; Chen, S.; Kong, F.; Dou, H.T.; Wang, H.; Xiao, Y.L.; Kang, M.; Sun, Z.Y.; Hu, Z.D.; et al. Predominance of Cryptococcus neoformans var. grubii multilocus sequence type 5 and emergence of isolates with non-wild-type minimum inhibitory concentrations to fluconazole: A multi-centre study in China. *Clin. Microbiol. Infect.* **2016**, *22*, 887.E1–887.E9. [CrossRef]
32. Ferreira-Paim, K.; Andrade-Silva, L.; Fonseca, F.M.; Ferreira, T.B.; Mora, D.J.; Andrade-Silva, J.; Khan, A.; Dao, A.; Reis, E.C.; Almeida, M.T.; et al. MLST-Based Population Genetic Analysis in a Global Context Reveals Clonality amongst Cryptococcus

neoformans var. grubii VNI Isolates from HIV Patients in Southeastern Brazil. *PLoS Negl. Trop. Dis.* **2017**, *11*, e0005223. [CrossRef] [PubMed]
33. Loftus, B.J.; Fung, E.; Roncaglia, P.; Rowley, D.; Amedeo, P.; Bruno, D.; Vamathevan, J.; Miranda, M.; Anderson, I.J.; Fraser, J.A.; et al. The genome of the basidiomycetous yeast and human pathogen Cryptococcus neoformans. *Science* **2005**, *307*, 1321–1324. [CrossRef] [PubMed]
34. Cuomo, C.A.; Rhodes, J.; Desjardins, C.A. Advances in Cryptococcus genomics: Insights into the evolution of pathogenesis. *Mem. Inst. Oswaldo Cruz.* **2018**, *113*, e170473. [CrossRef]
35. Gerstein, A.C.; Jackson, K.M.; McDonald, T.R.; Wang, Y.; Lueck, B.D.; Bohjanen, S.; Smith, K.D.; Akampurira, A.; Meya, D.B.; Xue, C.; et al. Identification of Pathogen Genomic Differences That Impact Human Immune Response and Disease during Cryptococcus neoformans Infection. *mBio* **2019**, *10*, e01440-19. [CrossRef] [PubMed]
36. Zaragoza, O. Basic principles of the virulence of Cryptococcus. *Virulence* **2019**, *10*, 490–501. [CrossRef]
37. Rhodes, J.; Desjardins, C.A.; Sykes, S.M.; Beale, M.A.; Vanhove, M.; Sakthikumar, S.; Chen, Y.; Gujja, S.; Saif, S.; Chowdhary, A.; et al. Tracing Genetic Exchange and Biogeography of Cryptococcus neoformans var. grubii at the Global Population Level. *Genetics* **2017**, *207*, 327–346. [CrossRef]
38. Huang, Y.M.; Tao, X.H.; Xu, D.F.; Yu, Y.; Teng, Y.; Xie, W.Q.; Fan, Y.B. HOG1 has an essential role in the stress response, virulence and pathogenicity of Cryptococcus gattii. *Exp. Ther. Med.* **2021**, *21*, 476. [CrossRef]
39. Baddley, J.W.; Chen, S.C.; Huisingh, C.; Benedict, K.; DeBess, E.E.; Galanis, E.; Jackson, B.R.; MacDougall, L.; Marsden-Haug, N.; Oltean, H.; et al. MSG07: An International Cohort Study Comparing Epidemiology and Outcomes of Patients With Cryptococcus neoformans or Cryptococcus gattii Infections. *Clin. Infect. Dis.* **2021**, *73*, 1133–1141. [CrossRef]
40. Cogliati, M. Global Molecular Epidemiology of Cryptococcus neoformans and Cryptococcus gattii: An Atlas of the Molecular Types. *Scientifica* **2013**, *2013*, 675213. [CrossRef]
41. Hagen, F.; Colom, M.F.; Swinne, D.; Tintelnot, K.; Iatta, R.; Montagna, M.T.; Torres-Rodriguez, J.M.; Cogliati, M.; Velegraki, A.; Burggraaf, A.; et al. Autochthonous and dormant Cryptococcus gattii infections in Europe. *Emerg. Infect. Dis.* **2012**, *18*, 1618–1624. [CrossRef]
42. Cogliati, M.; Puccianti, E.; Montagna, M.T.; De Donno, A.; Susever, S.; Ergin, C.; Velegraki, A.; Ellabib, M.S.; Nardoni, S.; Macci, C.; et al. Fundamental niche prediction of the pathogenic yeasts Cryptococcus neoformans and Cryptococcus gattii in Europe. *Environ. Microbiol.* **2017**, *19*, 4318–4325. [CrossRef] [PubMed]
43. Cogliati, M. Global warming impact on the expansion of fundamental niche of Cryptococcus gattii VGI in Europe. *Environ. Microbiol. Rep.* **2021**, *13*, 375–383. [CrossRef] [PubMed]
44. Chase, B.M.; Scott, L.; Meadows, M.E.; Gil-Romera, G.; Boom, A.; Carr, A.S.; Reimer, P.J.; Truc, L.; Valsecchi, V.; Quick, L.J. Rock hyrax middens: A palaeoenvironmental archive for southern African drylands. *Quat. Sci. Rev.* **2012**, *56*, 107–125. [CrossRef]
45. Chen, S.; Sorrell, T.; Nimmo, G.; Speed, B.; Currie, B.; Ellis, D.; Marriott, D.; Pfeiffer, T.; Parr, D.; Byth, K. Epidemiology and host- and variety-dependent characteristics of infection due to Cryptococcus neoformans in Australia and New Zealand. Australasian Cryptococcal Study Group. *Clin. Infect. Dis.* **2000**, *31*, 499–508. [CrossRef]
46. Harris, J.R.; Lockhart, S.R.; Debess, E.; Marsden-Haug, N.; Goldoft, M.; Wohrle, R.; Lee, S.; Smelser, C.; Park, B.; Chiller, T. Cryptococcus gattii in the United States: Clinical aspects of infection with an emerging pathogen. *Clin. Infect. Dis.* **2011**, *53*, 1188–1195. [CrossRef]
47. Galanis, E.; Macdougall, L.; Kidd, S.; Morshed, M. Epidemiology of Cryptococcus gattii, British Columbia, Canada, 1999-2007. *Emerg. Infect. Dis.* **2010**, *16*, 251–257. [CrossRef]
48. Saijo, T.; Chen, J.; Chen, S.C.; Rosen, L.B.; Yi, J.; Sorrell, T.C.; Bennett, J.E.; Holland, S.M.; Browne, S.K.; Kwon-Chung, K.J. Anti-granulocyte-macrophage colony-stimulating factor autoantibodies are a risk factor for central nervous system infection by Cryptococcus gattii in otherwise immunocompetent patients. *mBio* **2014**, *5*, e00912–e00914. [CrossRef]
49. Crum-Cianflone, N.F.; Lam, P.V.; Ross-Walker, S.; Rosen, L.B.; Holland, S.M. Autoantibodies to Granulocyte-Macrophage Colony-Stimulating Factor Associated With Severe and Unusual Manifestations of Cryptococcus gattii Infections. *Open Forum Infect. Dis.* **2017**, *4*, ofx211. [CrossRef]
50. Sim, B.Z.; Conway, L.; Smith, L.K.; Fairhead, L.; Der, Y.S.; Payne, L.; Binotto, E.; Smith, S.; Hanson, J. The aetiology and clinical characteristics of cryptococcal infections in Far North Queensland, tropical Australia. *PLoS ONE* **2022**, *17*, e0265739. [CrossRef]
51. Piffer, A.C.; Santos, F.M.D.; Thomé, M.P.; Diehl, C.; Garcia, A.W.A.; Kinskovski, U.P.; Schneider, R.O.; Gerber, A.; Feltes, B.C.; Schrank, A.; et al. Transcriptomic analysis reveals that mTOR pathway can be modulated in macrophage cells by the presence of cryptococcal cells. *Genet. Mol. Biol.* **2021**, *44*, e20200390. [CrossRef]
52. Holcomb, Z.E.; Steinbrink, J.M.; Zaas, A.K.; Betancourt, M.; Tenor, J.L.; Toffaletti, D.L.; Alspaugh, J.A.; Perfect, J.R.; McClain, M.T. Transcriptional Profiles Elucidate Differential Host Responses to Infection with Cryptococcus neoformans and Cryptococcus gattii. *J. Fungi* **2022**, *8*, 430. [CrossRef] [PubMed]
53. Elsegeiny, W.; Marr, K.A.; Williamson, P.R. Immunology of Cryptococcal Infections: Developing a Rational Approach to Patient Therapy. *Front. Immunol.* **2018**, *9*, 651. [CrossRef] [PubMed]
54. da Silva-Junior, E.B.; Firmino-Cruz, L.; Guimarães-de-Oliveira, J.C.; De-Medeiros, J.V.R.; de Oliveira Nascimento, D.; Freire-de-Lima, M.; de Brito-Gitirana, L.; Morrot, A.; Previato, J.O.; Mendonça-Previato, L.; et al. The role of Toll-like receptor 9 in a murine model of Cryptococcus gattii infection. *Sci. Rep.* **2021**, *11*, 1407. [CrossRef] [PubMed]

55. Kuo, P.H.; Wu, U.I.; Pan, Y.H.; Wang, J.T.; Wang, Y.C.; Sun, H.Y.; Sheng, W.H.; Chen, Y.C.; Chang, S.C. Neutralizing Anti-Granulocyte-Macrophage Colony-Stimulating Factor Autoantibodies in Patients With Central Nervous System and Localized Cryptococcosis: Longitudinal Follow-up and Literature Review. *Clin. Infect. Dis.* **2022**, *75*, 278–287. [CrossRef]
56. Huang, Y.; Zang, X.; Yang, C.; Deng, H.; Ma, X.; Xie, M.; Zhou, M.; Song, J.; Xue, X. Gene, virulence and related regulatory mechanisms in Cryptococcus gattii. *Acta Biochim. Biophys. Sin.* **2022**, *54*, 593–603. [CrossRef]
57. Wright, L.C.; Payne, J.; Santangelo, R.T.; Simpanya, M.F.; Chen, S.C.; Widmer, F.; Sorrell, T.C. Cryptococcal phospholipases: A novel lysophospholipase discovered in the pathogenic fungus Cryptococcus gattii. *Biochem. J.* **2004**, *384*, 377–384. [CrossRef]
58. Ueno, K.; Otani, Y.; Yanagihara, N.; Urai, M.; Nagamori, A.; Sato-Fukushima, M.; Shimizu, K.; Saito, N.; Miyazaki, Y. Cryptococcus gattii evades CD11b-mediated fungal recognition by coating itself with capsular polysaccharides. *Eur. J. Immunol.* **2021**, *51*, 2281–2295. [CrossRef]
59. Urai, M.; Kaneko, Y.; Ueno, K.; Okubo, Y.; Aizawa, T.; Fukazawa, H.; Sugita, T.; Ohno, H.; Shibuya, K.; Kinjo, Y.; et al. Evasion of Innate Immune Responses by the Highly Virulent Cryptococcus gattii by Altering Capsule Glucuronoxylomannan Structure. *Front. Cell. Infect. Microbiol.* **2015**, *5*, 101. [CrossRef]
60. Castelo-Branco, D.; da Rocha, M.G.; de Oliveira, J.S.; Araujo, G.D.S.; Martins, D.V.; Garcia, L.G.S.; Cordeiro, R.A.; Sidrim, J.J.C.; Pereira-Neto, W.A.; de Melo Guedes, G.M.; et al. The herbicide paraquat alters growth and melanin production on the Cryptococcus neoformans/Cryptococcus gattii species complex. *Can. J. Microbiol.* **2022**, *68*, 493–499. [CrossRef]
61. de Sousa, H.R.; de Oliveira, G.P., Jr.; Frazão, S.O.; Gorgonha, K.C.M.; Rosa, C.P.; Garcez, E.M.; Lucas, J., Jr.; Correia, A.F.; de Freitas, W.F.; Borges, H.M.; et al. Faster Cryptococcus Melanization Increases Virulence in Experimental and Human Cryptococcosis. *J. Fungi* **2022**, *8*, 393. [CrossRef]
62. Saidykhan, L.; Correia, J.; Romanyuk, A.; Peacock, A.F.A.; Desanti, G.E.; Taylor-Smith, L.; Makarova, M.; Ballou, E.R.; May, R.C. An in vitro method for inducing titan cells reveals novel features of yeast-to-titan switching in the human fungal pathogen Cryptococcus gattii. *PLoS Pathog.* **2022**, *18*, e1010321. [CrossRef] [PubMed]
63. Dos Santos, M.H.; Machado, M.P.; Kumaresan, P.R.; da Silva, T.A. Titan Cells and Yeast Forms of Cryptococcus neoformans and Cryptococcus gattii Are Recognized by GXMR-CAR. *Microorganisms* **2021**, *9*, 1886. [CrossRef] [PubMed]
64. Oliveira, L.S.S.; Pinto, L.M.; de Medeiros, M.A.P.; Toffaletti, D.L.; Tenor, J.L.; Barros, T.F.; Neves, R.P.; Neto, R.G.L.; Milan, E.P.; Padovan, A.C.B.; et al. Comparison of Cryptococcus gattii/neoformans Species Complex to Related Genera (Papiliotrema and Naganishia) Reveal Variances in Virulence Associated Factors and Antifungal Susceptibility. *Front. Cell. Infect. Microbiol.* **2021**, *11*, 642658. [CrossRef] [PubMed]
65. Freitas, G.J.C.; Santos, D.A. Cryptococcus gattii polysaccharide capsule: An insight on fungal-host interactions and vaccine studies. *Eur. J. Immunol.* **2021**, *51*, 2206–2209. [CrossRef]
66. Zang, X.; Ke, W.; Wang, L.; Wu, H.; Huang, Y.; Deng, H.; Zhou, M.; Wu, N.; Xue, X.; Shen, D. Molecular epidemiology and microbiological characteristics of Cryptococcus gattii VGII isolates from China. *PLoS Negl. Trop. Dis.* **2022**, *16*, e0010078. [CrossRef] [PubMed]
67. Arendrup, M.C.; Boekhout, T.; Akova, M.; Meis, J.F.; Cornely, O.A.; Lortholary, O. ESCMID and ECMM joint clinical guidelines for the diagnosis and management of rare invasive yeast infections. *Clin. Microbiol. Infect.* **2014**, *20* (Suppl. 3), 76–98. [CrossRef]
68. Baddley, J.W.; Forrest, G.N. Cryptococcosis in solid organ transplantation—Guidelines from the American Society of Transplantation Infectious Diseases Community of Practice. *Clin. Transplant.* **2019**, *33*, e13543. [CrossRef] [PubMed]
69. Kung, H.C.; Huang, P.Y.; Chen, W.T.; Ko, B.S.; Chen, Y.C.; Chang, S.C.; Chuang, Y.C. 2016 guidelines for the use of antifungal agents in patients with invasive fungal diseases in Taiwan. *J. Microbiol. Immunol. Infect.* **2018**, *51*, 1–17. [CrossRef]
70. Chang, C.C.; Blyth, C.C.; Chen, S.C.; Khanina, A.; Morrissey, C.O.; Roberts, J.A.; Thursky, K.A.; Worth, L.J.; Slavin, M.A. Introduction to the updated Australasian consensus guidelines for the management of invasive fungal disease and use of antifungal agents in the haematology/oncology setting, 2021. *Intern. Med. J.* **2021**, *51* (Suppl. 7), 3–17. [CrossRef]
71. Chen, S.C.; Sorrell, T.C.; Chang, C.C.; Paige, E.K.; Bryant, P.A.; Slavin, M.A. Consensus guidelines for the treatment of yeast infections in the haematology, oncology and intensive care setting, 2014. *Intern. Med. J.* **2014**, *44*, 1315–1332. [CrossRef]
72. WHO. *Guidelines for Diagnosing, Preventing and Managing Cryptococcal Disease among Adults, Adolescents and Children Living with HIV*; World Health Organization: Geneva, Switzerland, 2022; pp. 1–64.
73. Hoenigl, M.; Sprute, R.; Egger, M.; Arastehfar, A.; Cornely, O.A.; Krause, R.; Lass-Flörl, C.; Prattes, J.; Spec, A.; Thompson, G.R., 3rd; et al. The Antifungal Pipeline: Fosmanogepix, Ibrexafungerp, Olorofim, Opelconazole, and Rezafungin. *Drugs* **2021**, *81*, 1703–1729. [CrossRef] [PubMed]
74. Wang, L.; Zhang, M.; Guo, J.; Guo, W.; Zhong, N.; Shen, H.; Cai, J.; Zhu, Z.; Wu, W. In vitro activities of the tetrazole VT-1161 compared with itraconazole and fluconazole against Cryptococcus and non-albicans Candida species. *Mycologia* **2021**, *113*, 918–925. [CrossRef]
75. Qian, A.; Zheng, Y.; Wang, R.; Wei, J.; Cui, Y.; Cao, X.; Yang, Y. Design, synthesis, and structure-activity relationship studies of novel tetrazole antifungal agents with potent activity, broad antifungal spectrum and high selectivity. *Bioorg. Med. Chem. Lett.* **2018**, *28*, 344–350. [CrossRef] [PubMed]
76. Mitsuyama, J.; Nomura, N.; Hashimoto, K.; Yamada, E.; Nishikawa, H.; Kaeriyama, M.; Kimura, A.; Todo, Y.; Narita, H. In vitro and in vivo antifungal activities of T-2307, a novel arylamidine. *Antimicrob. Agents Chemother.* **2008**, *52*, 1318–1324. [CrossRef] [PubMed]

77. Wiederhold, N.P.; Najvar, L.K.; Jaramillo, R.; Olivo, M.; Patterson, H.; Connell, A.; Fukuda, Y.; Mitsuyama, J.; Catano, G.; Patterson, T.F. The Novel Arylamidine T-2307 Demonstrates In Vitro and In Vivo Activity against Candida auris. *Antimicrob. Agents Chemother.* **2020**, *64*, e02198-19. [CrossRef] [PubMed]
78. Pappas, P.G.; Bustamante, B.; Ticona, E.; Hamill, R.J.; Johnson, P.C.; Reboli, A.; Aberg, J.; Hasbun, R.; Hsu, H.H. Recombinant interferon- gamma 1b as adjunctive therapy for AIDS-related acute cryptococcal meningitis. *J. Infect. Dis.* **2004**, *189*, 2185–2191. [CrossRef]
79. Jarvis, J.N.; Meintjes, G.; Rebe, K.; Williams, G.N.; Bicanic, T.; Williams, A.; Schutz, C.; Bekker, L.G.; Wood, R.; Harrison, T.S. Adjunctive interferon-gamma immunotherapy for the treatment of HIV-associated cryptococcal meningitis: A randomized controlled trial. *AIDS* **2012**, *26*, 1105–1113. [CrossRef]
80. Thwaites, G.E.; Nguyen, D.B.; Nguyen, H.D.; Hoang, T.Q.; Do, T.T.; Nguyen, T.C.; Nguyen, Q.H.; Nguyen, T.T.; Nguyen, N.H.; Nguyen, T.N.; et al. Dexamethasone for the treatment of tuberculous meningitis in adolescents and adults. *N. Engl. J. Med.* **2004**, *351*, 1741–1751. [CrossRef]
81. Beardsley, J.; Wolbers, M.; Kibengo, F.M.; Ggayi, A.B.; Kamali, A.; Cuc, N.T.; Binh, T.Q.; Chau, N.V.; Farrar, J.; Merson, L.; et al. Adjunctive Dexamethasone in HIV-Associated Cryptococcal Meningitis. *N. Engl. J. Med.* **2016**, *374*, 542–554. [CrossRef]
82. Seaton, R.A.; Verma, N.; Naraqi, S.; Wembri, J.P.; Warrell, D.A. The effect of corticosteroids on visual loss in Cryptococcus neoformans var. gattii meningitis. *Trans. R. Soc. Trop. Med. Hyg.* **1997**, *91*, 50–52. [CrossRef]
83. Lane, M.; McBride, J.; Archer, J. Steroid responsive late deterioration in Cryptococcus neoformans variety gattii meningitis. *Neurology* **2004**, *63*, 713–714. [CrossRef]
84. Chang, C.C.; Hall, V.; Cooper, C.; Grigoriadis, G.; Beardsley, J.; Sorrell, T.C.; Heath, C.H. Consensus guidelines for the diagnosis and management of cryptococcosis and rare yeast infections in the haematology/oncology setting, 2021. *Intern. Med. J.* **2021**, *51* (Suppl. 7), 118–142. [CrossRef] [PubMed]
85. Rhein, J.; Huppler Hullsiek, K.; Tugume, L.; Nuwagira, E.; Mpoza, E.; Evans, E.E.; Kiggundu, R.; Pastick, K.A.; Ssebambulidde, K.; Akampurira, A.; et al. Adjunctive sertraline for HIV-associated cryptococcal meningitis: A randomised, placebo-controlled, double-blind phase 3 trial. *Lancet Infect. Dis.* **2019**, *19*, 843–851. [CrossRef] [PubMed]
86. Villanueva-Lozano, H.; Trevino-Rangel, R.J.; Gonzalez, G.M.; Hernandez-Rodriguez, P.A.; Camacho-Ortiz, A.; Castillo-Reyna, L.; Galindo-Alvarado, S.G.; Martinez-Resendez, M.F. Clinical evaluation of the antifungal effect of sertraline in the treatment of cryptococcal meningitis in HIV patients: A single Mexican center experience. *Infection* **2018**, *46*, 25–30. [CrossRef] [PubMed]
87. Boulware, D.R.; Nalintya, E.; Rajasingham, R.; Kirumira, P.; Naluyima, R.; Turya, F.; Namanda, S.; Rutakingirwa, M.K.; Skipper, C.P.; Nikweri, Y.; et al. Adjunctive sertraline for asymptomatic cryptococcal antigenemia: A randomized clinical trial. *Med. Mycol.* **2020**, *58*, 1037–1043. [CrossRef]
88. Ngan, N.T.T.; Thanh Hoang Le, N.; Vi Vi, N.N.; Van, N.T.T.; Mai, N.T.H.; Van Anh, D.; Trieu, P.H.; Lan, N.P.H.; Phu, N.H.; Chau, N.V.V.; et al. An open label randomized controlled trial of tamoxifen combined with amphotericin B and fluconazole for cryptococcal meningitis. *Elife* **2021**, *10*, e68929. [CrossRef]
89. Ueno, K.; Urai, M.; Takatsuka, S.; Abe, M.; Miyazaki, Y.; Kinjo, Y. Immunization with Antigen-Pulsed Dendritic Cells Against Highly Virulent Cryptococcus gattii Infection: Analysis of Cytokine-Producing T Cells. *Methods Mol. Biol.* **2017**, *1625*, 327–339. [CrossRef]
90. Ueno, K.; Yanagihara, N.; Shimizu, K.; Miyazaki, Y. Vaccines and Protective Immune Memory against Cryptococcosis. *Biol. Pharm. Bull.* **2020**, *43*, 230–239. [CrossRef]
91. Oliveira, L.V.N.; Wang, R.; Specht, C.A.; Levitz, S.M. Vaccines for human fungal diseases: Close but still a long way to go. *NPJ Vaccines* **2021**, *6*, 33. [CrossRef]
92. da Silva, T.A.; Hauser, P.J.; Bandey, I.; Laskowski, T.; Wang, Q.; Najjar, A.M.; Kumaresan, P.R. Glucuronoxylomannan in the Cryptococcus species capsule as a target for Chimeric Antigen Receptor T-cell therapy. *Cytotherapy* **2021**, *23*, 119–130. [CrossRef]
93. Perfect, J.R.; Bicanic, T. Cryptococcosis diagnosis and treatment: What do we know now. *Fungal Genet. Biol.* **2015**, *78*, 49–54. [CrossRef] [PubMed]
94. Hu, S.; Gu, F.; Chen, M.; Wang, C.; Li, J.; Yang, J.; Wang, G.; Zhou, Z.; Yang, Y. A novel method for identifying and distinguishing Cryptococcus neoformans and Cryptococcus gattii by surface-enhanced Raman scattering using positively charged silver nanoparticles. *Sci. Rep.* **2020**, *10*, 12480. [CrossRef] [PubMed]
95. Klein, K.R.; Hall, L.; Deml, S.M.; Rysavy, J.M.; Wohlfiel, S.L.; Wengenack, N.L. Identification of Cryptococcus gattii by use of L-canavanine glycine bromothymol blue medium and DNA sequencing. *J. Clin. Microbiol.* **2009**, *47*, 3669–3672. [CrossRef] [PubMed]
96. Cheng, Y.C.; Ling, J.F.; Chang, F.C.; Wang, S.J.; Fuh, J.L.; Chen, S.S.; Teng, M.M.; Chang, C.Y. Radiological manifestations of cryptococcal infection in central nervous system. *J. Chin. Med. Assoc.* **2003**, *66*, 19–26.
97. Duarte, S.B.L.; Oshima, M.M.; Mesquita, J.; do Nascimento, F.B.P.; de Azevedo, P.C.; Reis, F. Magnetic resonance imaging findings in central nervous system cryptococcosis: Comparison between immunocompetent and immunocompromised patients. *Radiol. Bras.* **2017**, *50*, 359–365. [CrossRef]
98. Song, K.D.; Lee, K.S.; Chung, M.P.; Kwon, O.J.; Kim, T.S.; Yi, C.A.; Chung, M.J. Pulmonary cryptococcosis: Imaging findings in 23 non-AIDS patients. *Korean J. Radiol.* **2010**, *11*, 407–416. [CrossRef]
99. Tintelnot, K.; Hagen, F.; Han, C.O.; Seibold, M.; Rickerts, V.; Boekhout, T. Pitfalls in Serological Diagnosis of Cryptococcus gattii Infections. *Med. Mycol.* **2015**, *53*, 874–879. [CrossRef]

100. Dhana, A. Diagnosis of Cryptococcosis and Prevention of Cryptococcal Meningitis Using a Novel Point-of-Care Lateral Flow Assay. *Case Rep. Med.* **2013**, *2013*, 640216. [CrossRef]
101. Vidal, J.E.; Boulware, D.R. Lateral flow assay for cryptococcal antigen: An important advance to improve the continuum of hiv care and reduce cryptococcal meningitis-related mortality. *Rev. Inst. Med. Trop. Sao Paulo* **2015**, *57* (Suppl. 19), 38–45. [CrossRef]
102. Tay, E.; Chen, S.C.; Green, W.; Lopez, R.; Halliday, C.L. Development of a Real-Time PCR Assay to Identify and Distinguish between Cryptococcus neoformans and Cryptococcus gattii Species Complexes. *J. Fungi* **2022**, *8*, 462. [CrossRef]
103. Ma, Q.; Yao, J.; Yuan, S.; Liu, H.; Wei, N.; Zhang, J.; Shan, W. Development of a lateral flow recombinase polymerase amplification assay for rapid and visual detection of Cryptococcus neoformans/*C. gattii* in cerebral spinal fluid. *BMC Infect. Dis.* **2019**, *19*, 108. [CrossRef] [PubMed]
104. Wang, L.; Wang, Y.; Wang, F.; Zhao, M.; Gao, X.; Chen, H.; Li, N.; Zhu, Q.; Liu, L.; Zhu, W.; et al. Development and Application of Rapid Clinical Visualization Molecular Diagnostic Technology for Cryptococcus neoformans/*C. gattii* Based on Recombinase Polymerase Amplification Combined With a Lateral Flow Strip. *Front. Cell. Infect. Microbiol.* **2021**, *11*, 803798. [CrossRef] [PubMed]
105. Firacative, C.; Trilles, L.; Meyer, W. MALDI-TOF MS enables the rapid identification of the major molecular types within the Cryptococcus neoformans/*C. gattii* species complex. *PLoS ONE* **2012**, *7*, e37566. [CrossRef] [PubMed]
106. Singhal, N.; Kumar, M.; Kanaujia, P.K.; Virdi, J.S. MALDI-TOF mass spectrometry: An emerging technology for microbial identification and diagnosis. *Front. Microbiol.* **2015**, *6*, 791. [CrossRef] [PubMed]

Journal of *Fungi*

MDPI

Review

Paracoccidioidomycosis: What We Know and What Is New in Epidemiology, Diagnosis, and Treatment

Paulo Mendes Peçanha [1], Paula Massaroni Peçanha-Pietrobom [2], Tânia Regina Grão-Velloso [3], Marcos Rosa Júnior [4], Aloísio Falqueto [1] and Sarah Santos Gonçalves [5,*]

1 Department of Medical Clinic, Division of Infectious Diseases, Federal University of Espírito Santo (UFES), Vitoria 29043900, Brazil
2 Department of Medicine, Division of Infectious Diseases, Federal University of São Paulo (UNIFESP), São Paulo 04039032, Brazil
3 Department of Dental Clinic, Federal University of Espírito Santo (UFES), Vitoria 29043900, Brazil
4 Department of Medical Clinic, Federal University of Espírito Santo (UFES), Vitoria 29043900, Brazil
5 Department of Pathology, Infectious Diseases Postgraduate Program, Center for Research in Medical Mycology, Federal University of Espírito Santo (UFES), Vitoria 29043900, Brazil
* Correspondence: sarah.tavares@ufes.br; Tel.: +55-27-33357290

Abstract: Paracoccidioidomycosis (PCM) is a systemic mycosis endemic to Latin America caused by thermodimorphic fungi of the genus *Paracoccidioides.* In the last two decades, enhanced understanding of the phylogenetic species concept and molecular variations has led to changes in this genus' taxonomic classification. Although the impact of the new species on clinical presentation and treatment remains unclear, they can influence diagnosis when serological methods are employed. Further, although the infection is usually acquired in rural areas, the symptoms may manifest years or decades later when the patient might be living in the city or even in another country outside the endemic region. Brazil accounts for 80% of PCM cases worldwide, and its incidence is rising in the northern part of the country (Amazon region), owing to new settlements and deforestation, whereas it is decreasing in the south, owing to agriculture mechanization and urbanization. Clusters of the acute/subacute form are also emerging in areas with major human intervention and climate change. Advances in diagnostic methods (molecular and immunological techniques and biomarkers) remain scarce, and even the reference center's diagnostics are based mainly on direct microscopic examination. Classical imaging findings in the lungs include interstitial bilateral infiltrates, and eventually, enlargement or calcification of adrenals and intraparenchymal central nervous system lesions are also present. Besides itraconazole, cotrimoxazole, and amphotericin B, new azoles may be an alternative when the previous ones are not tolerated, although few studies have investigated their use in treating PCM.

Keywords: paracoccidioidomycosis; *Paracoccidioides* spp.; endemic mycosis; thermodimorphic fungi

Citation: Peçanha, P.M.; Peçanha-Pietrobom, P.M.; Grão-Velloso, T.R.; Rosa Júnior, M.; Falqueto, A.; Gonçalves, S.S. Paracoccidioidomycosis: What We Know and What Is New in Epidemiology, Diagnosis, and Treatment. *J. Fungi* **2022**, *8*, 1098. https://doi.org/10.3390/jof8101098

Academic Editor: Alessandro C. Pasqualotto

Received: 1 September 2022
Accepted: 14 October 2022
Published: 18 October 2022

Publisher's Note: MDPI stays neutral with regard to jurisdictional claims in published maps and institutional affiliations.

1. Introduction

Paracoccidioidomycosis (PCM) is a systemic fungal infection caused by thermodimorphic microorganisms belonging to the genus *Paracoccidioides.* The disease is endemic to South and Central America, but imported cases have been reported in North America, Europe, Africa, and Asia. The fungus is a soil saprophyte; classically, humans get infected through agricultural activities. Thus, socioeconomic changes in Latin America in recent decades—namely, reduction in human labor in agriculture—have directly impacted the epidemiology of PCM [1–4].

Although PCM-related mortality is low, morbidity is high once the chronic form's sequelae are present in almost 50% of patients despite treatment [5]. Often, a lack of early clinical suspicion results in delayed treatment.

A recent review of endemic mycosis in America recognized that PCM had the fewest diagnostic tools available [6]. Microbiological studies on respiratory, ganglion, or mucocutaneous samples can be easily performed with a direct microscopic examination (DME). A limitation is that expertise is required to recognize the characteristic fungal structures [6].

Commercial kits for serological diagnosis have been developed, but countries still face production, distribution, and cost problems; thus, in most centers, the antigen is produced *in-house*. Some investigations have been conducted to determine triggers that detect PCM owing to all the species of *Paracoccidioides*. Novel molecular biomarkers have been studied, but warrant further investigation to validate their use and identify the best scenarios to employ them [6–9].

Treatment-wise, *Paracoccidioides* spp. are susceptible to sulfonamides, azoles, and amphotericin B (both conventional and lipidic formulations). Although itraconazole has been proven to be more effective than cotrimoxazole to treat mild and moderate PCM, in Latin America, the last is still largely used because of its lower cost. Similarly, although the lipid formulations of amphotericin B are less toxic than the conventional one, their high price limits its use in low-income countries [10–12].

This paper summarizes essential knowledge and updates in the epidemiology, diagnosis, and treatment of PCM. In addition, we aimed to highlight the need for advancement in diagnostic and therapeutic strategies that will reduce the burden of PCM in Latin America.

2. Epidemiology of Paracoccidioidomycosis

Species and Geographic Distribution

In recent years, the taxonomy of the genus *Paracoccidioides* has undergone notable changes. Until 2005, *Paracoccidioides brasiliensis* was considered the only species that caused PCM. In 2006, Matute et al. [13], through genotypic studies, revealed variations in *P. brasiliensis* encompassing four genetic variants (S1, PS2, PS3, and PS4) [12]. In 2009, Teixeira et al. [14] described the new species *Paracoccidioides lutzii* in the midwest region of Brazil. In 2017, Turissini et al. [15] proposed a new classification for the *P. brasiliensis* variants' phylogenetic species (S1, PS2, PS3, and PS4), comprising four new species: *P. brasiliensis sensu stricto*, for S1, the deep split of the S1 lineage into two clades named S1a and S1b [16]; *Paracoccidioides americana* for PS2; *Paracoccidioides restrepiensis* for PS3; and *Paracoccidioides venezuelensis* for PS4. In 2020, whole genome sequencing studies confirmed *P. brasiliensis* complex reclassification into new species [17].

Regarding geographic distribution, *P. brasiliensis sensu stricto* is widespread and predominant in the southern region of South America (Brazil, Argentina, Paraguay, Uruguay, and Bolivia). *P. americana* has been identified in the same regions, but in limited cases. *P. restrepiensis* is predominant in Colombia, with cases reported in Argentina, Peru, and Uruguay. *P. venezuelensis* is dominant in Venezuela and reported in Colombia as well. *P. lutzii* is prevalent in central-west Brazil, with scattered cases outside this area [14–18].

Recent developments in genetic studies of *Paracoccidioides'* different species and their epidemiology show that they co-exist in several regions of Latin America, with a clear overlapping distribution. Interestingly, when the magnitude of gene flow between species was analyzed, Mavengere et al. [19] confirmed that *Paracoccidioides* species rarely exchange genes despite extensive geographic overlap [16–19].

Although the different species have shown differences in conidial morphology and antigenic display, no recent studies have shown clear implications of species diversity on disease clinical manifestation and treatment [20–22]. Finally, recent genetic analyses of specimens from dolphins with lobomycosis placed the DNA sequences of *Lacazia loboi* within *Paracoccidioides* species, proposing the taxonomy of the dolphin pathogen as *Paracoccidioides cetii*, and the human pathogen as *Paracoccidioides loboi* [23].

In Latin America, Brazil has the highest number of PCM cases (80%), followed by Colombia, Venezuela, Ecuador, and Argentina [2]. The disease has been reported as far north as Mexico and south as Argentina [24,25]. No cases have been reported in Nicaragua, Belize, most Caribbean islands, Guyana, Surinam, or Chile [2,26].

Over 100 PCM cases have been reported in Europe, the United States of America, Canada, Africa, and Asia. The patients are usually non-autochthonous, mainly immigrants or travelers from endemic countries [27–29]. In a recent systematic review, Wagner et al. [4] identified 83 patients with PCM in 11 European countries, most of whom were from Spain, Italy, and Germany. The patients were mostly men aged 23–83 years, with a latency period of 6 days to 50 years [4]. In Asia, specifically in Japan, 16 imported cases of PCM—mainly from Brazil—have been reported [30–38]. Figure 1 shows all countries with endemic PCM and imported cases.

Figure 1. Geographic distribution of Paracoccidioidomycosis in relation to endemic areas and imported cases.

A major risk factor for PCM is soil exposure in rural areas, being an occupational disease for farmers in endemic regions [2,3]. Person-to-person transmission has not yet been documented [1]. In terms of personal and social history, smoking has been elicited in the histories of 90% of patients with chronic PCM; smokers have a 14-fold higher risk of developing the disease than nonsmokers [39,40]. Alcohol intake also increased the risk of chronic PCM by 3.5 times [2,39].

PCM's chronic form presents in adults from 30 to 60 years, with a male predominance (ratios of 15:1 to 22:1 in Brazil) [41,42]. The acute form of the disease shows an equal distribution between sexes, especially in childhood [41]. The predominance of chronic PCM in men can be explained by the protective effect of estrogen in women, which inhibits the transformation of conidia in yeast cells after menarche [43].

The incidence of PCM in Brazil ranges from 1 to 4 per 100,000 inhabitants per year in established endemic areas in the southeastern and southern regions. In areas of recent colonization in the northern region (Amazon), the incidence rises to 9–40 cases per 100,000 inhabitants, illustrating a hyperendemic pattern in the last few decades [2,44,45]. The prevalence rate of PCM infection was 45.8% in endemic rural areas in southeast Brazil according to surveys based on skin tests, with Colombia and Ecuador having rates of 45% and 41%, respectively. Conversely, in Argentina and Venezuela, the rates were as low as 7.8% and 10.2%, respectively [39,46–49]. Notably, since PCM does not require compulsory notification in most Latin American countries, its incidence and prevalence may be underreported [6].

The most extensive study on PCM mortality performed in Brazil by Coutinho et al. reported a mortality rate of 1.45 per million inhabitants, with a total of 3.181 attributed deaths per million inhabitants between 1980 and 1995 [50]. Currently, the mortality rate is decreasing in Brazil's southeastern and southern regions, and increasing in the north. The state of Paraná, which had the highest mortality rate for PCM in south Brazil (3–4.29 deaths per million inhabitants during 1980–1995), showed a reduction in the average annual mortality rate to 1.17 deaths per million inhabitants during 2007–2020 [51]. Contrastingly, the state of Rondônia in the northern region drastically progressed from a mortality rate of 3.65 per million inhabitants in 1980–1995 to 8.2 in 2002–2004 [44,52].

Death from PCM is due to disseminated disease, respiratory insufficiency, and adrenal insufficiency. These three are frequently associated with the chronic form of the disease and may develop long after antifungal treatment has been completed [41,53,54]. Despite the indolent course of the disease and the availability of curative medications, difficulty accessing diagnostics and treatment contributes to an unfavorable prognosis [3,50].

The epidemiology of PCM has significantly changed in terms of frequency, demographic characteristics, and geographic distribution, owing to human migration, environmental interventions, and climate change. The opening of new agricultural frontiers and deforestation have contributed to the rise in the incidence of PCM in northern Brazilian states. In contrast, the reduction in child labor in rural areas and mechanization of agriculture have decreased the number of new cases in the southern region [3,44,55,56]. An increase in *Paracoccidioides* spp. infections has been associated with the construction of the Yacyreta hydroelectric plant in northeastern Argentina [57]. Clusters of acute/subacute cases of PCM were reported after the El Niño phenomena in 1982/83 in southeast Brazil and 2009 in northeast Argentina [55,58], as well as after the massive land removal during the construction of the ring road in the Rio de Janeiro metropolitan area in 2016 [59]. These clusters underscore the urgent need for the surveillance of new cases in endemic regions undergoing climate change and human interventions, such as deforestation and massive constructions, to ensure early diagnosis and treatment [59].

3. Diagnosis of Paracoccidioidomycosis

3.1. Clinical Diagnosis

The first challenge in diagnosing PCM is to think about the disease when, even in endemic countries and more so in countries outside Latin America, physicians are unfamiliar with this systemic mycosis. Consequently, late diagnosis increases morbidity and mortality rates [4,40].

PCM infection is acquired by inhaling fungal propagules found in the environment, but only 1% to 2% of infected individuals will develop clinical manifestations during their lives [3]. Those who remain asymptomatic can control fungal replication through a robust Th-1 immune response pattern, characterized by cytokine release that activates macrophages and TCD4+ and TCD8+ cells, forming compact granulomas. This stage is named PCM infection [60]. Among those individuals who progress from infection to illness, 5–25% present with the acute/subacute clinical form of PCM, in which Th-2 and Th-9 immune response patterns activate B-lymphocytes that produce high levels of antigen-specific IgA, IgE, and IgG4 [61]. The other 75–95% of cases will evolve from the latent stage to the chronic form of the disease many years later, usually after the fourth decade of life [3].

Chronic PCM manifests gradually and may occur years after exposure to *Paracoccidioides* when the patient is already living in urban areas or outside endemic regions [3]. This form mainly affects the lungs; mucous membranes; skin; and, eventually, the adrenal and central nervous systems [2,41,53]. The main manifestation of chronic PCM in approximately 90% of patients is pulmonary, with symptoms of cough, dyspnea, and sputum expectoration [3,41]. A recent study by Dutra et al. [62] in a southeastern Brazilian hospital found that 59.6% of the patients with PCM had granulomatous ulcerated oral lesions. The oral lesions (Figure 2A,B) may be the first visible physical manifestation of the disease noticed by the patient, and may lead to a prompt diagnosis. However, even with adequate treatment, patients may develop the residual form of PCM, owing to fibrosis of the affected organs [3,5].

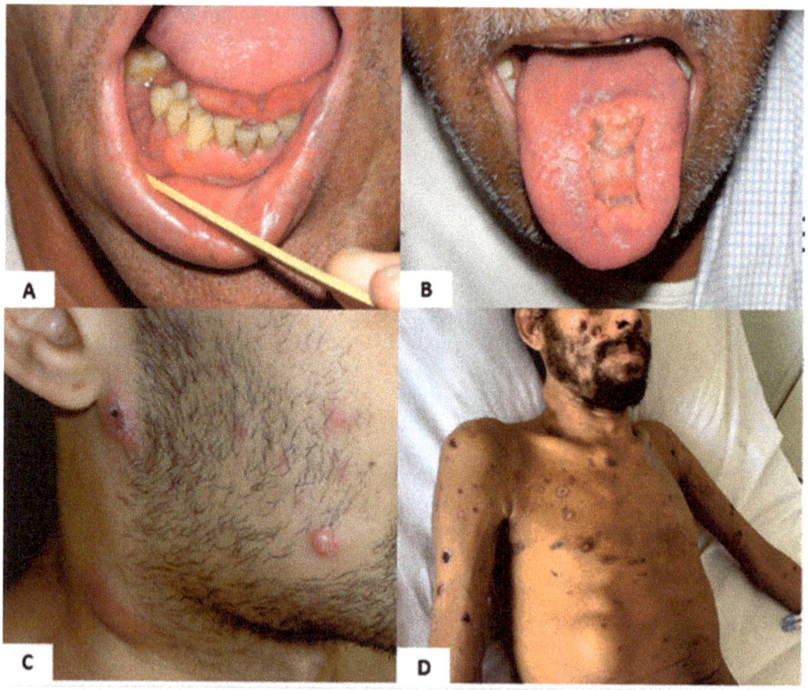

Figure 2. (**A**) Chronic form: gingival and lingual frenulus "mulberry-like" ulcers with hemorrhagic dots; (**B**) chronic form: deep ulcerative lesion on the tongue with infiltrative borders, hemorrhagic dots, covered with fibrin; (**C**) acute/subacute form: multiple polymorphic (nodular, papular, and ulcerated) skin lesions and cervical inflammatory lymphadenopathy; (**D**) disseminated form in a patient with PCM and HIV co-infection; the following may be seen: cervical lymphadenopathy, exuberant ulcerated, crusted skin lesions, and larges subcutaneous abscesses in the thorax and abdomen.

The acute/subacute form of the disease shows rapid and disseminated progression in the form of skin lesions; lymphadenopathy; and eventual suppuration, fever, and anorexia (Figure 2C) [3,41,63]. This form characteristically develops a few weeks or months after fungal exposure [2].

In immunocompromised patients, a mixed clinical form of PCM, with characteristics of both chronic and acute forms of the disease, has been observed. Pulmonary involvement can coincide with generalized adeno- and hepatosplenomegaly. In patients with the mixed form, multiple, exuberant skin involvement; lytic bone lesions; and central nervous system involvement can be present, indicative of severe disease (Figure 2D). Patients with HIV co-infected with PCM make up the bulk of immunosuppressed patients with PCM; however, cases of PCM in transplant patients and those receiving immunobiological therapy have also been reported [63–66].

PCM is often confused with tuberculosis in Latin America, owing to its high prevalence and the similar clinical presentations of both diseases [34]. Besides tuberculosis, the most relevant differential diagnoses of chronic pulmonary PCM are other fungal infections, such as coccidioidomycosis and histoplasmosis. Sarcoidosis, pneumoconiosis, and interstitial pneumonitis should also be considered. Moreover, it is essential to rule out concomitant diseases. Tuberculosis and PCM can occur simultaneously or sequentially in 5.5–19% of cases [41,53,67,68]. Additionally, PCM and solid cancers of the respiratory and gastrointestinal tracts share similar risk factors (male sex, smoking, and alcohol intake). Solid neoplasias and *Paracoccidioides* infection have been shown to co-exist in 0.16–11% of patients [69].

In patients with mucocutaneous PCM, the differential diagnosis should include leishmaniasis, tuberculosis, chromoblastomycosis, leprosy, syphilis, and neoplasia. In individuals with acute PCM, clinicians should be concerned about hematologic neoplasms, histoplasmosis, tuberculosis, and visceral leishmaniasis [3]. It is important to remember that many infectious diseases share the same endemic areas as PCM.

Finally, patients with chronic pulmonary PCM are at higher risk of more severe illness with COVID-19 coinfection. Despite that, during the pandemic, only one case of SARS-CoV-2 and *Paracoccidioides* spp. co-infection was described in an individual with the acute form of PCM. The scarcity of documented cases probably reflects underdiagnosis or underreporting in Latin American countries that had their health systems overwhelmed by COVID-19 [70–72].

3.2. Laboratory Diagnosis

Laboratory diagnosis via microscopy remains the gold standard method for diagnosing PCM. This may show the presence of the etiologic agent in biological fluids and tissue sections or the isolation of the fungus from clinical specimens, owing to the characteristic appearance of typical *Paracoccidioides* spp. yeast forms [73]. Other tools, such as cultures, immunodiffusion assays, and polymerase chain reaction (PCR) tests [7–9,74,75], are also used. Different types of clinical samples may be collected for testing. Mucocutaneous scrapings, sputum, bronchoalveolar lavage (BAL), cerebrospinal fluid (CSF), lymph node aspirate, biopsy, and tissue samples are those most frequently collected [76,77]. Prior processing of some samples is needed to increase their sensitivity to detection methods, including centrifugation (for sputum, BAL, CSF, and lymph node aspirate) and maceration of fragmented, biopsied tissues [7]. The sputum sample should be prepared with potassium hydroxide, sodium hydroxide, and N-acetyl-L-cysteine before being added to a suitable culture medium at 25 °C [73]. Figure 3 shows the main diagnostic tools employed for laboratory diagnosis of PCM.

3.2.1. Mycological Diagnosis

This method includes visualizing fungal elements through DME, followed by isolation of the agent in culture media [25]. DME of the sputum, BAL, CSF, lymph node aspirate, and mucocutaneous scraping are prepared with the addition of 10–20% KOH or calcofluor, making it possible to visualize *Paracoccidioides* spp. in their parasitic form with multiple budding cells (blastoconidia) surrounding it, connected by short cellular bridges (Figure 3B,D) [78–82]. However, it is important to mention that for biopsy and tissue samples, the slides are mounted using 40% KOH [78]. *Paracoccidioides* cells have a thick mucopolysaccharide wall with a double-contour appearance that is birefringent under light microscopy [83,84]. *Paracoccidioides* structures resembling a "ship's wheel" or "Mickey Mouse" are deemed pathognomonic findings [9,79,82].

The size and multiple budding distinguish *Paracoccidioides* spp. from other fungi. Nevertheless, *Paracoccidioides* isolates can be mistaken for *Histoplasma capsulatum* and *Cryptococcus* spp. when it produces small, non-budding cells, and when *Cryptococcus* does not produce its capsule efficiently [84,85].

Moreto et al. [7] evaluated the diagnostic methods for PCM at a university hospital between 1976 and 2004. They observed that in the DME of 51 different tissue specimens and 112 sputum samples, the sensitivity was 75% and 63%, respectively. For 483 sputum cell blocks, the values found were 55%. Since the PCM chronic form is the most frequent form encountered, sputum is the biological material most commonly evaluated under DME, and the sensitivity will depend on the processing method of that material [40,53,86]. Although DME is a simple, fast, and low-cost technique implemented in small laboratories, its sensitivity is low [40,45,53,86,87]. Thus, DME cannot provide a conclusive diagnosis in cases of negative results. Owing to the heterogeneity of the sample fractions, DME can mistakenly lead to the assumption that the fungus does not exist in the entire sample [75].

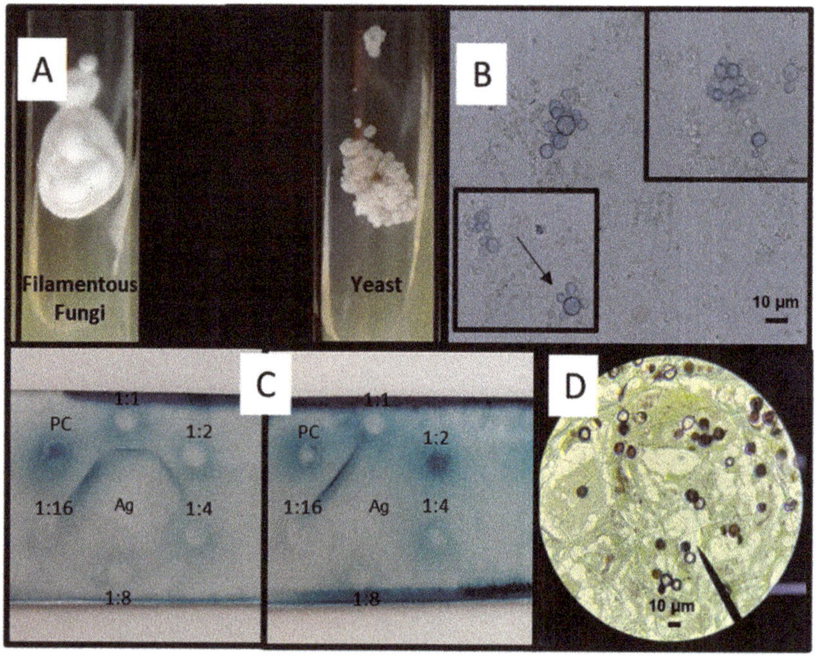

Figure 3. Laboratory diagnosis of Paracoccidioidomycosis. (**A**) *Paracoccidioides* spp. on Sabouraud agar slants for 14 days at 25 °C (left, filamentous phase) and 37 °C (right, yeast phase); (**B**) yeast form showing multiple budding (5 to 15 μm) seen upon direct examination of a lymph node aspirate, stained with KOH and Parker ink; Bars 10 μm (**C**) double immunodiffusion assay: *Paracoccidioides brasiliensis* exoantigen (Ag) is in the center well; sera sample from a patient with paracoccidioidomycosis is used in different titrations (wells 1:1 to 1:16) and positive control (PC); positive 1:2 (**left**) and negative (**right**); (**D**) Grocott's methenamine silver stain showing *Paracoccidioides* yeast cells. Bars 10 μm.

3.2.2. Cultures

Paracoccidioides take an average of 3–6 weeks to grow on fungal culture media [88–90]. Nonetheless, the growth of this fungus varies in the different studies, with a sensitivity of 80 to 97% of cases [21,90–92]. The results should be evaluated about 4 weeks after the cultivation of the sample at 25 °C. To reduce the running cultivation time, samples must be cultured simultaneously at 25 °C and 37 °C. The culture media most frequently used are Sabouraud, Mycosel, and Fava-Netto agar. Other media, such as mycobiotic agar, brain heart infusion agar (BHI), Sabouraud dextrose plus BHI broth (SABHI), agar-yeast extract-phosphate, agar-yeast extract, agar-yeast extract-penicillin plus streptomycin and cycloheximide, and Kelley medium with hemoglobin, are also employed [75,78,79,84,93]. According to Hahn et al. [78], the successful recovery of *Paracoccidioides* from clinical specimens depends on several factors, including the culture media and the number of tubes or plates seeded, besides the decontamination with antibiotics of sputum and bronchial lavage fluid. Despite its slowness, culture is always recommended because it allows species identification using molecular biology [9,94–96].

3.2.3. Histopathological Diagnosis

Histopathological examination is a valuable tool (≥95% sensitivity) for PCM diagnosis. It can also determine disease severity [7,53,97]. The sections can be stained with hematoxylin/eosin (H & E), Grocott's methenamine silver, and periodic acid–Schiff (PAS). These last two are specific stains that increase sensitivity [62]. When stained with H&E,

an inflammatory response can be observed in the parasite–host interaction. Organizing granulomas or a combination of suppurative and granulomatous infiltrates can also be seen [7,75,98]. If the samples are sufficient, DME and tissue culture can be performed [86]. It is worth mentioning that biological samples for histological examination are fixed in formaldehyde solution. For culture, the biopsy and tissue samples must be placed in a sterile container under a sterile physiological saline solution.

3.2.4. Immunological Diagnosis: Antibody Detection

In 1916, Arthur Moses, an assistant physician at the Oswaldo Cruz Foundation, isolated an antigen from *P. brasiliensis*-infected cells used in a complement fixation (CF) test to diagnose PCM [20]. Since then, numerous antigens and antibody-based serological assays have been developed as alternative methods for detecting fungal structures in biological fluids/tissues and disease monitoring [99,100]. Double immunodiffusion (DID), counterimmunoelectrophoresis (CIE), immunofluorescence, radioimmunoassay, enzyme-based immunoassays (ELISA), immunoblotting, dot immunoassay, western blotting, and latex particle agglutination (LA) are some important techniques in use [9,99–108]. Though many validated methods for detecting anti-*Paracoccidioides* serum antibodies exist, most are conducted only in research centers. There also remains significant limitations, owing to cross-reactivity with other infectious fungi, such as *Histoplasma* and *Aspergillus* [25,99,109].

DID is the most widely used method for detecting anti-*Paracoccidioides* serum antibodies in endemic countries [99,110]. Its advantages include the ease of performing the quantitative techniques, low cost, and high specificity (85–100%) [34,53,89,90]. However, its sensitivity can range from 80 to 95% depending on the antigen applied [111,112]. CIE has similar specificity to DID (>95%) with slightly higher sensitivity (77–100%). Both techniques are recommended as serological screening tests for patients with suspected PCM because of their faster turnaround time than microbiological methods [25,99,111,113,114].

By applying the immunoblotting technique, de Camargo et al. [115] assessed several exoantigens produced by *P. brasiliensis* isolates against serum from PCM-positive patients. IgG anti-*P. brasiliensis* was discovered in some cell surface components, but the most promising were glycoproteins gp70 (70 KDa) and gp43 (43 KDa). The latter is commonly recognized by IgG antibodies, and is reactive in 100% of patients with PCM caused by *P. brasiliensis sensu stricto*.

Several components with antigenic ability to distinguish circulating antibodies in patients with suspected PCM have been tested. Glycoprotein gp43 is the most commonly used component that can be presented as a cell-free antigen (CFA), exo-antigen (ExoAg), and recombinant or purified antigen [9,110,115,116]. Throughout the exponential growth phase of *P. brasiliensis*, its cells secrete this antigen, found in almost all isolates [75]. However, there is decreased expression of this antigen in patients infected with *P. lutzii*, and false-negative results may result [9,110,115,116]. These differences in antigenic composition are probably related to phylogenetic peculiarities [117]. In addition, gp43 may trigger cross-reactivity in patients with histoplasmosis or lobomycosis because its epitope is a galactose-containing carbohydrate, common among pathogenic fungi [75,118].

By evaluating different antigenic preparations from *P. lutzii* using the immunodiffusion technique, Gegembauer et al. [112] demonstrated that tests employing *P. brasiliensis* antigens might yield false-negative results when *P. lutzii* is the causative agent [109,119]. Maifrede et al. [109] showed that 7 of 21 sera samples negative for *P. brasiliensis* antigen were positive for *P. lutzii* when the Pb339 exoantigen and PIEPM208 CFA were applied. We can infer that the frequency of *P. lutzii* may be higher than reported in endemic areas because gp43 is the most commonly used antigen in routine laboratory examinations [9,77,109].

In 2021, an American company began commercializing a DID-based test to detect *Paracoccidioides* serum antibodies (ID Antigen®). In the same year, Cocio and Martinez [110] used CIE and DID to evaluate the sensitivity and specificity of the antigen in ID Antigen®. They found that of the 24 PCM-positive serum samples of patients with active PCM, 100% were reactive in CIE methodology using ID Antigen®, including 11 cases of infection by

P. brasiliensis sensu stricto, one by *P. americana* and one by *P. lutzii*. The test's specificity was 100%, with negative results for histoplasmosis, aspergillosis, and other diseases, and an overall 75% sensitivity with PCM sera. Therefore, the antigen available in the commercial test could diagnose PCM caused by three different species.

Considering that five *Paracoccidioides* species have been recognized as PCM agents in endemic areas, new antigen preparations must be, and are, being investigated to expand the use of PCM serology with increased sensitivity and specificity.

3.2.5. Antigen Detection

Antibody detection is not the best choice for all patients with PCM. To illustrate, in many studies, immunocompromised patients and those with severe forms of acute/subacute disease showed decreased antibodies, with half showing none [84,120,121]. Therefore, identifying antigens instead in these cases would be more suitable.

In 1997, using the inhibition ELISA technique (inh-ELISA), Colombian researchers discovered a monoclonal antibody to detect the 87 kDa antigen in patients with PCM. In patients with the disease, acute, multifocal, and unifocal forms were detected in 100%, 83.3%, and 60% of patients, respectively [122]. Since then, several antigenic molecules and tools have been characterized and evaluated [74,78,105–108]. Notably, anti-gp43 and anti-gp70 monoclonal antibodies are still the most frequently investigated glycoproteins [86]. However, cross-reactions have been obtained with heterologous sera, such as sera from patients with aspergillosis, cryptococcosis, and histoplasmosis [96].

Xavier et al. [106] analyzed the Platelia™ *Aspergillus* enzyme immunoassay (EIA) (Bio-Rad, Marnes-la-Coquette, France) as a diagnostic tool for 30 PCM patients and found a positivity rate of 50%. This method is widely used to detect galactomannan in patients suspected of having invasive aspergillosis [123].

Recently, Melo et al. [108] investigated the performance of (1,3)-β-D-glucan assays (BDG), a test used to diagnose invasive fungal infections in patients with PCM. Fifty-two serum samples from 29 patients with acute and chronic PCM were evaluated. Despite its excellent diagnostic sensitivity (96.5%), it did not contribute to disease monitoring.

Some commercial methods have been validated and are currently available for detecting specific fungal antigens in patients with cryptococcosis, histoplasmosis, coccidioidomycosis, blastomycosis, aspergillosis, and candidiasis. However, progress has not been made on making commercial tests for detecting *Paracoccidioides* antigens available. It is essential to highlight that if the antigen tests were commercially available, they could make the serological diagnosis of PCM more accessible to patients who live far from referral centers.

3.2.6. Molecular Detection

In the last century, molecular tools have provided crucial information for the taxonomic classification and epidemiological, diagnostic, and therapeutic management of pathogenic fungi. Several molecular methods, including PCR-derived techniques, have opened doors for the early diagnosis of fungal diseases and the identification of etiologic agents [9]. PCR, loop-mediated isothermal amplification (LAMP), quantitative real-time PCR (qPCR), nested and semi-nested PCR, and duplex PCR-assay have been found to detect *Paracoccidioides* genetic material directly from clinical samples [7,9,77,124–128]. Most assays are based on primary markers, such as the *GP43* gene and the internal transcribed spacer (ITS) region of ribosomal DNA [77].

In 2021, Pinheiro et al. [77] developed a duplex PCR single-assay capable of detecting and differentiating members of the *P. brasiliensis* complex and *P. lutzii* from paraffin-embedded tissue blocks [62]. This methodology became vital in clinical laboratory practice, particularly in diagnosing atypical cases, such as those with seronegative yet positive DME results, and in examining patients with co-infections.

Despite having similar sensitivity and specificity as DME and histopathological techniques, molecular methods might have a better yield with materials with a low burden of infection (serum, BAL, CSF), and may be more sensitive than DID. However, the molecular approach is performed based on *in-house* tests, for which currently external quality assessments are lacking [7,77,128]. Moreover, using molecular techniques for diagnosing disease-causing fungi directly from the clinical sample is challenging because of the complexity of DNA extraction. In addition, databases with genome sequences for these microorganisms are under construction. The very nomenclature of these agents requires continuous updating in the laboratory. In other words, the molecular tools, to be successfully used, have to adapt to the objectives of the study. Knowledge is constantly evolving in the study of fungi, and these techniques are not yet validated for use in the routine diagnosis of PCM. Furthermore, it is worth emphasizing that molecular tests are expensive, and most of the population affected by PCM belongs to developing countries.

4. Diagnostic Imaging

The conventional imaging of different systems affected by the fungus can contribute to PCM diagnosis and help identify the disease's acute, chronic, or sequel forms. The most common findings in PCM are pulmonary lesions, for which chest radiography or computed tomography (CT) can be used, with the latter being more sensitive to abnormalities. Ultrasonography, CT, or magnetic resonance imaging (MRI) can be used to examine the abdominal region. For the head and neck, CT or MRI are used; and for examining the osteoarticular system, radiography, CT, or MRI are acceptable options [115]. Recently, Cunha et al. [129] studied the effectiveness of F-fluorodeoxyglucose-positron emission tomography (FDG-PET/CT)/CT) in evaluating the extent of active disease in patients with PCM under antifungal treatment, and demonstrated that FDG-PET/CT could help detect active lesions and is more sensitive than conventional imaging methods.

In the lungs, which are more affected in the chronic form of the disease, the involvement is usually bilateral and symmetric, with lesions occurring mainly in the periphery and middle-third regions (Figure 4). There are multiple CT presentations, including consolidations, ground-glass opacities, nodules, masses, and cavitations (Figure 5); interlobular septal thickening is the most common finding. In patients from endemic areas, the so-called "butterfly wing" pattern and bilateral symmetric opacities in the middle region of the lungs are suggestive of PCM [130]. The "reversed halo" sign on CT, defined as a focal and round area of ground-glass opacity surrounded by a complete or nearly complete consolidation ring, can be observed in up to 10% of cases [34]. However, it is not specific to this disease. It also occurs in patients with tuberculosis, mucormycosis, invasive pulmonary aspergillosis, *Pneumocystis carinii* pneumonia, organizing pneumonia, granulomatous polyangiitis (Wegner), lymphomatoid granulomatosis, sarcoidosis, and lepidic predominant adenocarcinoma, among others [130]. Pleural effusion and pneumothorax, also manifestations of PCM, are rare findings in patients with the chronic form. There is significant heterogeneity in pulmonary presentations. Fibrotic lesions can develop in patients and be visualized on CT as architectural distortions, traction bronchiectasis, honeycomb lesions, thickening of the alveolar and interlobular septum, paracicatricial emphysema, and parenchymal bands [130]. There are usually no pulmonary parenchymal lesions in the acute form of the disease, frequently seen with pleural effusion or lymph node enlargement (37%) [130].

The abdomen is less frequently affected than the chest. Abdominal involvement is mainly observed in patients with the acute form of PCM, which may primarily involve the adrenal glands, liver, spleen, lymph nodes, and intestinal loops. The adrenal gland is the most commonly affected abdominal organ. There may be diffuse enlargements of the gland with heterogeneous attenuation on CT and MRI, with peripheral contrast enhancement in the acute form of the disease. Adrenal atrophy and calcification are usually observed during the chronic phase (Figure 6). Differential diagnoses should include other granulomatous infections, such as histoplasmosis and tuberculosis; old hemorrhage; and neoplasms, such as lymphoma, primary tumors, and metastases [131].

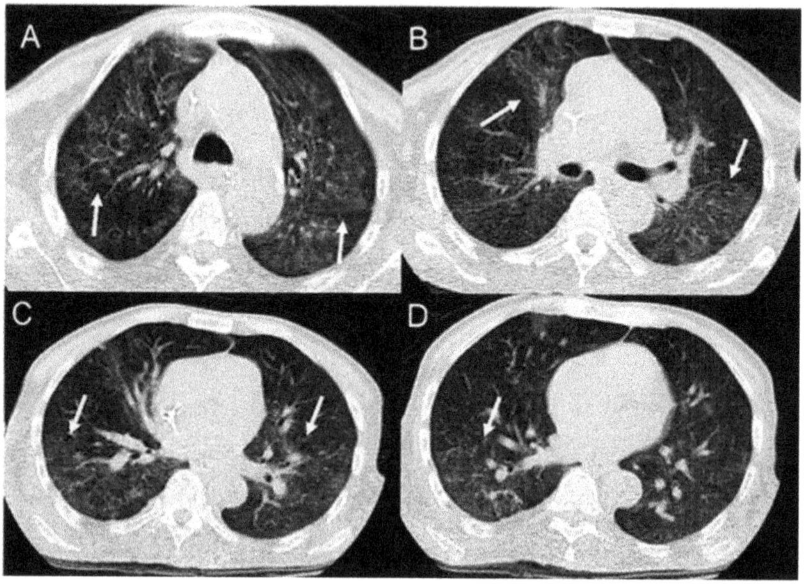

Figure 4. Computed tomography showing ground-glass opacities (arrows) with bilateral and symmetric distribution, occurring mainly in the periphery and the middle-third of the lungs ("butterfly wing" pattern), as shown from top to bottom in (**A–D**).

Figure 5. Computed tomography showing nodular opacities bilaterally distributed (arrows in **A–D**), with cavitation (arrow in **C**), as shown from top to bottom in figures (**A–D**).

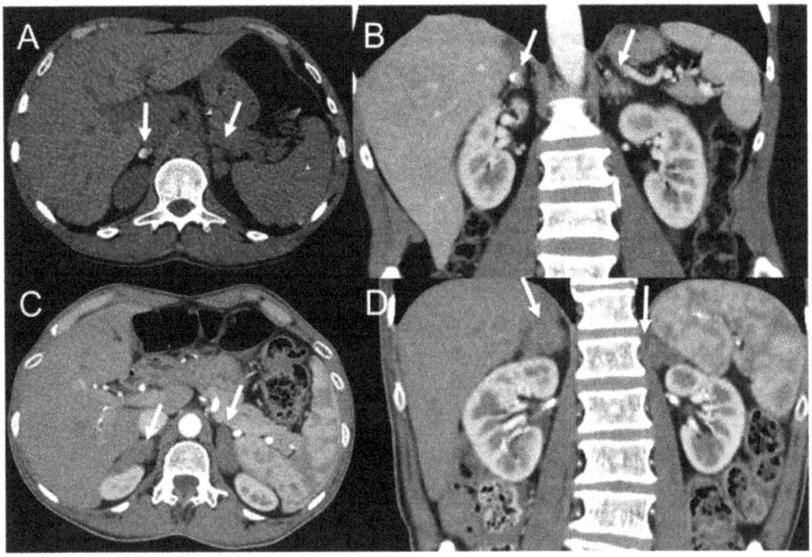

Figure 6. Computed tomography axial without contrast (**A**) and coronal post-contrast (**B**) images showing asymmetric adrenal thickening without significant contrast enhancement. The right adrenal is calcified, whereas the left is enlarged. Computed tomography axial post-contrast (**C**) and coronal post-contrast (**D**) from another patient showing adrenal thickening.

The literature demonstrates a wide range of involvement of the central nervous system, ranging from 1 to 27% in different case series. The involvement can be intraparenchymal, (rarely) meningeal, or a mix of both. Parenchymal involvement usually presents as hypointense lesions on T2WI (T2-weighted imaging) MRI sequences, with peripheral enhancement after contrast (Figure 7). Occasionally, the lesions may emit iso- or hyperintense signals on T2WI, and variable, but predominantly hypointense, signals on T1. Diffusion restriction mimicking a brain abscess may also be observed.

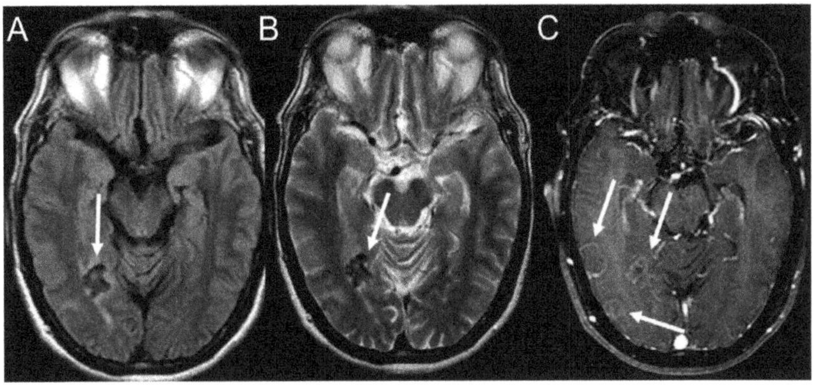

Figure 7. Magnetic resonance imaging showing a hypointense lesion on fluid-attenuated inversion recovery (FLAIR) (**A**) and T2-weighted imaging (T2WI) (**B**), with an annular enhancement on T1 post-contrast (**C**) located in the right occipital and temporal lobes.

Contrast enhancement can be ring-shaped, nodular, or heterogeneous [132]. Rosa et al. [133] demonstrated the presence of a "double halo" sign on a susceptibility-weighted imaging (SWI) MRI sequence. This finding can contribute to imaging-based diagnosis for patients with suspected PCM. These investigators also demonstrated the association between PCM and mesial temporal sclerosis, known as dual pathology, in patients with epilepsy. The meningeal form commonly presents as leptomeningeal contrast enhancement. Spinal cord involvement is rare and difficult to differentiate from neoplastic lesions [134]. Isolated involvement of the osteoarticular system is rare and, when present, is part of a multisystem process. Osteolytic lesions with cortical destruction are generally observed, especially in the metaphyses and epiphyses of long bones, without periosteal reactions [135].

In summary, the most recent advances in diagnostic imaging were FDG-PET/CT for evaluating the extent of active disease in patients with PCM undergoing antifungal treatment, demonstrating that it is more sensitive than conventional imaging. Additionally, the description of the presence of the "double halo" sign on SWI MRI sequences can help diagnose neuroparacoccidioidomycosis.

5. Treatment

The treatment of PCM has evolved in the last decades, and the current antifungal options are cotrimoxazole, triazoles, and amphotericin B. However, there is a lack of robust trials, and current guidelines are primarily based on non-comparative studies, expert opinions, and a couple of studies that compared itraconazole and cotrimoxazole [3,10,136].

Itraconazole (ITZ) is the first therapeutic choice for mild-to-moderate presentations of PCM, with response rates of 85–90% and an adequate tolerance profile [3,10,11,36]. Currently, the drug is available in almost all Latin American countries, but with a limited presence in non-referral hospitals and at a high cost [6]. Moreover, intravenous formulations of itraconazole are not easily available, and the oral absorption of capsules is erratic, leading to unpredictable supratherapeutic or subtherapeutic plasma levels [12]. To overcome this limitation, a novel formulation of ITZ, labeled SUper BioAvailable (SUBA)-itraconazole, was developed, with a relative bioavailability of 180% when compared with conventional itraconazole (C-ITZ), and an absolute bioavailability up to 90%. Indeed, some studies suggested less variability of SUBA compared with C-ITZ capsules, especially under fasted conditions, but a recent open-label comparative trial of 160 mg SUBA bid versus 200 mg C-ITZ bid for the treatment of endemic mycosis showed almost identical serum levels with similar specific adverse events. Of note, PCM cases were not included in the trial [137,138].

Sulfanilamides were the first agents used to treat PCM in the 1940s, and, until now, cotrimoxazole has been widely used in South America to treat mild and moderate forms of PCM, owing to its lower cost than that of itraconazole [11]. It also has the advantages of oral and venous formulations, good absorption with predictable serum levels, and fewer drug–drug interactions than azoles [10].

The duration of treatment may vary from 9 to 18 months with itraconazole, and 18 to 24 months with cotrimoxazole [3]. No in vitro or in vivo evidence suggests that the different species of *Paracoccidioides* require modifications in doses of antifungal agents or the duration of treatment [21].

Severe and disseminated forms should be treated with amphotericin B for 2–4 weeks until clinical stabilization and then transitioned to the maintenance of oral treatment. Lipid formulations are preferred (3–5 mg/kg/day) because they are less toxic than deoxycholate. Unfortunately, its availability is limited in Latin America [6,73,139]. If amphotericin B is contraindicated, high-dose intravenous cotrimoxazole (800 mg/160 mg every eight hours) may be an alternative [11].

Second-generation triazoles (voriconazole, posaconazole, and isavuconazole) have not been used extensively to treat PCM. Their high cost still prevents their large-scale use, but they are expected to be potential substitutes to itraconazole when it is not tolerated [140,141]. Notably, most endemic countries from Central and South America reported limited access to posaconazole and isavuconazole [6].

A significant challenge in effectively treating PCM is the need for prolonged antifungal use, leading to poor compliance [12]. Additionally, despite long-term treatment, the sequelae, due to chronic inflammatory processes and fibrosis, may profoundly impact organ function, and are not entirely resolved after antifungal medication [136,142]. Persistent fungal antigen stimulation and immune system activation can lead to fibrosis through excessive extracellular matrix component deposition and alterations in the tissue-scarring process, despite appropriate treatment [67,143].

Strategies have been developed to reduce lung damage. Vaccination with an immunogenic recombinant antigen of *P. brasiliensis* has shown the potential to attenuate fibrosis [144,145]. Experimental animal models treated with antifungal drugs and other antibiotics or immunomodulatory compounds (itraconazole + pentoxifylline and cotrimoxazole + azithromycin) have also achieved good results [146,147]. A monoclonal antibody specific to neutrophils (mAb-anti-Ly6G), associated with itraconazole, reduced pro-inflammatory cells, fungal load, and pro-inflammatory cytokines in mice [143,148]. In addition, a recent biological therapy comprising a *P. brasiliensis* cell wall glycoconjugate monoclonal antibody (mAbF1.4) combined with cotrimoxazole showed promising results in reducing the pulmonary fungal burden in mice [149]. Additionally, mesenchymal stem cell transplantation in combination with antifungals has been found to attenuate the inflammatory response and fibrosis induced by *P. brasiliensis* [150]. Further trials in human subjects are needed to determine the effectiveness of the new therapies for PCM.

6. Conclusions

PCM is undoubtedly a public health problem in Latin America, where it is an occupational disease in rural populations. Indeed, despite the mechanization of agriculture reducing human exposure to the fungus in classically endemic regions, the expansion of agriculture frontiers has led to the rise of new cases of PCM among rural workers, especially in the Amazon region. Simultaneously, reports of clusters of the acute form of the disease suggest that climate changes and anthropic environmental interventions may be risk factors for PCM development once they expose individuals to high fungal burdens, even away from rural areas. Given the increase in the number of transplants performed in endemic areas, and in the indications for the use of immunobiological therapy, a growing number of immunosuppressed people are at risk of developing PCM. In non-endemic countries, when there is a clinical suspicion of infection with *Paracoccidioides* spp., it is crucial to consider individual travel and migration history up to decades ago if the chronic form of the disease is to be considered.

Improving patients' quality of life after PCM treatment depends on early diagnosis and new therapeutic strategies to reduce pulmonary sequelae. Public health policies aiming to spread knowledge of the disease among general practitioners in rural areas of Latin America are crucial.

In summary, efforts must be combined to achieve advances in new biomarkers for diagnosing PCM and determining the causative species. This will allow us to know more about these microorganisms' molecular epidemiology, since culture media recovery is slow and not always possible. Moreover, this process reinforces the necessity to develop and standardize new antigens for serology diagnosis when differences in antigenic composition are probably related to phylogenetic peculiarities.

In addition, efforts must be made to ensure access to itraconazole throughout Latin America, besides optimizing antifibrotic treatments and pulmonary rehabilitation.

Author Contributions: Conceptualization, P.M.P.-P. and S.S.G.; P.M.P., P.M.P.-P. and T.R.G.-V. contributed to writing the 'Epidemiology and diagnosing of Paracoccidioidomycosis and sections' (Species and Geographic Distribution and Clinical diagnosis, respectively). S.S.G. contributed to writing the 'diagnosing of Paracoccidioidomycosis section' (Laboratory Diagnosis, Mycological Diagnosis, Cultures, Histopathological Diagnosis, Immunological Diagnosis: Antibody Detection, Antigen Detection, Molecular Detection). M.R.J. contributed to writing the 'Diagnostic imaging' section. P.M.P.-P. contributed to writing the 'Treatment' section. P.M.P.-P. and S.S.G. contributed to writing the conclusions. P.M.P., M.R.J., S.S.G. and A.F. prepared all figures. S.S.G. and T.R.G.-V. were responsible for editing the manuscript. All authors have read and agreed to the published version of the manuscript.

Funding: This research received no external funding.

Institutional Review Board Statement: Not applicable.

Informed Consent Statement: The authors confirm that the ethical policies of the journal have been strictly followed and the appropriate ethical review committee approval has been received. This study was approved by the Research Ethics Committee (under the CAAE number: 15734713.0.0000.5060) of the Federal University of Espírito Santo (UFES).

Data Availability Statement: Not applicable.

Acknowledgments: We are grateful to Patricio Godoy-Martinez and Simone B. Maifrede for the donations of images included in Figure 3. All individuals included in this section have consented to acknowledgment.

Conflicts of Interest: The authors declare no conflict of interest.

References

1. Bocca, A.L.; Amaral, A.C.; Teixeira, M.M.; Sato, P.K.; Shikanai-Yasuda, M.A.; Soares Felipe, M.S. Paracoccidioidomycosis: Eco-epidemiology, taxonomy and clinical and therapeutic issues. *Future Microbiol.* **2013**, *8*, 1177–1191. [CrossRef] [PubMed]
2. Martinez, R. New Trends in Paracoccidioidomycosis Epidemiology. *J. Fungi* **2017**, *3*, 1. [CrossRef] [PubMed]
3. Shikanai-Yasuda, M.A.; Mendes, R.P.; Colombo, A.L.; Queiroz-Telles, F.; Kono, A.S.G.; Paniago, A.M.M.; Nathan, A.; Valle, A.; Bagagli, E.; Benard, G.; et al. Brazilian guidelines for the clinical management of paracoccidioidomycosis. *Rev. Soc. Bras. Med. Trop.* **2017**, *50*, 715–740. [CrossRef] [PubMed]
4. Wagner, G.; Moertl, D.; Glechner, A.; Mayr, V.; Klerings, I.; Zachariah, C.; Van den Nest, M.; Gartlehner, G.; Willinger, B. Paracoccidioidomycosis Diagnosed in Europe-A Systematic Literature Review. *J. Fungi* **2021**, *7*, 157. [CrossRef]
5. Costa, A.N.; Benard, G.; Albuquerque, A.L.; Fujita, C.L.; Magri, A.S.; Salge, J.M.; Shikanai-Yasuda, M.A.; Carvalho, C.R. The lung in paracoccidioidomycosis: New insights into old problems. *Clinics* **2013**, *68*, 441–448. [CrossRef]
6. Caceres, D.H.; Echeverri Tirado, L.C.; Bonifaz, A.; Adenis, A.; Gomez, B.L.; Flores, C.L.B.; Canteros, C.E.; Santos, D.W.; Arathoon, E.; Soto, E.R.; et al. Current situation of endemic mycosis in the Americas and the Caribbean: Proceedings of the first international meeting on endemic mycoses of the Americas (IMEMA). *Mycoses* **2022**. [CrossRef]
7. Moreto, T.C.; Marques, M.E.; de Oliveira, M.L.; Moris, D.V.; de Carvalho, L.R.; Mendes, R.P. Accuracy of routine diagnostic tests used in paracoccidioidomycosis patients at a university hospital. *Trans. R. Soc. Trop. Med. Hyg.* **2011**, *105*, 473–478. [CrossRef]
8. da Silva, J.F.; de Oliveira, H.C.; Marcos, C.M.; Assato, P.A.; Fusco-Almeida, A.M.; Mendes-Giannini, M.J.S. Advances and challenges in paracoccidioidomycosis serology caused by Paracoccidioides species complex: An update. *Diagn. Microbiol. Infect. Dis.* **2016**, *84*, 87–94. [CrossRef]
9. Pinheiro, B.G.; Hahn, R.C.; Camargo, Z.P.; Rodrigues, A.M. Molecular Tools for Detection and Identification of Paracoccidioides Species: Current Status and Future Perspectives. *J. Fungi* **2020**, *6*, 293. [CrossRef]
10. Cavalcante Rde, S.; Sylvestre, T.F.; Levorato, A.D.; de Carvalho, L.R.; Mendes, R.P. Comparison between itraconazole and cotrimoxazole in the treatment of paracoccidiodomycosis. *PLoS Negl. Trop. Dis.* **2014**, *8*, e2793. [CrossRef]
11. Shikanai-Yasuda, M.A. Paracoccidioidomycosis Treatment. *Rev. Inst. Med. Trop. Sao Paulo* **2015**, *57* (Suppl. 19), 31–37. [CrossRef] [PubMed]
12. Andrade, U.V.; Oliveira, S.; Chang, M.R.; Pereira, E.F.; Marques, A.; Carvalho, L.R.; Mendes, R.P.; Paniago, A.M.M. Treatment compliance of patients with paracoccidioidomycosis in Central-West Brazil. *J. Bras. Pneumol.* **2019**, *45*, e20180167. [CrossRef] [PubMed]
13. Matute, D.R.; McEwen, J.G.; Puccia, R.; Montes, B.A.; San-Blas, G.; Bagagli, E.; Rauscher, J.T.; Restrepo, A.; Morais, F.; Nino-Vega, G.; et al. Cryptic speciation and recombination in the fungus Paracoccidioides brasiliensis as revealed by gene genealogies. *Mol. Biol. Evol.* **2006**, *23*, 65–73. [CrossRef] [PubMed]
14. Teixeira, M.M.; Theodoro, R.C.; de Carvalho, M.J.; Fernandes, L.; Paes, H.C.; Hahn, R.C.; Mendoza, L.; Bagagli, E.; San-Blas, G.; Felipe, M.S. Phylogenetic analysis reveals a high level of speciation in the Paracoccidioides genus. *Mol. Phylogenet. Evol.* **2009**, *52*, 273–283. [CrossRef] [PubMed]
15. Turissini, D.A.; Gomez, O.M.; Teixeira, M.M.; McEwen, J.G.; Matute, D.R. Species boundaries in the human pathogen Paracoccidioides. *Fungal Genet. Biol.* **2017**, *106*, 9–25. [CrossRef] [PubMed]
16. Munoz, J.F.; Farrer, R.A.; Desjardins, C.A.; Gallo, J.E.; Sykes, S.; Sakthikumar, S.; Misas, E.; Whiston, E.A.; Bagagli, E.; Soares, C.M.; et al. Genome Diversity, Recombination, and Virulence across the Major Lineages of Paracoccidioides. *mSphere* **2016**, *1*, e00213-16. [CrossRef]
17. Teixeira, M.M.; Cattana, M.E.; Matute, D.R.; Munoz, J.F.; Arechavala, A.; Isbell, K.; Schipper, R.; Santiso, G.; Tracogna, F.; Sosa, M.L.A.; et al. Genomic diversity of the human pathogen Paracoccidioides across the South American continent. *Fungal Genet. Biol.* **2020**, *140*, 103395. [CrossRef]

18. Nery, A.F.; de Camargo, Z.P.; Rodrigues, A.M.; Portela, T.F.; Hoffmann-Santos, H.D.; Pinheiro, B.G.; Possa, A.P.; Cavalcante, L.; Hagen, F.; Hahn, R.C. Puzzling paracoccidioidomycosis: Factors associated with the severity of Paracoccidioides lutzii infections. *Int. J. Infect. Dis.* **2021**, *107*, 284–290. [CrossRef]
19. Mavengere, H.; Mattox, K.; Teixeira, M.M.; Sepulveda, V.E.; Gomez, O.M.; Hernandez, O.; McEwen, J.; Matute, D.R. Paracoccidioides Genomes Reflect High Levels of Species Divergence and Little Interspecific Gene Flow. *mBio* **2020**, *11*, e01999-20. [CrossRef]
20. Pereira, E.F.; Gegembauer, G.; Chang, M.R.; Camargo, Z.P.; Nunes, T.F.; Ribeiro, S.M.; Carvalho, L.R.; Maldonado, B.M.; Mendes, R.P.; Paniago, A.M.M. Comparison of clinico-epidemiological and radiological features in paracoccidioidomycosis patients regarding serological classification using antigens from Paracoccidioides brasiliensis complex and Paracoccidioides lutzii. *PLoS Negl. Trop. Dis.* **2020**, *14*, e0008485. [CrossRef]
21. Hahn, R.C.; Rodrigues, A.M.; Della Terra, P.P.; Nery, A.F.; Hoffmann-Santos, H.D.; Gois, H.M.; Fontes, C.J.F.; de Camargo, Z.P. Clinical and epidemiological features of paracoccidioidomycosis due to Paracoccidioides lutzii. *PLoS Negl. Trop. Dis.* **2019**, *13*, e0007437. [CrossRef] [PubMed]
22. de Macedo, P.M.; Teixeira, M.M.; Barker, B.M.; Zancope-Oliveira, R.M.; Almeida-Paes, R.; Francesconi do Valle, A.C. Clinical features and genetic background of the sympatric species Paracoccidioides brasiliensis and Paracoccidioides americana. *PLoS Negl. Trop. Dis.* **2019**, *13*, e0007309. [CrossRef]
23. Vilela, R.; Huebner, M.; Vilela, C.; Vilela, G.; Pettersen, B.; Oliveira, C.; Mendoza, L. The taxonomy of two uncultivated fungal mammalian pathogens is revealed through phylogeny and population genetic analyses. *Sci. Rep.* **2021**, *11*, 18119. [CrossRef] [PubMed]
24. Lopez-Martinez, R.; Hernandez-Hernandez, F.; Mendez-Tovar, L.J.; Manzano-Gayosso, P.; Bonifaz, A.; Arenas, R.; Padilla-Desgarennes Mdel, C.; Estrada, R.; Chavez, G. Paracoccidioidomycosis in Mexico: Clinical and epidemiological data from 93 new cases (1972-2012). *Mycoses* **2014**, *57*, 525–530. [CrossRef] [PubMed]
25. Cordova, L.A.; Torres, J. Paracoccidioidomycosis. In *StatPearls*; StatPearls Publishing: Treasure Island, FL, USA, 2022.
26. Theodoro, R.C.; Teixeira Mde, M.; Felipe, M.S.; Paduan Kdos, S.; Ribolla, P.M.; San-Blas, G.; Bagagli, E. Genus paracoccidioides: Species recognition and biogeographic aspects. *PLoS ONE* **2012**, *7*, e37694. [CrossRef]
27. Linares, G.; Baker, R.D.; Linares, L. Paracoccidioidomycosis in the United States (South American Blastomycosis). *Arch. Otolaryngol.* **1971**, *93*, 514–518. [CrossRef]
28. Murray, H.W.; Littman, M.L.; Roberts, R.B. Disseminated paracoccidioidomycosis (South American blastomycosis) in the United States. *Am. J. Med.* **1974**, *56*, 209–220. [CrossRef]
29. Rahman, R.; Davies, L.; Mohareb, A.M.; Pecanha-Pietrobom, P.M.; Patel, N.J.; Solomon, I.H.; Meredith, D.M.; Tsai, H.K.; Guenette, J.P.; Bhattacharyya, S.; et al. Delayed Relapse of Paracoccidioidomycosis in the Central Nervous System: A Case Report. *Open Forum Infect. Dis.* **2020**, *7*, ofaa077. [CrossRef]
30. Chikamori, T.; Saka, S.; Nagano, H.; Saeki, S.; Lacaz Cda, S.; Rodrigues, M.C.; Cassaguerra, C.M.; Braccialli, M.L. Paracoccidioidomycosis in Japan. Report of a case. *Rev. Inst. Med. Trop. Sao Paulo* **1984**, *26*, 267–271. [CrossRef]
31. Kawayama, T.; Sawa, A.; Sueyasu, Y.; Arikawa, K.; Shiraishi, T.; Ichikawa, Y.; Oizumi, K. [Chronic pulmonary paracoccidioidomyosis in a Japanese Adult]. *Nihon Kyobu Shikkan Gakkai Zasshi* **1996**, *34*, 911–915.
32. Fujio, J.; Nishimura, K.; Miyaji, M. Epidemiological survey of the imported mycoses in Japan. *Nihon Ishinkin Gakkai Zasshi* **1999**, *40*, 103–109. [CrossRef] [PubMed]
33. Kamei, K.; Sano, A.; Kikuchi, K.; Makimura, K.; Niimi, M.; Suzuki, K.; Uehara, Y.; Okabe, N.; Nishimura, K.; Miyaji, M. The trend of imported mycoses in Japan. *J. Infect. Chemother.* **2003**, *9*, 16–20. [CrossRef] [PubMed]
34. Miyaji, M.; Kamei, K. Imported mycoses: An update. *J. Infect. Chemother.* **2003**, *9*, 107–113. [CrossRef] [PubMed]
35. Igarashi, T.; Kurose, T.; Itabashi, K.; Nakano, I.; Okamoto, K.; Sano, A.; Kimura, K.; Kaji, H. [A case of chronic pulmonary paracoccidioidomycosis]. *Nihon Kokyuki Gakkai Zasshi* **2004**, *42*, 629–633. [PubMed]
36. Onda, H.; Komine, M.; Murata, S.; Ohtsuki, M. Letter: Imported paracoccidioidomycosis in Japan. *Dermatol. Online J.* **2011**, *17*, 11. [CrossRef] [PubMed]
37. Kurai, H.; Ohmagari, N.; Ito, K.; Kawamura, I.; Suzuki, J.; Hadano, Y.; Endo, M.; Iida, Y.; Okinaka, K.; Kamei, K. [Case of oral paracoccidioidomycosis suspected to be pharyngeal cancer]. *Med. Mycol. J.* **2012**, *53*, 49–52. [CrossRef]
38. Yoshimura, Y.; Tachikawa, N.; Oosawa, T.; Kosuge, Y.; Kamei, K. [A case of paracoccidioidomycosis with severe adrenal insufficiency]. *Kansenshogaku Zasshi* **2012**, *86*, 291–294. [CrossRef]
39. dos Santos, W.A.; da Silva, B.M.; Passos, E.D.; Zandonade, E.; Falqueto, A. [Association between smoking and paracoccidioidomycosis: A case-control study in the State of Espirito Santo, Brazil]. *Cad. Saude Publica* **2003**, *19*, 245–253. [CrossRef]
40. Queiroz-Telles, F.V.; Pecanha Pietrobom, P.M.; Rosa Junior, M.; Baptista, R.M.; Pecanha, P.M. New Insights on Pulmonary Paracoccidioidomycosis. *Semin. Respir. Crit. Care Med.* **2020**, *41*, 53–68. [CrossRef]
41. Pecanha, P.M.; Batista Ferreira, M.E.; Massaroni Pecanha, M.A.; Schmidt, E.B.; Lamas de Araujo, M.; Zanotti, R.L.; Potratz, F.F.; Delboni Nunes, N.E.; Goncalves Ferreira, C.U.; Delmaestro, D.; et al. Paracoccidioidomycosis: Epidemiological and Clinical Aspects in 546 Cases Studied in the State of Espirito Santo, Brazil. *Am. J. Trop. Med. Hyg.* **2017**, *97*, 836–844. [CrossRef]
42. Bellissimo-Rodrigues, F.; Bollela, V.R.; Da Fonseca, B.A.; Martinez, R. Endemic paracoccidioidomycosis: Relationship between clinical presentation and patients' demographic features. *Med. Mycol.* **2013**, *51*, 313–318. [CrossRef] [PubMed]

43. Restrepo, A.; Salazar, M.E.; Cano, L.E.; Stover, E.P.; Feldman, D.; Stevens, D.A. Estrogens inhibit mycelium-to-yeast transformation in the fungus Paracoccidioides brasiliensis: Implications for resistance of females to paracoccidioidomycosis. *Infect. Immun.* **1984**, *46*, 346–353. [CrossRef] [PubMed]
44. Vieira Gde, D.; Alves Tda, C.; Lima, S.M.; Camargo, L.M.; Sousa, C.M. Paracoccidioidomycosis in a western Brazilian Amazon State: Clinical-epidemiologic profile and spatial distribution of the disease. *Rev. Soc. Bras. Med. Trop.* **2014**, *47*, 63–68. [CrossRef] [PubMed]
45. Krakhecke-Teixeira, A.G.; Yamauchi, D.H.; Rossi, A.; de Sousa, H.R.; Garces, H.G.; Junior, J.L.; Junior, A.O.S.; Felipe, M.S.S.; Bagagli, E.; de Andrade, H.F., Jr.; et al. Clinical and Eco-Epidemiological Aspects of a Novel Hyperendemic Area of Paracoccidioidomycosis in the Tocantins-Araguaia Basin (Northern Brazil), Caused by *Paracoccidioides* sp. *J. Fungi* **2022**, *8*, 502. [CrossRef] [PubMed]
46. Goncalves, A.J.; Londero, A.T.; Terra, G.M.; Rozenbaum, R.; Abreu, T.F.; Nogueira, S.A. Paracoccidioidomycosis in children in the state of Rio de Janeiro (Brazil). Geographic distribution and the study of a "reservarea". *Rev. Inst. Med. Trop. Sao Paulo* **1998**, *40*, 11–13. [CrossRef]
47. Cadavid, D.; Restrepo, A. Factors associated with Paracoccidiodes brasiliensis infection among permanent residents of three endemic areas in Colombia. *Epidemiol. Infect.* **1993**, *111*, 121–133. [CrossRef]
48. van Gelderen de Komaid, A.; Duran, E.; Borges de Kestelman, I. Histoplasmosis and Paracoccidioidomycosis in northwestern Argentina III. Epidemiological survey in Vipos, La Toma, and Choromoro-Trancas, Tucuman, Argentina. *Eur. J. Epidemiol.* **1999**, *15*, 383–388. [CrossRef]
49. Cermeno, J.; Cermeno, J.; Godoy, G.; Hernandez, I.; Orellan, Y.; Blanco, Y.; Penna, S.; Garcia, L.; Mender, T.; Gonsalvez, M.; et al. Epidemiological study of paracoccidioidomycosis and histoplasmosis in a suburb of San Felix city, Bolivar state, Venezuela. *Investig. Clin.* **2009**, *50*, 213–220.
50. Coutinho, Z.F.; Silva, D.; Lazera, M.; Petri, V.; Oliveira, R.M.; Sabroza, P.C.; Wanke, B. Paracoccidioidomycosis mortality in Brazil (1980–1995). *Cad. Saude Publica* **2002**, *18*, 1441–1454. [CrossRef]
51. de Suguiura, I.M.S.; Ono, M.A. Compulsory notification of paracoccidioidomycosis: A 14-year retrospective study of the disease in the state of Parana, Brazil. *Mycoses* **2022**, *65*, 354–361. [CrossRef]
52. Prado, M.; Silva, M.B.; Laurenti, R.; Travassos, L.R.; Taborda, C.P. Mortality due to systemic mycoses as a primary cause of death or in association with AIDS in Brazil: A review from 1996 to 2006. *Mem. Inst. Oswaldo Cruz* **2009**, *104*, 513–521. [CrossRef] [PubMed]
53. Paniago, A.M.; Aguiar, J.I.; Aguiar, E.S.; da Cunha, R.V.; Pereira, G.R.; Londero, A.T.; Wanke, B. [Paracoccidioidomycosis: A clinical and epidemiological study of 422 cases observed in Mato Grosso do Sul]. *Rev. Soc. Bras. Med. Trop.* **2003**, *36*, 455–459. [CrossRef] [PubMed]
54. Tobon, A.M.; Agudelo, C.A.; Osorio, M.L.; Alvarez, D.L.; Arango, M.; Cano, L.E.; Restrepo, A. Residual pulmonary abnormalities in adult patients with chronic paracoccidioidomycosis: Prolonged follow-up after itraconazole therapy. *Clin. Infect. Dis.* **2003**, *37*, 898–904. [CrossRef]
55. Barrozo, L.V.; Benard, G.; Silva, M.E.; Bagagli, E.; Marques, S.A.; Mendes, R.P. First description of a cluster of acute/subacute paracoccidioidomycosis cases and its association with a climatic anomaly. *PLoS Negl. Trop. Dis.* **2010**, *4*, e643. [CrossRef]
56. Arantes, T.D.; Theodoro, R.C.; Teixeira Mde, M.; Bosco Sde, M.; Bagagli, E. Environmental Mapping of Paracoccidioides spp. in Brazil Reveals New Clues into Genetic Diversity, Biogeography and Wild Host Association. *PLoS Negl. Trop. Dis.* **2016**, *10*, e0004606. [CrossRef]
57. Mangiaterra, M.L.; Giusiano, G.E.; Alonso, J.M.; Gorodner, J.O. [Paracoccidioides brasiliensis infection in a subtropical region with important environmental changes]. *Bull. Soc. Pathol. Exot.* **1999**, *92*, 173–176.
58. Giusiano, G.; Aguirre, C.; Vratnica, C.; Rojas, F.; Corallo, T.; Cattana, M.E.; Fernandez, M.; Mussin, J.; de Los Angeles Sosa, M. Emergence of acute/subacute infant-juvenile paracoccidioidomycosis in Northeast Argentina: Effect of climatic and anthropogenic changes? *Med. Mycol.* **2019**, *57*, 30–37. [CrossRef] [PubMed]
59. do Valle, A.C.F.; Marques de Macedo, P.; Almeida-Paes, R.; Romao, A.R.; Lazera, M.D.S.; Wanke, B. Paracoccidioidomycosis after Highway Construction, Rio de Janeiro, Brazil. *Emerg. Infect. Dis.* **2017**, *23*, 1917–1919. [CrossRef] [PubMed]
60. Benard, G.; Romano, C.C.; Cacere, C.R.; Juvenale, M.; Mendes-Giannini, M.J.; Duarte, A.J. Imbalance of IL-2, IFN-gamma and IL-10 secretion in the immunosuppression associated with human paracoccidioidomycosis. *Cytokine* **2001**, *13*, 248–252. [CrossRef]
61. Juvenale, M.; Negro, G.; Duarte, A.J.S.; Benard, G. Antibody isotypes to a Paracoccidioides brasiliensis somatic antigen in sub-acute and chronic form paracoccidioidomycosis. *J. Med. Microbiol.* **2001**, *50*, 127–134. [CrossRef]
62. Dutra, L.M.; Silva, T.H.M.; Falqueto, A.; Pecanha, P.M.; Souza, L.R.M.; Goncalves, S.S.; Velloso, T.R.G. Oral paracoccidioidomycosis in a single-center retrospective analysis from a Brazilian southeastern population. *J. Infect. Public Health* **2018**, *11*, 530–533. [CrossRef]
63. Almeida, F.A.; Neves, F.F.; Mora, D.J.; Reis, T.A.; Sotini, D.M.; Ribeiro, B.M.; Andrade-Silva, L.E.; Nascentes, G.N.; Ferreira-Paim, K.; Silva-Vergara, M.L. Paracoccidioidomycosis in Brazilian Patients With and Without Human Immunodeficiency Virus Infection. *Am. J. Trop. Med. Hyg.* **2017**, *96*, 368–372. [CrossRef]
64. Covre, L.C.P.; Hombre, P.M.; Falqueto, A.; Pecanha, P.M.; Valim, V. Pulmonary paracoccidioidomycosis: A case report of reactivation in a patient receiving biological therapy. *Rev. Soc. Bras. Med. Trop.* **2018**, *51*, 249–252. [CrossRef] [PubMed]

65. de Almeida, J.N., Jr.; Pecanha-Pietrobom, P.M.; Colombo, A.L. Paracoccidioidomycosis in Immunocompromised Patients: A Literature Review. *J. Fungi* **2018**, *5*, 2. [CrossRef] [PubMed]
66. Falcao, E.M.; de Macedo, P.M.; Freitas, D.F.S.; Freitas, A.D.; Grinsztejn, B.; Veloso, V.G.; Almeida-Paes, R.; do Valle, A.C.F. Paracoccidioidomycosis in people living with HIV/AIDS: A historical retrospective cohort study in a national reference center for infectious diseases, Rio de Janeiro, Brazil. *PLoS Negl. Trop. Dis.* **2022**, *16*, e0010529. [CrossRef]
67. Benard, G. An overview of the immunopathology of human paracoccidioidomycosis. *Mycopathologia* **2008**, *165*, 209–221. [CrossRef]
68. Jasenosky, L.D.; Scriba, T.J.; Hanekom, W.A.; Goldfeld, A.E. T cells and adaptive immunity to Mycobacterium tuberculosis in humans. *Immunol. Rev.* **2015**, *264*, 74–87. [CrossRef] [PubMed]
69. Shikanai-Yasuda, M.A.; Conceicao, Y.M.; Kono, A.; Rivitti, E.; Campos, A.F.; Campos, S.V. Neoplasia and paracoccidioidomycosis. *Mycopathologia* **2008**, *165*, 303–312. [CrossRef]
70. Baldo, M.E. A diagnostic challenge in an individual with Paracoccidioidomycosis during hospitalization in times of COVID-19. *Int. J. Innov. Educ. Res.* **2022**, *10*, 149–158. [CrossRef]
71. de Macedo, P.M.; Freitas, D.F.S.; Varon, A.G.; Lamas, C.D.C.; Ferreira, L.C.F.; Freitas, A.D.; Ferreira, M.T.; Nunes, E.P.; Siqueira, M.M.; Veloso, V.G.; et al. COVID-19 and acute juvenile paracoccidioidomycosis coinfection. *PLoS Negl. Trop. Dis.* **2020**, *14*, e0008559. [CrossRef]
72. Messina, F.A.; Giusiano, G.; Santiso, G. Endemic Mycoses and COVID-19: A Review. *Curr. Fungal Infect. Rep.* **2022**, *16*, 98–106. [CrossRef]
73. Thompson, G.R., 3rd; Le, T.; Chindamporn, A.; Kauffman, C.A.; Alastruey-Izquierdo, A.; Ampel, N.M.; Andes, D.R.; Armstrong-James, D.; Ayanlowo, O.; Baddley, J.W.; et al. Global guideline for the diagnosis and management of the endemic mycoses: An initiative of the European Confederation of Medical Mycology in cooperation with the International Society for Human and Animal Mycology. *Lancet Infect. Dis.* **2021**, *21*, e364–e374. [CrossRef]
74. Lauand, F. [Contribution to the study of the morphology of Paracoccidioides brasiliensis in oral tissues]. *Rev. Inst. Med. Trop. Sao Paulo* **1966**, *8*, 69–78. [PubMed]
75. Teles, F.R.; Martins, M.L. Laboratorial diagnosis of paracoccidioidomycosis and new insights for the future of fungal diagnosis. *Talanta* **2011**, *85*, 2254–2264. [CrossRef]
76. Talhari, C.; de Souza, J.V.; Parreira, V.J.; Reinel, D.; Talhari, S. Oral exfoliative cytology as a rapid diagnostic tool for paracoccidioidomycosis. *Mycoses* **2008**, *51*, 177–178. [CrossRef]
77. Pinheiro, B.G.; Possa, A.P.; Della Terra, P.P.; de Carvalho, J.A.; Ricci, G.; Nishikaku, A.S.; Hahn, R.C.; Camargo, Z.P.; Rodrigues, A.M. A New Duplex PCR-Assay for the Detection and Identification of Paracoccidioides Species. *J. Fungi* **2021**, *7*, 169. [CrossRef] [PubMed]
78. Hahn, R.C.; Hagen, F.; Mendes, R.P.; Burger, E.; Nery, A.F.; Siqueira, N.P.; Guevara, A.; Rodrigues, A.M.; de Camargo, Z.P. Paracoccidioidomycosis: Current Status and Future Trends. *Clin. Microbiol. Rev.* **2022**, e0023321. [CrossRef]
79. Queiroz-Telles, F.; Escuissato, D.L. Pulmonary paracoccidioidomycosis. *Semin. Respir. Crit. Care Med.* **2011**, *32*, 764–774. [CrossRef]
80. Brummer, E.; Castaneda, E.; Restrepo, A. Paracoccidioidomycosis: An update. *Clin. Microbiol. Rev.* **1993**, *6*, 89–117. [CrossRef]
81. Negro, G.; Pereira, C.N.; Andrade, H.F.; Palacios, S.A.; Vidal, M.M.S.; Charbel, C.E.; Benard, G. Evaluation of tests for antibody response in the follow-up of patients with acute and chronic forms of paracoccidioidomycosis. *J. Med. Microbiol.* **2000**, *49*, 37–46. [CrossRef]
82. Colombo, A.; Queiroz-Telles, F. Paracoccidioidomycosis. In *Atlas of Infections Diseases*, 2nd ed.; Mandell, G.D., Kauffman, C.A., Eds.; Current Medicine: Philadelphia, PA, USA, 2007; pp. 53–70.
83. Ameen, M.; Talhari, C.; Talhari, S. Advances in paracoccidioidomycosis. *Clin. Exp. Dermatol.* **2010**, *35*, 576–580. [CrossRef] [PubMed]
84. Gaviria, M.; Rivera, V.; Munoz-Cadavid, C.; Cano, L.E.; Naranjo, T.W. Validation and clinical application of a nested PCR for paracoccidioidomycosis diagnosis in clinical samples from Colombian patients. *Braz. J. Infect. Dis.* **2015**, *19*, 376–383. [CrossRef] [PubMed]
85. Londero, A.T.; Severo, L.C.; Ramos, C.D. Small forms and hyphae of paracoccidioides brasiliensis in human tissue. *Mycopathologia* **1980**, *72*, 17–19. [CrossRef] [PubMed]
86. Mendes, R.P.; Cavalcante, R.S.; Marques, S.A.; Marques, M.E.A.; Venturini, J.; Sylvestre, T.F.; Paniago, A.M.M.; Pereira, A.C.; da Silva, J.F.; Fabro, A.T.; et al. Paracoccidioidomycosis: Current Perspectives from Brazil. *Open Microbiol. J.* **2017**, *11*, 224–282. [CrossRef]
87. Zancopé-Oliveira, R.M.; Pizzini, C.V.; Muniz, M.d.M.; Valle, A.C.F.d.; Almeida-Paes, R. Diagnostic aspects of Paracoccidioidomycosis. *Curr. Trop. Med. Rep.* **2014**, *1*, 111–118. [CrossRef]
88. Restrepo, A.; Robledo, M.; Giraldo, R.; Hernandez, H.; Sierra, F.; Gutierrez, F.; Londono, F.; Lopez, R.; Calle, G. The gamut of paracoccidioidomycosis. *Am. J. Med.* **1976**, *61*, 33–42. [CrossRef]
89. Montenegro, M.R.; Miyaji, M.; Franco, M.; Nishimura, K.; Coelho, K.I.; Horie, Y.; Mendes, R.P.; Sano, A.; Fukushima, K.; Fecchio, D. Isolation of fungi from nature in the region of Botucatu, state of Sao Paulo, Brazil, an endemic area of paracoccidioidomycosis. *Mem. Inst. Oswaldo Cruz* **1996**, *91*, 665–670. [CrossRef]
90. Restrepo, A.; Tobón, A.M.; Cano, L.E. Paracoccidioides brasiliensis. In *Principles and Practice of Infectious Diseases*, 7th ed.; Mendell, G.D.B., Bennett, J.E., Dolin, R., Eds.; Elsevier: Philadelphia, PA, USA, 2010; pp. 3336–3357.

91. Restrepo, A.; Tobón, A.M.; Agudelo, C.A. Paracoccidioidomycosis. In *Diagnosis and Treatment of Human Mycoses. Infectious Disease*; Hospenthal, D.R., Rinaldi, M.G., Eds.; Humana Press: Totowa, NJ, USA, 2008.
92. Alvarado, P.; Teixeira, M.M.; Cavallera, E.; Paes, H.C.; Guerra, G.; Santander, G.; Merino-Alado, R. Epidemiology of paracoccidioidomycosis in Venezuela: A retrospective study from 1954 to 2019. *Mem. Inst. Oswaldo Cruz* **2021**, *116*, e210203. [CrossRef]
93. Sifuentes-Osornio, J.; Corzo-Leon, D.E.; Ponce-de-Leon, L.A. Epidemiology of Invasive Fungal Infections in Latin America. *Curr. Fungal Infect. Rep.* **2012**, *6*, 23–34. [CrossRef]
94. Gomez, B.L.; Figueroa, J.I.; Hamilton, A.J.; Diez, S.; Rojas, M.; Tobon, A.M.; Hay, R.J.; Restrepo, A. Antigenemia in patients with paracoccidioidomycosis: Detection of the 87-kilodalton determinant during and after antifungal therapy. *J. Clin. Microbiol.* **1998**, *36*, 3309–3316. [CrossRef]
95. da Silva, S.H.; Colombo, A.L.; Blotta, M.H.; Queiroz-Telles, F.; Lopes, J.D.; de Camargo, Z.P. Diagnosis of neuroparacoccidioidomycosis by detection of circulating antigen and antibody in cerebrospinal fluid. *J. Clin. Microbiol.* **2005**, *43*, 4680–4683. [CrossRef] [PubMed]
96. Marques da Silva, S.H.; Colombo, A.L.; Blotta, M.H.; Lopes, J.D.; Queiroz-Telles, F.; Pires de Camargo, Z. Detection of circulating gp43 antigen in serum, cerebrospinal fluid, and bronchoalveolar lavage fluid of patients with paracoccidioidomycosis. *J. Clin. Microbiol.* **2003**, *41*, 3675–3680. [CrossRef]
97. Blotta, M.H.; Mamoni, R.L.; Oliveira, S.J.; Nouer, S.A.; Papaiordanou, P.M.; Goveia, A.; Camargo, Z.P. Endemic regions of paracoccidioidomycosis in Brazil: A clinical and epidemiologic study of 584 cases in the southeast region. *Am. J. Trop. Med. Hyg.* **1999**, *61*, 390–394. [CrossRef] [PubMed]
98. Wanke, B.; Londero, A.T. Epidemiology and paracoccidioidomycosis infection. In *Paracoccidioidomycosis*; Franco, M.F., Lacaz, C.S., Restrepo-Moreno, A., Del Negro, G., Eds.; CRC Press: Boca Raton, FL, USA, 1993; pp. 109–120.
99. de Camargo, Z.P. Serology of paracoccidioidomycosis. *Mycopathologia* **2008**, *165*, 289–302. [CrossRef] [PubMed]
100. Dos Santos, P.O.; Rodrigues, A.M.; Fernandes, G.F.; da Silva, S.H.; Burger, E.; de Camargo, Z.P. Immunodiagnosis of paracoccidioidomycosis due to Paracoccidioides brasiliensis using a latex test: Detection of specific antibody anti-gp43 and specific antigen gp43. *PLoS Negl. Trop. Dis.* **2015**, *9*, e0003516. [CrossRef]
101. Restrepo, A.; Moncada, L.H. [A slide latex test for the diagnosis of paracoccidiodomycosis]. *Bol. Oficina Sanit. Panam.* **1978**, *84*, 520–532.
102. Mendes-Giannini, M.J.; Bueno, J.P.; Shikanai-Yasuda, M.A.; Ferreira, A.W.; Masuda, A. Detection of the 43,000-molecular-weight glycoprotein in sera of patients with paracoccidioidomycosis. *J. Clin. Microbiol.* **1989**, *27*, 2842–2845. [CrossRef]
103. Blotta, M.H.; Altemani, A.M.; Amaral, E.; Silva, L.J.; Camargo, Z.P. Placental involvement in paracoccidioidomycosis. *J. Med. Vet. Mycol.* **1993**, *31*, 249–257. [CrossRef]
104. Taborda, C.P.; Camargo, Z.P. Diagnosis of paracoccidioidomycosis by dot immunobinding assay for antibody detection using the purified and specific antigen gp43. *J. Clin. Microbiol.* **1994**, *32*, 554–556. [CrossRef]
105. Camargo, Z.P.; Berzaghi, R.; Amaral, C.C.; Silva, S.H. Simplified method for producing Paracoccidioides brasiliensis exoantigens for use in immunodiffusion tests. *Med. Mycol.* **2003**, *41*, 539–542. [CrossRef]
106. Xavier, M.O.; Pasqualotto, A.C.; Cardoso, I.C.; Severo, L.C. Cross-reactivity of Paracoccidioides brasiliensis, Histoplasma capsulatum, and Cryptococcus species in the commercial Platelia Aspergillus enzyme immunoassay. *Clin. Vaccine Immunol.* **2009**, *16*, 132–133. [CrossRef] [PubMed]
107. Silveira-Gomes, F.; Sarmento, D.N.; Pinto, T.M.; Pimentel, R.F.; Nepomuceno, L.B.; Espirito Santo, E.P.; Mesquita-da-Costa, M.; Camargo, Z.P.; Marques-da-Silva, S.H. Development and evaluation of a latex agglutination test for the serodiagnosis of paracoccidioidomycosis. *Clin. Vaccine Immunol.* **2011**, *18*, 604–608. [CrossRef] [PubMed]
108. Melo, A.S.A.; Santos, D.; Lima, S.L.; Rodrigues, A.M.; de Camargo, Z.P.; Finkelman, M.; Colombo, A.L. Evaluation of (1 -> 3)-beta-D-glucan assay for diagnosing paracoccidioidomycosis. *Mycoses* **2020**, *63*, 38–42. [CrossRef]
109. Maifrede, S.B.; Kruschewsky, W.L.L.; Patricio, S.A.; Falqueto, A.; Pecanha, P.M.; Malaquias, L.C.C.; Possa, A.P.; de Camargo, Z.P.; Rodrigues, A.M.; Goncalves, S.S.; et al. Screening paracoccidioidomycosis by double immunodiffusion test in a referred diagnostic center in Brazilian southeastern: An accessible tool. *Infection* **2021**, *49*, 1257–1264. [CrossRef]
110. Cocio, T.A.; Martinez, R. Serological diagnosis of paracoccidioidomycosis using a Paracoccidioides spp. comercial antigen and the counterimmunoelectrophoresis method. *Braz. J. Infect. Dis.* **2021**, *25*, 101607. [CrossRef] [PubMed]
111. Silva Ferreira, C.; de Castro Ribeiro, E.M.; Miranda Goes, A.; Mello Silva, B. Current strategies for diagnosis of paracoccidioidomycosis and prospects of methods based on gold nanoparticles. *Future Microbiol.* **2016**, *11*, 973–985. [CrossRef]
112. Gegembauer, G.; Araujo, L.M.; Pereira, E.F.; Rodrigues, A.M.; Paniago, A.M.; Hahn, R.C.; de Camargo, Z.P. Serology of paracoccidioidomycosis due to Paracoccidioides lutzii. *PLoS Negl. Trop. Dis.* **2014**, *8*, e2986. [CrossRef]
113. Mendes-Giannini, M.J.; Ricci, L.C.; Uemura, M.A.; Toscano, E.; Arns, C.W. Infection and apparent invasion of Vero cells by Paracoccidioides brasiliensis. *J. Med. Vet. Mycol.* **1994**, *32*, 189–197. [CrossRef]
114. Cano, L.E.; Restrepo, A. Predictive value of serologic tests in the diagnosis and follow-up of patients with paracoccidioidomycosis. *Rev. Inst. Med. Trop. Sao Paulo* **1987**, *29*, 276–283. [CrossRef]
115. Camargo, Z.P.; Unterkircher, C.; Travassos, L.R. Identification of antigenic polypeptides of Paracoccidioides brasiliensis by immunoblotting. *J. Med. Vet. Mycol.* **1989**, *27*, 407–412. [CrossRef]

116. Puccia, R.; Schenkman, S.; Gorin, P.A.; Travassos, L.R. Exocellular components of Paracoccidioides brasiliensis: Identification of a specific antigen. *Infect. Immun.* **1986**, *53*, 199–206. [CrossRef] [PubMed]
117. Batista, J., Jr.; de Camargo, Z.P.; Fernandes, G.F.; Vicentini, A.P.; Fontes, C.J.; Hahn, R.C. Is the geographical origin of a Paracoccidioides brasiliensis isolate important for antigen production for regional diagnosis of paracoccidioidomycosis? *Mycoses* **2010**, *53*, 176–180. [CrossRef] [PubMed]
118. Albuquerque, C.F.; da Silva, S.H.; Camargo, Z.P. Improvement of the specificity of an enzyme-linked immunosorbent assay for diagnosis of paracoccidioidomycosis. *J. Clin. Microbiol.* **2005**, *43*, 1944–1946. [CrossRef] [PubMed]
119. Rodrigues, A.M.; Kubitschek-Barreira, P.H.; Pinheiro, B.G.; Teixeira-Ferreira, A.; Hahn, R.C.; de Camargo, Z.P. Immunoproteomic Analysis Reveals Novel Candidate Antigens for the Diagnosis of Paracoccidioidomycosis Due to Paracoccidioides lutzii. *J. Fungi* **2020**, *6*, 357. [CrossRef] [PubMed]
120. Freitas-da-Silva, G.; Roque-Barreira, M.C. Antigenemia in paracoccidioidomycosis. *J. Clin. Microbiol.* **1992**, *30*, 381–385. [CrossRef]
121. Silva, M.F.; Bocca, A.L.; Ferracini, R., Jr.; Figueiredo, F.; Silva, C.L. Cellular requirements for immunomodulatory effects caused by cell wall components of Paracoccidioides brasiliensis on antibody production. *Clin. Exp. Immunol.* **1997**, *109*, 261–271. [CrossRef]
122. Gomez, B.L.; Figueroa, J.I.; Hamilton, A.J.; Ortiz, B.; Robledo, M.A.; Hay, R.J.; Restrepo, A. Use of monoclonal antibodies in diagnosis of paracoccidioidomycosis: New strategies for detection of circulating antigens. *J. Clin. Microbiol.* **1997**, *35*, 3278–3283. [CrossRef]
123. Xavier, M.O.; Araujo, J.S.; Aquino, V.R.; Severo, C.B.; Guazzelli, L.S.; Severo, L.C.; Pasqualotto, A.C. Variability in Galactomannan detection by Platelia Aspergillus EIA according to the Aspergillus species. *Rev. Inst. Med. Trop. Sao Paulo* **2013**, *55*. [CrossRef]
124. Gomes, G.M.; Cisalpino, P.S.; Taborda, C.P.; de Camargo, Z.P. PCR for diagnosis of paracoccidioidomycosis. *J. Clin. Microbiol.* **2000**, *38*, 3478–3480. [CrossRef]
125. Endo, S.; Komori, T.; Ricci, G.; Sano, A.; Yokoyama, K.; Ohori, A.; Kamei, K.; Franco, M.; Miyaji, M.; Nishimura, K. Detection of gp43 of Paracoccidioides brasiliensis by the loop-mediated isothermal amplification (LAMP) method. *FEMS Microbiol. Lett.* **2004**, *234*, 93–97. [CrossRef]
126. Buitrago, M.J.; Merino, P.; Puente, S.; Gomez-Lopez, A.; Arribi, A.; Zancope-Oliveira, R.M.; Gutierrez, M.C.; Rodriguez-Tudela, J.L.; Cuenca-Estrella, M. Utility of real-time PCR for the detection of Paracoccidioides brasiliensis DNA in the diagnosis of imported paracoccidioidomycosis. *Med. Mycol.* **2009**, *47*, 879–882. [CrossRef] [PubMed]
127. Koishi, A.C.; Vituri, D.F.; Dionizio Filho, P.S.; Sasaki, A.A.; Felipe, M.S.; Venancio, E.J. A semi-nested PCR assay for molecular detection of Paracoccidioides brasiliensis in tissue samples. *Rev. Soc. Bras. Med. Trop.* **2010**, *43*, 728–730. [CrossRef] [PubMed]
128. Dias, L.; de Carvalho, L.F.; Romano, C.C. Application of PCR in serum samples for diagnosis of paracoccidioidomycosis in the southern Bahia-Brazil. *PLoS Negl. Trop. Dis.* **2012**, *6*, e1909. [CrossRef]
129. Cunha, P.G.e.a. Annual Congress of the European Association of Nuclear Medicine October 13–17, 2018 Dusseldorf, Germany. *Eur. J. Nucl. Med. Mol. Imaging* **2018**, *45*, 1–844. [CrossRef]
130. Rosa Junior, M.; Baldon, I.V.; Amorim, A.F.C.; Fonseca, A.P.A.; Volpato, R.; Lourenco, R.B.; Baptista, R.M.; de Mello, R.A.F.; Pecanha, P.; Falqueto, A. Imaging paracoccidioidomycosis: A pictorial review from head to toe. *Eur. J. Radiol.* **2018**, *103*, 147–162. [CrossRef]
131. Vermelho, M.B.; Correia, A.S.; Michailowsky, T.C.; Suzart, E.K.; Ibanes, A.S.; Almeida, L.A.; Khoury, Z.; Barba, M.F. Abdominal alterations in disseminated paracoccidioidomycosis: Computed tomography findings. *Radiol. Bras.* **2015**, *48*, 81–85. [CrossRef]
132. Rosa Junior, M.; Amorim, A.C.; Baldon, I.V.; Martins, L.A.; Pereira, R.M.; Campos, R.P.; Goncalves, S.S.; Velloso, T.R.G.; Pecanha, P.; Falqueto, A. Paracoccidioidomycosis of the Central Nervous System: CT and MR Imaging Findings. *AJNR Am. J. Neuroradiol.* **2019**, *40*, 1681–1688. [CrossRef]
133. Rosa, J.M.; Grenfell, M.L.R.; PeCanha, P.M. Hippocampal sclerosis in paracoccidioidomycosis. *Arq. Neuropsiquiatr.* **2020**, *78*, 384. [CrossRef]
134. Colli, B.O.; Assirati Junior, J.A.; Machado, H.R.; Figueiredo, J.F.; Chimelli, L.; Salvarani, C.P.; Dos Santos, F. Intramedullary spinal cord paracoccidioidomycosis. Report of two cases. *Arq. Neuropsiquiatr.* **1996**, *54*, 466–473. [CrossRef]
135. Savarese, L.G.; Monsignore, L.M.; de Andrade Hernandes, M.; Martinez, R.; Nogueira-Barbosa, M.H. Magnetic resonance imaging findings of paracoccidioidomycosis in the musculoskeletal system. *Trop. Med. Int. Health* **2015**, *20*, 1346–1354. [CrossRef]
136. Griffiths, J.; Lopes Colombo, A.; Denning, D.W. The case for paracoccidioidomycosis to be accepted as a neglected tropical (fungal) disease. *PLoS Negl. Trop. Dis.* **2019**, *13*, e0007195. [CrossRef] [PubMed]
137. Rauseo, A.M.; Mazi, P.; Lewis, P.; Burnett, B.; Mudge, S.; Spec, A. Bioavailability of Single-Dose SUBA-Itraconazole Compared to Conventional Itraconazole under Fasted and Fed Conditions. *Antimicrob. Agents Chemother.* **2021**, *65*, e0013421. [CrossRef] [PubMed]
138. Pappas, P.G.; Spec, A.; Miceli, M.; McGwin, G.; McMullen, R.; Thompson, G.R.R. An open-label comparative trial of SUBA-itraconazole (SUBA) versus conventional itraconazole (c-itra) for treatment of proven and probable endemic mycoses (MSG-15): A pharmacokinetic (PK) and adverse Event (AE) analysis. *Open Forum Infect. Dis.* **2021**, *8*, S72. [CrossRef]
139. Pecanha, P.M.; de Souza, S.; Falqueto, A.; Grao-Veloso, T.R.; Lirio, L.V.; Ferreira, C.U., Jr.; Santos, A.R.; Costa, H.G.; de Souza, L.R.; Tuon, F.F. Amphotericin B lipid complex in the treatment of severe paracoccidioidomycosis: A case series. *Int. J. Antimicrob. Agents* **2016**, *48*, 428–430. [CrossRef]

140. Queiroz-Telles, F.; Goldani, L.Z.; Schlamm, H.T.; Goodrich, J.M.; Espinel-Ingroff, A.; Shikanai-Yasuda, M.A. An open-label comparative pilot study of oral voriconazole and itraconazole for long-term treatment of paracoccidioidomycosis. *Clin. Infect. Dis.* **2007**, *45*, 1462–1469. [CrossRef]
141. Thompson, G.R., 3rd; Rendon, A.; Ribeiro Dos Santos, R.; Queiroz-Telles, F.; Ostrosky-Zeichner, L.; Azie, N.; Maher, R.; Lee, M.; Kovanda, L.; Engelhardt, M.; et al. Isavuconazole Treatment of Cryptococcosis and Dimorphic Mycoses. *Clin. Infect. Dis.* **2016**, *63*, 356–362. [CrossRef]
142. Gonzalez, A. The Therapy of Pulmonary Fibrosis in Paracoccidioidomycosis: What Are the New Experimental Approaches? *J. Fungi* **2020**, *6*, 217. [CrossRef]
143. Puerta-Arias, J.D.; Pino-Tamayo, P.A.; Arango, J.C.; Gonzalez, A. Depletion of Neutrophils Promotes the Resolution of Pulmonary Inflammation and Fibrosis in Mice Infected with Paracoccidioides brasiliensis. *PLoS ONE* **2016**, *11*, e0163985. [CrossRef]
144. Magalhaes, A.; Ferreira, K.S.; Almeida, S.R.; Nosanchuk, J.D.; Travassos, L.R.; Taborda, C.P. Prophylactic and therapeutic vaccination using dendritic cells primed with peptide 10 derived from the 43-kilodalton glycoprotein of Paracoccidioides brasiliensis. *Clin. Vaccine Immunol.* **2012**, *19*, 23–29. [CrossRef]
145. Taborda, C.P.; Uran, M.E.; Nosanchuk, J.D.; Travassos, L.R. Paracoccidioidomycosis: Challenges in the Development of a Vaccine against an Endemic Mycosis in the Americas. *Rev. Inst. Med. Trop. Sao Paulo* **2015**, *57* (Suppl. 19), 21–24. [CrossRef]
146. Naranjo, T.W.; Lopera, D.E.; Diaz-Granados, L.R.; Duque, J.J.; Restrepo, A.M.; Cano, L.E. Combined itraconazole-pentoxifylline treatment promptly reduces lung fibrosis induced by chronic pulmonary paracoccidioidomycosis in mice. *Pulm. Pharmacol. Ther.* **2011**, *24*, 81–91. [CrossRef] [PubMed]
147. Finato, A.C.; Almeida, D.F.; Dos Santos, A.R.; Nascimento, D.C.; Cavalcante, R.S.; Mendes, R.P.; Soares, C.T.; Paniago, A.M.M.; Venturini, J. Evaluation of antifibrotic and antifungal combined therapies in experimental pulmonary paracoccidioidomycosis. *Med. Mycol.* **2020**, *58*, 667–678. [CrossRef] [PubMed]
148. Puerta-Arias, J.D.; Pino-Tamayo, P.A.; Arango, J.C.; Salazar-Pelaez, L.M.; Gonzalez, A. Itraconazole in combination with neutrophil depletion reduces the expression of genes related to pulmonary fibrosis in an experimental model of paracoccidioidomycosis. *Med. Mycol.* **2018**, *56*, 579–590. [CrossRef]
149. Boniche-Alfaro, C.; Kischkel, B.; Thomaz, L.; Carvalho-Gomes, M.M.; Lopes-Bezerra, L.M.; Nosanchuk, J.D.; Taborda, C.P. Antibody- Based Immunotherapy Combined With Antimycotic Drug TMP-SMX to Treat Infection With Paracoccidioides brasiliensis. *Front. Immunol.* **2021**, *12*, 725882. [CrossRef] [PubMed]
150. Arango, J.C.; Puerta-Arias, J.D.; Pino-Tamayo, P.A.; Salazar-Pelaez, L.M.; Rojas, M.; Gonzalez, A. Impaired anti-fibrotic effect of bone marrow-derived mesenchymal stem cell in a mouse model of pulmonary paracoccidioidomycosis. *PLoS Negl. Trop. Dis.* **2017**, *11*, e0006006. [CrossRef] [PubMed]

Journal of *Fungi*

MDPI

Review

Blastomycosis—Some Progress but Still Much to Learn

Matthew F. Pullen [1], Jonathan D. Alpern [2,3] and Nathan C. Bahr [4,*]

1 Division of Infectious Diseases and International Medicine, Department of Medicine, University of Minnesota, Minneapolis, MN 55455, USA
2 Division of Infectious Diseases, HealthPartners, Bloomington, MN 55425, USA
3 Department of Medicine, University of Minnesota, Minneapolis, MN 55455, USA
4 Division of Infectious Diseases, Department of Medicine, University of Kansas Medical Center, Kansas City, KS 66160, USA
* Correspondence: nbahr@kumc.edu

Abstract: Blastomycosis, caused by *Blastomyces* spp., is an endemic mycosis capable of causing significant disease throughout the body. Higher rates of infection are seen in the Mississippi and Ohio River valleys, the Great Lakes region of the United States and Canada, much of Africa, and, to a lesser extent, in India and the Middle East. Limited reporting inhibits our true understanding of the geographic distribution of blastomycosis. An estimated 50% of those infected remain asymptomatic. Of those who present with symptomatic disease, pulmonary involvement is most common, while the most common extrapulmonary sites are the skin, bones, genitourinary system, and central nervous system. Itraconazole is the standard therapy for mild–moderate disease. Data for other azoles are limited. Amphotericin is used for severe disease, and corticosteroids are occasionally used in severe disease, but evidence for this practice is limited. Despite increasing incidence and geographic reach in recent years, there are still significant knowledge gaps in our understanding of blastomycosis. Here, we provide an updated review of the epidemiology, clinical presentations, and diagnostic and therapeutic approaches for this infection. We also discuss areas needing further research.

Keywords: *Blastomyces*; *Blastomyces dermatitidis*; blastomycosis; mycosis; diseases; fungus

Citation: Pullen, M.F.; Alpern, J.D.; Bahr, N.C. Blastomycosis—Some Progress but Still Much to Learn. *J. Fungi* **2022**, *8*, 824. https://doi.org/10.3390/jof8080824

Academic Editor: Alessandro C. Pasqualotto

Received: 19 July 2022
Accepted: 3 August 2022
Published: 7 August 2022

1. Introduction

Endemic mycoses are fungal pathogens known to cause human disease and that have characteristic geographic distributions [1]. These fungi are increasingly important in aging and vulnerable populations, such as post-transplant or otherwise immunocompromised persons [2,3]. Among the endemic fungi, *Blastomyces* is relatively rare, but it has potentially devastating consequences. Mortality data are fairly limited due to gaps in disease reporting for blastomycosis, yet, case fatality rates of 4–22% have been reported, depending on patient demographics, such as advanced age, male sex, ethnicity, and living in hyperendemic regions [4]. Thus, knowledge of blastomycosis is incredibly important for clinicians in hyperendemic areas and those seeing patients who have traveled to such areas. In addition, we must continue to be aware that areas of endemicity are shifting for blastomycosis and other endemic mycoses [5]. The primary etiologic agents of blastomycosis are *Blastomyces dermatitidis* species complex (comprised of *B. dermatitidis* and the more recently described *Blastomyces gilchristii*) and *Blastomyces helicus*. In the Middle East and Africa, *Blastomyces percusus* and *Blastomyces emzantsi* are uncommon causes of blastomycosis; a newer member of the *Blastomyces* genus, *Blastomyces parvus*, has also been described as a rare cause of atypical, granulomatous pulmonary blastomycosis [6–8].

While most commonly seen as a pulmonary pathogen, *Blastomyces* spp. can disseminate throughout the body, often mimicking other infections and/or malignant processes [9,10]. Given the syndromic similarity of *Blastomyces* infection to other disease processes, a high index of clinical suspicion is crucial to ensure the prompt recognition and treatment of blastomycosis. There are disproportionately little recent data on blastomycosis, particularly regarding the increasing geographic spread of these fungi and the utilization of new antifungal agents [2,5,11]. In this review, we examine the current data on the epidemiology, clinical presentation, and treatment of blastomycosis and comment on particularly important areas where additional research is urgently needed.

2. Epidemiology

The natural environment of *Blastomyces* is moist, acidic soil, particularly in vegetation-dense areas near rivers or other water sources. Early efforts to isolate *Blastomyces* in soil samples were largely based on culture or recovery from intravenous injection of soil samples into mice, often a difficult process with few successes. With the advent of PCR technology, nucleic acid detection has successfully identified *Blastomyces* in the expected ecologic niche, that is, moist soil with dense woodland or vegetation nearby, or from tissues of animals exposed to these environments [12–18]. Thus, *Blastomyces* is often reported in persons who work outdoors, hunt or fish, or have recent exposures to areas where soil and vegetation have been disturbed (construction, excavation, etc.) [5,19]. Reinforcing this soil–spore connection, the risk of blastomycosis has been shown to be higher in canines, particularly sporting or hunting dogs, exposed to these endemic regions, which is likely due to their closer proximity to the soil and digging/sniffing behaviors [20,21]. *Blastomyces*, being a thermally dimorphic fungus, grows as a filamentous, mycelial mold form in the natural environment, undergoing a phase transition into a pathogenic budding yeast form in warmer environments, such as the human (or other mammals) body, capable of evading host immune defenses to cause both pulmonary and disseminated infection [22].

Within the United States, *Blastomyces* is typically considered endemic within the states along the Ohio and Mississippi River Valleys and Great Lakes region, with extension into several central and eastern provinces of Canada (Ontario, Quebec, Manitoba, and Saskatchewan), which has been observed in both earlier and more recent epidemiologic studies (Figure 1) [5,23–26]. Presently, blastomycosis is a reportable infection in only five states within the United States (Arkansas, Louisiana, Michigan, Minnesota, and Wisconsin) [27]. While these states do represent regions with a historically higher incidence of blastomycosis, the relative lack of data from neighboring states raises questions about the true incidence outside of known endemic regions, particularly in light of case reports of blastomycosis in non-endemic areas dating back decades [26,28–30]. The bulk of epidemiologic data predate the identification of *B. gilchristii*, which is included in the *B. dermatitidis* complex designation, though some studies suggest that *B. gilchristii* is more geographically restricted to Canada and the northern United States region [31]. By comparison, *B. helicus* has been more commonly isolated in the western United States (Colorado, Idaho, Montana, California, Nebraska, Texas, and Utah) and western Canada (Alberta and Saskatchewan) [5,32].

Outside of North America, blastomycosis has been widely reported across Africa, though as is the case in much of North America, it is not a commonly reportable infection, so epidemiologic data are often limited to case reports [5,8,33–40]. Though less often reported, blastomycosis has also been seen in the Middle East and India, but differences in disease patterns and a recent molecular examination of several cases suggest that, in these regions, two separate species (*Blastomyces percusus* and *Blastomyces emzantsi*) may represent a significant portion of the disease burden [8,41,42].

The non-reportable status of blastomycosis in much of the world, as well as a dearth of broader epidemiologic studies of this infection, make it difficult to truly assess the emerging geographic spread of *Blastomyces*. Couple this with the possible expansion of endemic fungi domains due to climate change, and it is easy to see that more study is needed on the true incidence and geography of these infections [43].

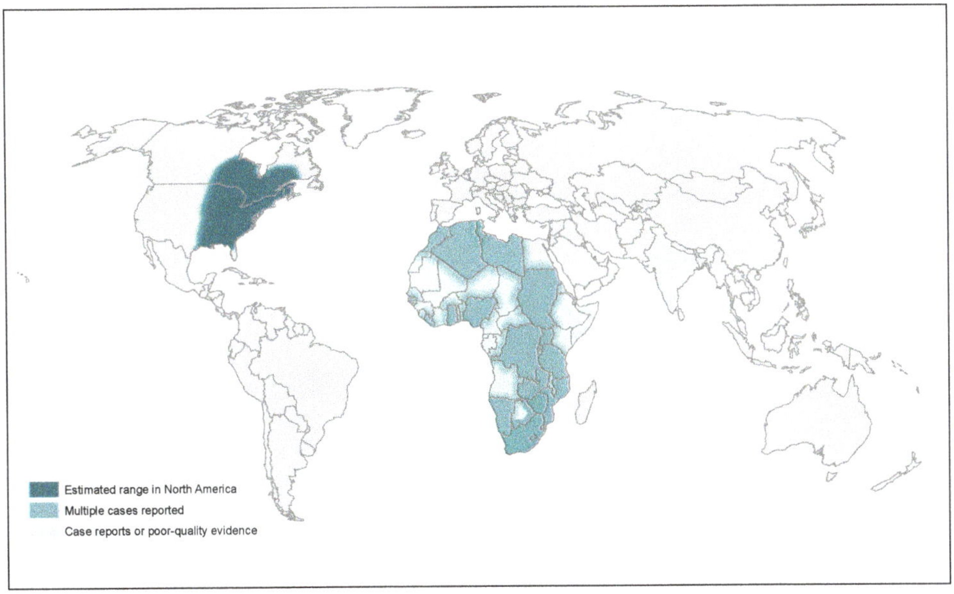

Figure 1. Map of the estimated distribution of *Blastomyces*. Figure used with permission from Ashraf et al. [5].

3. Clinical Presentation

Prior studies of point-source outbreaks of blastomycosis have suggested that an estimated 50% of persons infected with *Blastomyces* after an exposure experience minimal to no symptoms [44,45]. In those who go on to develop symptomatic infection, the typical incubation period is 4–6 weeks [46]. As expected, due to the inhalational route of exposure, pulmonary infection is the most common symptomatic presentation, accounting for at least three-quarters of diagnosed blastomycosis cases [47]. Amongst the *Blastomyces* spp., those within the *B. dermatidis complex* infection are more common in immunocompetent individuals and more often cause pulmonary symptoms compared to *B. helicus* infection, which is seen more often in immunocompromised persons and as a systemic disease [32]. Those species more common outside of the United States (*B. percusus* and *B. emzantsi*) seem to have a proclivity for dissemination from the lungs to cutaneous and osseous sites, though further studies are needed to solidify this pathogen–presentation relationship [7].

3.1. Pulmonary Blastomycosis

The most common presentation of symptomatic pulmonary blastomycosis is an indolent, chronic process, with patients typically complaining of weight loss, intermittent low-grade fever, chest pain, fatigue/malaise, dyspnea, and cough (often with scant sputum production and occasionally associated with hemoptysis) [9,10,47–49]. Less commonly, patients present with (or progress to) an acute form of pulmonary infection. In its acute form, pulmonary blastomycosis presentations can range in severity from subclinical pneumonia to acute respiratory distress syndrome (ARDS) [9,47,50]. Typical symptoms include fever, night sweats, dyspnea, cough (productive or non-productive), hemoptysis, fatigue, malaise, anorexia, and weight loss, often leading to an initial misdiagnosis of bacterial pneumonia [10,48,49]. Of those who rapidly develop ARDS due to acute pulmonary blastomycosis, mortality approaches 50% and is particularly severe if there was a diagnostic delay [46,47].

Imaging findings in these syndromes are variable and nonspecific. The most common presentation on chest imaging is airspace consolidation, most often described as patchy opacities. Less commonly, in fulminant pulmonary blastomycosis, large and/or bilateral consolidations can be found [10,50,51]. Other less common radiographic findings include mass-like lesions [52,53], granulomatous or cavitary lesions similar to tuberculosis [54,55], nodules [10,51], diffuse interstitial "tree-in-bud" nodularities, or, rarely, miliary disease with endobronchial extension. Many of these findings are more common when focal airspace opacity is also present. Figure 2 represents the more common, non-specific chest X-ray findings seen in pulmonary blastomycosis. The wide spectrum of radiographic findings in pulmonary blastomycosis often leads to misdiagnosis as other pulmonary infections or syndromes, such as bacterial pneumonia, sarcoidosis, tuberculosis, and malignancy [9,10]. Slow or no response to typical therapy for these conditions should raise suspicion for blastomycosis, particularly with the patient history.

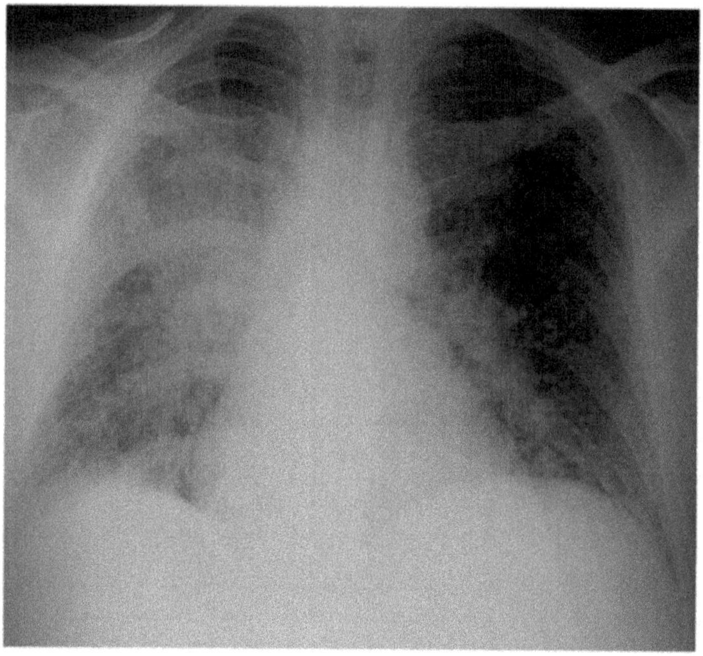

Figure 2. Pulmonary blastomycosis chest X-ray image. Chest X-ray seen in a patient with pulmonary blastomycosis demonstrating a right lower lobe consolidation and bilateral military nodules. Image sourced from Sarkar et al. under a creative commons license (CC BY 3.0) [56].

Of those who develop symptomatic infection, an estimated 25–40% develop extrapulmonary disease, most often disseminated from pulmonary disease (though direct inoculation can rarely occur) [46,47]. The most commonly involved extrapulmonary sites are the skin (40–80% of disseminated disease), bones (5–25%), and the genitourinary system (less than 10%) [46]. Central nervous system infection is rare, occurring in 5–10% of disseminated disease, with immunocompromised populations at higher risk [47,57]. Figure 3 summarizes the common sites of dissemination and the frequencies of dissemination to these sites.

3.2. *Cutaneous Blastomycosis*

Skin is the most common site of extrapulmonary blastomycosis, typically presenting initially as papulopustular lesions. These commonly progress to warty, verrucous plaques with heaped margins or, less often, lesions with central ulceration, abscesses, or violaceous nodules [58,59]. Severe cutaneous blastomycosis can expand to hundreds of lesions (typically not the scalp) and may also cause osteomyelitis by direct extension [47,58,59]. Mucus membranes are less commonly affected, though laryngeal, oral, and nasal lesions have been reported [60]. Similar to the pulmonary form of blastomycosis, the cutaneous form is often misdiagnosed as malignancy (basal or squamous cell carcinoma); bacterial infection (tuberculosis); or other skin processes, such as pyoderma gangrenosum or keratoacanthoma [60,61]. Rarely, isolated cutaneous blastomycosis without concurrent pulmonary infection is thought to occur related to a resolved prior pulmonary infection or direct inoculation; these infections are usually more limited [59,62]. Figure 4 shows several different cutaneous presentations of blastomycosis.

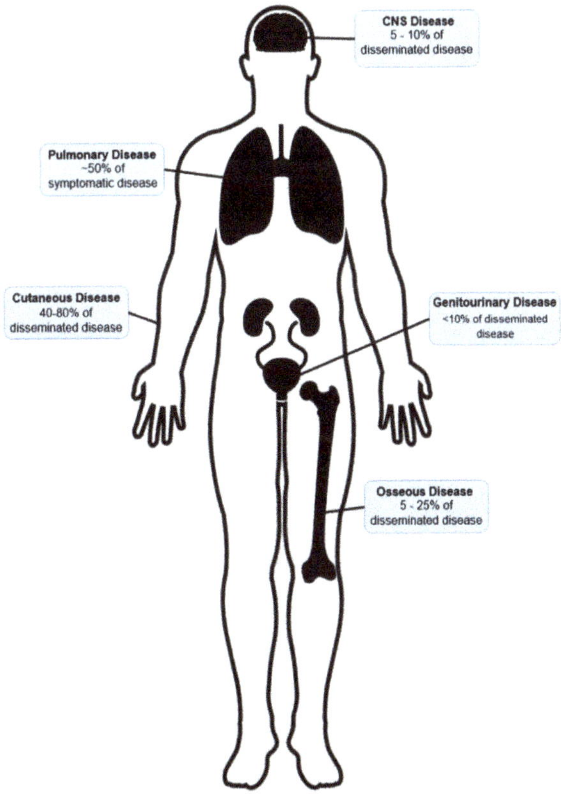

Figure 3. Sites of disseminated blastomycosis. CNS: central nervous system.

3.3. *Osseous Blastomycosis*

Osteomyelitis due to blastomycosis is typically associated with extension into the surrounding tissues causing abscesses, sinus tracts, and septic arthritis of adjacent joints [47,63]. Prior case series have shown a predilection for lower extremities and the lumbar or thoracic spine, though virtually any bone can be involved [47,61,64–66]. Lesions may or may not cause pain and may mimic cancers either as bone masses or lytic lesions [63,67,68]. Concurrent pulmonary disease is common but not universal [47,61,64].

3.4. *Genitourinary Blastomycosis*

Blastomycosis can also affect the genitourinary system via dissemination (though it has been rarely described in isolation) [69,70]. In men, the two classic syndromes are prostatitis +/−abscess and epididymitis [47,71]. Typical symptoms of *Blastomyces* prostatitis include urinary obstruction, dysuria, hematuria, hematospermia, and perineal or suprapubic pain [70,72,73]. Similarly, epididymitis typically presents with pain in the affected region, along with scrotal and/or testicular swelling, with rare sinus tract formation [47]. Genitourinary disease in female patients has been described less commonly but most often presents as tubo-ovarian abscesses, endometritis, and salpingitis [47,74–76].

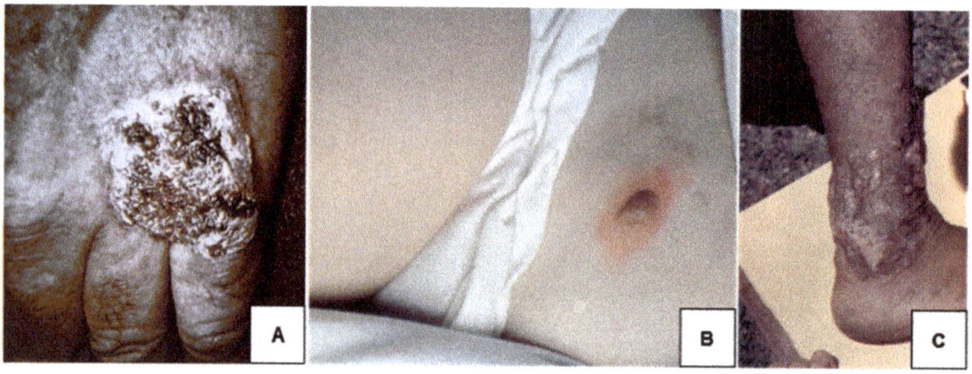

Figure 4. Cutaneous blastomycosis. (**A**) Verrucous blastomycosis; (**B**) nodular cutaneous blastomycosis with bulla formation; (**C**) keloidal blastomycosis. All images were obtained through the CDC Public Health Image Library (https://phil.cdc.gov (accessed on 1 August 2022)).

3.5. Central Nervous System Blastomycosis

CNS blastomycosis typically presents in the setting of multisystem disease (isolated CNS disease has also been rarely reported) [77–79]. Common presentations include a typical meningitis pattern (headache, nuchal rigidity, or other signs of meningism). Space-occupying lesions can occur within the cranium or spine and may cause abscesses [61,77,80,81]. True to its "great mimic" reputation, there are several case reports highlighting the frequent misdiagnosis of *Blastomyces* meningitis as tuberculous meningitis [82–84].

In all forms of blastomycosis, a common theme is the mimicry of other conditions [9,10,82–84]. Thus, particularly after treatment failure for more common diseases, blastomycosis should be strongly considered. In a patient with proper geographic- or exposure-related risk factors, blastomycosis should be considered immediately. Though pulmonary disease is most common (and is typically present in those who also have extrapulmonary disease), extrapulmonary forms are highly morbid and can also be fatal [46,47].

4. Diagnosis

The epidemiology outlined and history taking (for example, outdoor activities, such as chopping wood or digging into dirt, or even exposure to a construction site that disrupts soil or decaying wood) are crucial considerations that must play a part in the diagnosis of blastomycosis [5,19]. Interestingly, a patient's dog having been diagnosed with *Blastomyces* infection may be another helpful clue [85]. Of course, when ordering diagnostic tests, the symptoms being experienced by the patient affect which test(s) might be helpful. Regardless of whether sputum or other tissues are used, histopathologic techniques are crucial to the prompt diagnosis of blastomycosis, and early use may allow for the avoidance of diagnostic delays [86].

Blastomyces yeast is typically described as broad-based budding, and it is 8–20 μM in diameter with a doubly refractile cell wall [87]. Interestingly, giant forms have been described occasionally that may be confused with *Coccidioides* [88]. Stains such as Gomori methenamine silver, calcofluor white, periodic acid–Schiff, and 10% potassium hydroxide are commonly used to visualize *Blastomyces* yeast [47,89,90]. Sputum staining with potassium hydroxide is of particular interest in that it is a rapid method (15–30 min) with a fairly high sensitivity (36–90%) [89,91]. Obviously, this method does not identify all cases, but when used up front, it may allow for prompt diagnosis and treatment. Figure 5 demonstrates the common microbiologic findings of *Blastomyces*, notably broad-based budding yeast cells. In disseminated blastomycosis, examination of skin tissue samples may be quite useful, and these techniques can also be utilized on bone, blood, cerebrospinal fluid (CSF), or prostatic tissue samples depending on the site(s) involved.

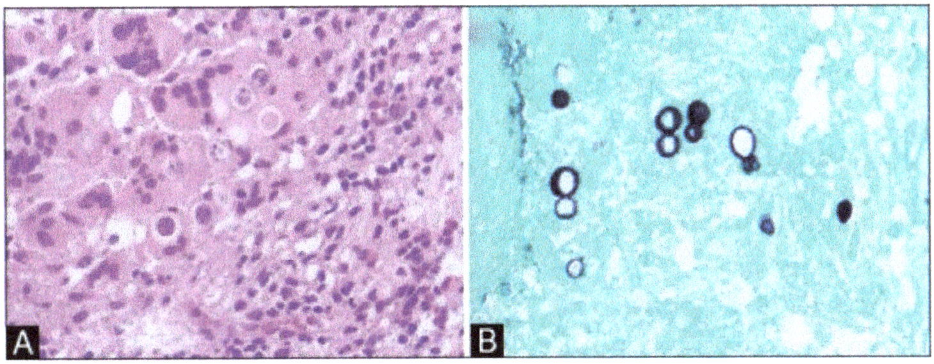

Figure 5. Characteristic *Blastomyces* microscopy. Hematoxylin and eosin staining of a *Blastomyces*-containing cerebellar mass at ×400 magnification (**A**) with large, round, thick-walled yeast cells. Grocott's methenamine-silver-stained section of the same mass at ×400 magnification (**B**) with broad-based budding yeast cells. Image source is Kochar et al., 2016, used under a creative commons license (CC BY-NC-SA 3.0) [92].

Fungal culture is another traditional method of diagnosis. While often more sensitive, culture may take up to five weeks to grow *Blastomyces* (though one or two weeks is more common) [87]. One interesting study found cultures from different respiratory fluids to have different yields, although they were limited to some degree by small numbers. Generally, respiratory cultures seem to have sensitivities >70%, with exact numbers depending on the fluid type. Culturing multiple specimen types (tracheal secretions, sputum, and bronchoalveolar lavage (BAL) fluid) seems to be additive [87,91]. A DNA probe (AccuProbe, Hologic, Inc., San Diego, CA, USA) is also available to be used on cultures growing fungi that have not yet been identified; however, this may cross-react with *Emergomyces* spp. [93].

Blastomyces spp. galactomannan antigen detection by an enzyme immunoassay is also commercially available and can be used on urine, serum, BAL fluid, and CSF [93]. This technology is relatively rapid (though for some assays, shipping slows turn-around time). Urine antigen testing is generally the most sensitive (75–90% depending on disease severity, the location of infection, assay type, and the laboratory performing testing), and levels may correlate with disease severity [94–98]. BAL fluid may have similar sensitivity (80%), although one study found a sensitivity closer to serum, wherein the EIA generally performs more poorly (50–60%) [94–96,99]. Similar to the potassium hydroxide staining of pulmonary fluid samples, if positive, these tests are quite helpful, but a negative test cannot rule out blastomycosis. One study found higher sensitivity in immunocompromised persons than in those without immune deficits, presumably because of higher antigen loads in such patients [100]. Importantly, broad cross-reactivity occurs with members of the *Histoplasma*, *Talaromyces*, and *Paracoccidiodes* genera [89,95,96,101].

Complement fixation and immunodiffusion antibody detection methods have been available for some time but are not frequently used given variable (and generally poor) reported performance characteristics [87,98]. However, a new EIA detecting antibodies against BAD-1 (a cell wall adhesion antigen) has shown higher sensitivity (88%) and specificity (94–99%) and, thus, may have some role. Importantly, this test is not currently available commercially despite the initial study having been published in 2014 [102].

Nucleic acid amplification tests, such as polymerase chain reaction (PCR), have been developed for *B. dermatitidis* but are not commercially available. Potential advantages could be use on tissue (possibly including paraffin-embedded tissue) or body fluids and without cross-reactions with other fungi [93,103]. Broad-range fungal PCR and metagenomic next-generation sequencing (mNGS) certainly can detect *Blastomyces*, and the use of mNGS has been reported in a patient suspected to have tuberculosis or lung cancer. Ultimately, that patient was found (via tissue mNGS) to have *B. dermatitidis* infection [104]. However, the role of these technologies for blastomycosis is not clear at this time.

5. Treatment

Updated Clinical Practice Guidelines for the treatment of blastomycosis were published by the Infectious Disease Society of America (IDSA) in 2008 [46]. No updates to this guideline have been published in over a decade, which is likely due to the uncommon occurrence of blastomycosis and the resultant dearth of large, well-run clinical trials. With the rare exception of asymptomatic infection or mild pulmonary blastomycosis in immunocompetent hosts that improve clinically and radiographically prior to diagnosis [105,106], all cases of blastomycosis should be treated to prevent disease progression [46]. The approach to treatment generally depends on the site of infection, disease severity, and the immune status of the patient, as summarized in Table 1. Additional factors that should be weighed include drug-related toxicities and patient co-morbidities.

Table 1. Management of blastomycosis.

Type of Infection	Drug(s) of Choice	Duration of Therapy
Pulmonary, mild to moderate	Itraconazole [a]	6–12 months
Pulmonary, severe	Induction: Liposomal amphotericin (or amphotericin B deoxycholate) [b] Step-down: Itraconazole [a]	Induction therapy × 1–2 weeks (or until clinical improvement), then oral therapy × 6–12 months.
Disseminated, mild–moderate, no CNS involvement	Itraconazole [a]	At least 12 months
Disseminated, severe or with CNS involvement	Induction: Liposomal amphotericin (or amphotericin B deoxycholate) [b] Step-down: Itraconazole [a]	Induction therapy × 1–2 weeks (or until clinical improvement), then oral therapy × 6–12 months.
Immunosuppressed, any form of blastomycosis	Induction: Liposomal amphotericin (or amphotericin B deoxycholate) [b] Step-down: Itraconazole [a]	Induction therapy × 1–2 weeks (or until clinical improvement), then oral therapy × 6–12 months (can be continued as lifelong suppression if ongoing immunosuppression) [c]

CNS: central nervous system. Treatment recommendations based on 2008 IDSA guidelines [46]. [a] 200 mg three times daily × 3 days, followed by 200 mg twice daily for remainder of therapy. [b] Liposomal formulation dosed at 3–5 mg/kg/day; deoxycholate dosed at 0.7–1 mg/kg/day. [c] Lifelong suppression dosed at 200 mg daily.

5.1. *Pulmonary Blastomycosis*

The choice of initial therapy for pulmonary blastomycosis depends on the severity of disease. Unfortunately, there are limited data available to guide clinicians in the assessment of disease severity, which is often left to clinical judgement [46]. This remains an important area of further study.

5.1.1. Mild-to-Moderate Pulmonary Blastomycosis

Mild-to-moderate disease is usually treated with itraconazole 200 mg three times per day for three days, followed by 200 mg once or twice daily for 6–12 months. Treatment is often continued for a few months after clinical and radiographic resolution, though more data are needed to support this practice and to determine the optimal duration of therapy [46]. The disadvantages of itraconazole include extensive drug–drug interactions, drug-related side effects, and absorption issues. The oral suspension formulation of itraconazole should be taken on an empty stomach and is preferred due to a more predictable absorption and improved bioavailability compared to the capsule formulation [107]. However, gastrointestinal side effects and the current high cost of treatment may limit its use. By contrast, the capsule formulation of itraconazole should be taken with a full meal or with acidic beverages to improve absorption. Acid-blocking agents should be avoided. To ensure adequate drug levels and tissue penetration, serum levels should be obtained after two weeks of therapy [46]. In 2018, a novel formulation of itraconazole, super bioavailable ("SUBA")-itraconazole was approved by the Food and Drug Administration (FDA) as a 65 mg capsule for the treatment of systemic fungal infections, including pulmonary and extrapulmonary blastomycosis. A recent study showed that SUBA-itraconazole has less variable absorption as compared to conventional itraconazole under fasted conditions. Whether the use of SUBA-itraconazole is associated with superior outcomes in blastomycosis is unknown [108,109].

For patients who do not tolerate itraconazole, alternative options exist; however, these are less effective and/or have limited data to support their use. High-dose fluconazole (400–800 mg daily) and ketoconazole (400–800 mg daily) have been used successfully, though ketoconazole-related side effects are common at high doses, and ketoconazole should generally be avoided [46,110,111]. The roles of voriconazole and posaconazole in the treatment of pulmonary blastomycosis remain poorly understood, with data limited to case reports [53,112].

5.1.2. Moderately Severe to Severe Pulmonary Blastomycosis

For severe disease, the recommended initial treatment is intravenous amphotericin B (lipid formulation of amphotericin B at 3–5 mg/kg per day; or amphotericin B deoxycholate 0.7–1 mg/kg per day) for 1–2 weeks or until clinical improvement, followed by step-down therapy to oral itraconazole (200 mg three times per day for three days, followed by 200 mg twice daily for 6–12 months) [46]. The lipid formulation of amphotericin is preferred due to the lower rates of nephrotoxicity [113]. Patients who require mechanical ventilatory support and develop adult respiratory distress syndrome (ARDS) have a poor prognosis [114]. The use of corticosteroids in this setting remains controversial and unproven, with data limited to case reports [115]. Although not recommended by the IDSA guidelines, adjunctive corticosteroids are recommended on a case-by-case basis for patients with life-threatening pulmonary blastomycosis during the first two weeks of therapy by the American Thoracic Society and some experts [116].

5.2. Disseminated/Extrapulmonary Blastomycosis

Extrapulmonary blastomycosis always requires treatment, even if complete tissue resection is anticipated, such as with surgical resection of cutaneous blastomycosis lesions [46]. Identifying whether there is central nervous system (CNS) involvement is a key step to ensure that a treatment regimen is used with adequate CNS penetration. Itraconazole is thought to achieve poor concentrations in the CNS and should generally be avoided [117]. Interestingly, it is used for step-down therapy in CNS histoplasmosis [118]. Similarly, determining the presence of osteoarticular disease is important, as osteoarticular blastomycosis can be more difficult to treat. The IDSA guidelines recommend at least a 12-month treatment duration due to a higher risk of relapse; however, there are little data to support this practice, and most recommendations are primarily based on expert opinion [46,119]. Disseminated disease without CNS involvement is treated the same as pulmonary disease, with severe disease starting with amphotericin upfront and mild–moderate disease focusing on itraconazole alone. High-dose fluconazole (400–800 mg per day) can be used as an alternative option for mild–moderate disease or step-down therapy, but it is less effective [46,110]. The roles of voriconazole and posaconazole are poorly understood.

5.3. CNS Blastomycosis

The recommended treatment of CNS blastomycosis is a lipid formulation of amphotericin B (higher doses are more commonly used but with incomplete evidence to support this practice) for 4–6 weeks followed by an azole for at least 12 months and until CNS abnormalities have resolved [46]. Liposomal amphotericin B has been shown in animal models to have superior CNS penetration as compared to alternative amphotericin B formulations, although whether it is the same in humans is not clear [46,120]. The optimal choice of azole following the completion of amphotericin B is poorly understood, with clinical data primarily limited to case reports [46]. The most experience appears to be with the off-label use of voriconazole, which has known intrinsic activity against *Blastomyces* and achieves adequate CSF concentrations [117]. Itraconazole has poor CNS concentration and, thus, is less effective at treating disease at this site, but in some settings, it is commonly used. By contrast, fluconazole has excellent CNS penetration but has less intrinsic activity against blastomycosis [107]. Surgical debridement may be needed in some cases, though more research is needed to define the optimal role and timing of debridement [121].

6. Immunosuppressed Patients with Blastomycosis

Immunocompromised patients (patients with HIV/AIDS, recipients of solid organ transplants, and patients with hematologic malignancies) appear to be at greater risk of developing severe disease and respiratory failure, and of dying from blastomycosis [122–124]. Intravenous liposomal amphotericin B is used in the same way as described above and is similarly followed by itraconazole. Life-long suppression with itraconazole 200 mg per day should be considered in patients for whom immunosuppression cannot be withheld and/or in patients experiencing relapse despite recommended treatment [46]. Although induction therapy with itraconazole is not recommended, a recent retrospective analysis of patients with proven blastomycosis identified multiple patients with solid organ transplant and malignancies on chemotherapy who were treated successfully with itraconazole monotherapy for the full duration of treatment [124]. Though promising, clinical trial data are needed to determine whether itraconazole, or other azoles, have a role in the initial treatment of blastomycosis in immunocompromised hosts. Furthermore, the close monitoring of drug (itraconazole and immune suppression medications) levels is particularly important in patients who have undergone solid organ transplant.

7. Blastomycosis in Pregnancy and Newborns

Intravenous liposomal amphotericin is the mainstay of therapy in blastomycosis affecting pregnant patients and should be continued until after the delivery or resolution of the infection, whichever occurs first. This is based on the possible teratogenic effects of azoles seen in animal studies [46]. There are limited data suggesting that fetal risk is higher than maternal risk in blastomycosis of pregnancy, owing to the potential for trans-placental transmission of *Blastomyces* [125]. Thus, if a mother is confirmed to have blastomycosis during pregnancy, it is recommended that the placenta be examined after delivery for signs of *Blastomyces* infection and that the newborn be monitored for emerging signs and symptoms of blastomycosis. Amphotericin deoxycholate is recommended to treat the newborn if they do become infected [46].

8. Alternative Azoles

There are emerging data supporting isavuconazole as an option for blastomycosis in immunocompromised patients, including those with CNS blastomycosis. This drug is appealing due to it having a smaller drug–drug interaction profile than other azoles. A recent retrospective case series of 14 patients with blastomycosis (half of whom were immunocompromised, and half of whom had CNS involvement) treated with isavuconazole demonstrated a 79% cure rate, suggesting that this drug could potentially be used in a broad profile of patients, particularly when other azoles would cause safety or drug interaction issues [126]. Similarly, a retrospective case series of blastomycosis in recipients of solid organ transplants at a single medical center in Wisconsin included a subset of patients treated with voriconazole, posaconazole, or fluconazole. Though not powered to examine differences in outcomes between these therapies, it offers a glimpse into the real-world use of these azoles in patients with blastomycosis, and it suggests that, in the right context, they may be useful alternatives to itraconazole [127]. That being said, there is still a significant need for large-scale clinical trials to strengthen this data.

9. Future Research

Significant knowledge gaps exist in our understanding of the evolving epidemiology of blastomycosis, as well as the effectiveness and appropriateness of newer azole medications for this infection. In the United States, blastomycosis is not a reportable disease outside of five states within the Mississippi River valley region [27]. Due to this, epidemiologic data on this infection outside of those states are piecemeal and often limited to case reports. Those epidemiologic studies that have been performed are typically focused on patients presenting to large academic medical centers, potentially skewing the geolocation data to the site of diagnosis rather than the site of exposure [28]. A large, multisite study that decouples the diagnosis and exposure sites for those with symptomatic blastomycosis would provide significant insight into the growing geographic reach of this endemic mycosis. Newer diagnostic methods, notably next-generation sequencing and targeted fungal PCR assay approaches similar to 13 s, are also largely unexplored in blastomycosis (and endemic mycoses in general); these technologies have the potential to provide a low-cost, rapid diagnosis with high specificity and sensitivity, but, unfortunately, they have not had much funding or attention to date.

Though itraconazole is the foundation of treatment for blastomycosis, there are occasionally situations where it is not a viable therapeutic option (intolerance, drug–drug interactions, unavailability, cost, etc.). Experience with other azoles as primary therapy, or as a transition from amphotericin in severe or CNS blastomycosis, is data-poor and mostly based on case reports or studies involving only a very small number of patients, leaving clinicians to make difficult therapeutic decisions without much of a data-driven foothold. Additionally, even when itraconazole is appropriate and available, there are significant questions that remain regarding the ideal length of therapy and the role of adjunctive steroids. Future studies, either focused broadly on the outcomes of patients with endemic mycoses treated with a variety of azole-based regimens or, more narrowly, on a specific azole, would help establish the non-inferiority of these alternative agents and could provide more therapeutic options for clinicians treating difficult blastomycosis cases.

10. Conclusions

Blastomyces is a significant fungal pathogen endemic to a large swath of the United States, Africa, and potentially India and the Middle East. Though primarily a pulmonary pathogen, *Blastomyces* is capable of causing severe disease throughout much of the body and is of particular concern in the immunocompromised population. As summarized above, diagnosis requires keen attention to the patient's personal exposure risk epidemiologic factors, and their clinical presentation, as well as microbiologic and other laboratory data, as available. Itraconazole is often the foundation of treatment for mild-to-moderate blastomycosis, with more severe or CNS disease requiring an initial amphotericin phase. Experience with alternative azoles is limited, with voriconazole perhaps having the best evidence. However, other azoles may still be viable options in cases where itraconazole is contraindicated or unavailable. Despite increasing incidence, there remain significant knowledge gaps in the epidemiology and management of this infection—areas where future research needs to be carried out.

Author Contributions: Conceptualization: M.F.P., J.D.A. and N.C.B., Methodology: M.F.P., J.D.A. and N.C.B., Writing–original draft preparation and editing: M.F.P., J.D.A. and N.C.B. All authors have read and agreed to the published version of the manuscript.

Funding: Bahr receives funding from the National Institutes of Neurological Disorders and Stroke, K23 NS110470.

Institutional Review Board Statement: Not applicable.

Informed Consent Statement: Not applicable.

Data Availability Statement: Not applicable.

Conflicts of Interest: The authors declare no conflict of interest.

References

1. Pfaller, M.A.; Diekema, D. Epidemiology of Invasive Mycoses in North America. *Crit. Rev. Microbiol.* **2010**, *36*, 1–53. [CrossRef]
2. Vallabhaneni, S.; Mody, R.K.; Walker, T.; Chiller, T. The Global Burden of Fungal Diseases. *Infect. Dis. Clin. N. Am.* **2016**, *30*, 1–11. [CrossRef]
3. Wheat, J. Endemic mycoses in AIDS: A clinical review. *Clin. Microbiol. Rev.* **1995**, *8*, 146–159. [CrossRef] [PubMed]
4. Khuu, D.; Shafir, S.; Bristow, B.; Sorvillo, F. Blastomycosis Mortality Rates, United States, 1990–2010. *Emerg. Infect. Dis.* **2014**, *20*, 1789–1794. [CrossRef]
5. Ashraf, N.; Kubat, R.C.; Poplin, V.; Adenis, A.A.; Denning, D.W.; Wright, L.; McCotter, O.; Schwartz, I.S.; Jackson, B.R.; Chiller, T.; et al. Re-drawing the Maps for Endemic Mycoses. *Mycopathologia* **2020**, *185*, 843–865. [CrossRef]
6. Linder, K.A.; Kauffman, C.A. Current and New Perspectives in the Diagnosis of Blastomycosis and Histoplasmosis. *J. Fungi* **2020**, *7*, 12. [CrossRef] [PubMed]
7. Maphanga, T.G.; Birkhead, M.; Muñoz, J.F.; Allam, M.; Zulu, T.G.; Cuomo, C.A.; Schwartz, I.S.; Ismail, A.; Naicker, S.D.; Mpembe, R.S.; et al. Human Blastomycosis in South Africa Caused by *Blastomyces percursus* and *Blastomyces emzantsi* sp. nov., 1967 to 2014. *J. Clin. Microbiol.* **2020**, *58*, e01661-19. [CrossRef]
8. Schwartz, I.S.; Muñoz, J.F.; Kenyon, C.R.; Govender, N.P.; McTaggart, L.; Maphanga, T.G.; Richardson, S.; Becker, P.; Cuomo, C.A.; McEwen, J.G.; et al. Blastomycosis in Africa and the Middle East: A Comprehensive Review of Reported Cases and Reanalysis of Historical Isolates Based on Molecular Data. *Clin. Infect. Dis.* **2020**, *73*, e1560–e1569. [CrossRef]
9. Kuzo, R.S.; Goodman, L.R. Blastomycosis. *Semin. Roentgenol.* **1996**, *31*, 45–51. [CrossRef]
10. Fang, W.; Washington, L.; Kumar, N. Imaging Manifestations of Blastomycosis: A Pulmonary Infection with Potential Dissemination. *RadioGraphics* **2007**, *27*, 641–655. [CrossRef]
11. Caceres, D.H.; Echeverri Tirado, L.C.; Bonifaz, A.; Adenis, A.; Gomez, B.L.; Lizette Banda Flores, C.; Canteros, C.E.; Santos, D.W.; Arathoon, E.; Soto, E.R.; et al. Current Situation of Endemic Mycosis in the Americas and the Caribbean: Proceedings of the First International Meeting on Endemic Mycoses of the Americas (IMEMA). *Myocses* **2022**, in press. [CrossRef]
12. Pfister, J.R.; Archer, J.R.; Hersil, S.; Boers, T.; Reed, K.D.; Meece, J.K.; Anderson, J.L.; Burgess, J.W.; Sullivan, T.D.; Klein, B.S.; et al. Non-Rural Point Source Blastomycosis Outbreak Near a Yard Waste Collection Site. *Clin. Med. Res.* **2010**, *9*, 57–65. [CrossRef] [PubMed]
13. Burgess, J.W.; Schwan, W.R.; Volk, T.J. PCR-based detection of DNA from the human pathogen *Blastomyces dermatitidis* from natural soil samples. *Med. Mycol.* **2006**, *44*, 741–748. [CrossRef]
14. Denton, J.F.; McDonough, E.S.; Ajello, L.; Ausherman, R.J. Isolation of Blastomyces dermatitidis from Soil. *Science* **1961**, *133*, 1126–1127. [CrossRef] [PubMed]
15. Tenenbaum, M.J.; Greenspan, J.; Kerkering, T.M.; Utz, J.P. Blastomycosis. *CRC Crit. Rev. Microbiol.* **1982**, *9*, 139–163. [CrossRef]
16. Sarosi, G.A.; Serstock, D.S. Isolation of Blastomyces dermatitidis from pigeon manure. *Am. Rev. Respir. Dis.* **1976**, *114*, 1179–1183. [CrossRef]
17. Bialek, R.; Cirera, A.C.; Herrmann, T.; Aepinus, C.; Shearn-Bochsler, V.I.; Legendre, A.M. Nested PCR Assays for Detection of *Blastomyces dermatitidis* DNA in Paraffin-Embedded Canine Tissue. *J. Clin. Microbiol.* **2003**, *41*, 205–208. [CrossRef]
18. Jackson, K.M.; Pelletier, K.C.; Scheftel, J.; Kerkaert, J.D.; Robinson, S.L.; McDonald, T.; Bender, J.B.; Knight, J.F.; Ireland, M.; Nielsen, K. Blastomyces dermatitidis Environmental Prevalence in Minnesota: Analysis and Modeling Using Soil Collected at Basal and Outbreak Sites. *Appl. Environ. Microbiol.* **2021**, *87*, e01922-20. [CrossRef]
19. Baumgardner, D.J.; Burdick, J.S. An Outbreak of Human and Canine Blastomycosis. *Clin. Infect. Dis.* **1991**, *13*, 898–905. [CrossRef]
20. Chen, T.; Legendre, A.M.; Bass, C.; Mays, S.E.; Odoi, A. A case-control study of sporadic canine blastomycosis in Tennessee, USA. *Med. Mycol.* **2008**, *46*, 843–852. [CrossRef]
21. Furcolow, M.L.; Busey, J.F.; Menges, R.W.; Chick, E.W. Prevalence and incidence studies of human and canine blastomycosis. II. Yearly incidence studies in three selected states, 1960–1967. *Am. J. Epidemiol.* **1970**, *92*, 121–131. [CrossRef] [PubMed]
22. McBride, J.A.; Gauthier, G.M.; Klein, B.S. Turning on virulence: Mechanisms that underpin the morphologic transition and pathogenicity of *Blastomyces. Virulence* **2018**, *10*, 801–809. [CrossRef] [PubMed]
23. Manetti, A.C. Hyperendemic urban blastomycosis. *Am. J. Public Health* **1991**, *81*, 633–636. [CrossRef] [PubMed]
24. Seitz, A.E.; Adjemian, J.; Steiner, C.A.; Prevots, D.R. Spatial epidemiology of blastomycosis hospitalizations: Detecting clusters and identifying environmental risk factors. *Med. Mycol.* **2015**, *53*, 447–454. [CrossRef]

25. Brown, E.; McTaggart, L.R.; Dunn, D.; Pszczolko, E.; Tsui, K.G.; Morris, S.K.; Stephens, D.; Kus, J.V.; Richardson, S.E. Epidemiology and Geographic Distribution of Blastomycosis, Histoplasmosis, and Coccidioidomycosis, Ontario, Canada, 1990–2015. *Emerg. Infect. Dis.* **2018**, *24*, 1257–1266. [CrossRef]
26. Baddley, J.W. Geographic Distribution of Endemic Fungal Infections among Older Persons, United States. *Emerg. Infect. Dis.* **2011**, *17*, 1664–1669. [CrossRef]
27. Centers for Disease Control and Prevention. Reportable Fungal Diseases by State. 2021. Available online: https://www.cdc.gov/fungal/fungal-disease-reporting-table.html (accessed on 17 June 2022).
28. Benedict, K.; Thompson, G.R.; Deresinski, S.; Chiller, T. Mycotic Infections Acquired outside Areas of Known Endemicity, United States. *Emerg. Infect. Dis.* **2015**, *21*, 1935–1941. [CrossRef]
29. Zampogna, J.C.; Hoy, M.J.; Ramos-Caro, F.A. Primary Cutaneous North American Blastomycosis in an Immunosuppressed Child. *Pediatr. Dermatol.* **2003**, *20*, 128–130. [CrossRef]
30. Lehman, W.L.; Ray, L.F. Blastomycosis in Oregon. *Northwest Med.* **1951**, *50*, 39–40.
31. McTaggart, L.R.; Brown, E.; Richardson, S.E. Phylogeographic Analysis of Blastomyces dermatitidis and Blastomyces gilchristii Reveals an Association with North American Freshwater Drainage Basins. *PLoS ONE* **2016**, *11*, e0159396. [CrossRef]
32. Schwartz, I.S.; Wiederhold, N.P.; E Hanson, K.; Patterson, T.F.; Sigler, L. *Blastomyces helicus*, a New Dimorphic Fungus Causing Fatal Pulmonary and Systemic Disease in Humans and Animals in Western Canada and the United States. *Clin. Infect. Dis.* **2018**, *68*, 188–195. [CrossRef] [PubMed]
33. A Frean, J.; Carman, W.F.; Crewe-Brown, H.H.; A Culligan, G.; Young, C.N. Blastomyces dermatitidis infections in the RSA. *S. Afr. Med. J.* **1989**, *76*, 13–16. [PubMed]
34. Fragoyannis, S.; Van Wyk, G.; De Beer, M. North American blastomycosis in South Africa: A case report. *S. Afr. Med. J.* **1977**, *51*, 169–171. [PubMed]
35. Emmons, C.; Murray, I.; Lurie, H.; King, M.; Tulloch, J.; Connor, D. North American blastomycosis: Two autochthonous cases from Africa. *Med. Mycol.* **1964**, *3*, 306–311. [CrossRef] [PubMed]
36. Emerson, P.A.; Higgins, E.; Branfoot, A. North American blastomycosis in Africans. *Br. J. Dis. Chest* **1984**, *78*, 286–291. [CrossRef]
37. Carman, W.F.; Frean, J.A.; Crewe-Brown, H.H.; Culligan, G.A.; Young, C.N. Blastomycosis in Africa. *Mycopathologia* **1989**, *107*, 25–32. [CrossRef] [PubMed]
38. Baily, G.G.; Robertson, V.J.; Neill, P.; Garrido, P.; Levy, L.F. Blastomycosis in Africa: Clinical Features, Diagnosis, and Treatment. *Clin. Infect. Dis.* **1991**, *13*, 1005–1008. [CrossRef]
39. Jordaan, H.F. Blastomycosis in Namibia-report of a case successfully treated with itraconazole. *Clin. Exp. Dermatol.* **1989**, *14*, 347–351. [CrossRef]
40. Ibrahim, T.M.; Edinol, S.T. Pleural effusion from blastomycetes in an adult Nigerian: A case report. *Niger. Postgrad. Med. J.* **2001**, *8*, 148–149.
41. Malak, J.A.; Farah, F.S. Blastomycosis in the Middle East. Report of a suspected case of North American blastomycosis. *Br. J. Dermatol.* **1971**, *84*, 161–166. [CrossRef]
42. Randhawa, H.S.; Chowdhary, A.; Kathuria, S.; Roy, P.; Misra, D.S.; Jain, S.; Chugh, T.D. Blastomycosis in India: Report of an imported case and current status. *Med. Mycol.* **2013**, *51*, 185–192. [CrossRef] [PubMed]
43. Silversides, A.; Greer, A.; Ng, V.; Fisman, D. CMA opposes gender discrimination against doctors. *Can. Med. Assoc. J.* **2008**, *178*, 715–722. [CrossRef] [PubMed]
44. Klein, B.S.; Vergeront, J.M.; Weeks, R.J.; Kumar, U.N.; Mathai, G.; Varkey, B.; Kaufman, L.; Bradsher, R.W.; Stoebig, J.F.; Davis, J.P.; et al. Isolation of *Blastomyces dermatitidis* in Soil Associated with a Large Outbreak of Blastomycosis in Wisconsin. *N. Engl. J. Med.* **1986**, *314*, 529–534. [CrossRef] [PubMed]
45. Tosh, F.E.; Hammerman, K.J.; Weeks, R.J.; Sarosi, G.A. A common source epidemic of North American blastomycosis. *Am. Rev. Respir. Dis.* **1974**, *109*, 525–529. [CrossRef] [PubMed]
46. Chapman, S.W.; Dismukes, W.E.; Proia, L.A.; Bradsher, R.W.; Pappas, P.G.; Threlkeld, M.G.; Kauffman, C.A. Clinical Practice Guidelines for the Management of Blastomycosis: 2008 Update by the Infectious Diseases Society of America. *Clin. Infect. Dis.* **2008**, *46*, 1801–1812. [CrossRef]
47. McBride, J.A.; Gauthier, G.M.; Klein, B.S. Clinical Manifestations and Treatment of Blastomycosis. *Clin. Chest Med.* **2017**, *38*, 435–449. [CrossRef]
48. Crampton, T.L.; Light, R.B.; Berg, G.M.; Meyers, M.P.; Schroeder, G.C.; Hershfield, E.S.; Embil, J.M. Epidemiology and Clinical Spectrum of Blastomycosis Diagnosed at Manitoba Hospitals. *Clin. Infect. Dis.* **2002**, *34*, 1310–1316. [CrossRef]
49. Fanella, S.; Skinner, S.; Trepman, E.; Embil, J.M. Blastomycosis in children and adolescents: A 30-year experience from Manitoba. *Med. Mycol.* **2011**, *49*, 627–632. [CrossRef]
50. Maini, R.; Ranjha, S.; Tandan, N.; Regmi, M.R.; Ullah, S.; Stone, T.; Nayeemuddin, F.; Pervin, N. Pulmonary Blastomycosis: A case series and review of unique radiological findings. *Med. Mycol. Case Rep.* **2020**, *28*, 49–54. [CrossRef]
51. Brown, L.R.; Swensen, S.J.; Van Scoy, R.E.; Prakash, U.B.; Coles, D.T.; Colby, T.V. Roentgenologic Features of Pulmonary Blastomycosis. *Mayo Clin. Proc.* **1991**, *66*, 29–38. [CrossRef]
52. Tobar, P.; Sanchez-Nadales, A.; Caldeira, E.; Saad, P. Unusual presentation of pulmonary blastomycosis complicated by amphotericin-induced refractory electrolyte abnormalities. *Oxf. Med. Case Rep.* **2020**, *2020*, omaa093. [CrossRef]

53. Hussaini, S.M.Q.; Madut, D.; Tong, B.C.; Pavlisko, E.N.; Schell, W.A.; Perfect, J.R.; Thielman, N.M. Pulmonary blastomycosis presenting as primary lung cancer. *BMC Infect. Dis.* **2018**, *18*, 336. [CrossRef] [PubMed]
54. Kumar, A.; Kunoor, A.; Eapen, M.; Singh, P.K.; Chowdhary, A. Blastomycosis Misdiagnosed as Tuberculosis, India. *Emerg. Infect. Dis.* **2019**, *25*, 1776–1777. [CrossRef] [PubMed]
55. Koroscil, U.M.T.; Skabelund, U.A. Chronic Pulmonary Blastomycosis Mimicking Pulmonary Tuberculosis. *Mil. Med.* **2018**, *183*, e332–e333. [CrossRef]
56. Sarkar, P.K.; Malhotra, P.; Sriram, P.S. Rapid Progression of Pulmonary Blastomycosis in an Untreated Patient of Chronic Lymphocytic Leukemia. *Case Rep. Med.* **2014**, *2014*, 514382. [CrossRef] [PubMed]
57. Bariola, J.R.; Perry, P.; Pappas, P.G.; Proia, L.; Shealey, W.; Wright, P.W.; Sizemore, J.M.; Robinson, M.; Bradsher, J.R.W. Blastomycosis of the Central Nervous System: A Multicenter Review of Diagnosis and Treatment in the Modern Era. *Clin. Infect. Dis.* **2010**, *50*, 797–804. [CrossRef]
58. Smith, J.A.; Riddell, J.; Kauffman, C.A. Cutaneous Manifestations of Endemic Mycoses. *Curr. Infect. Dis. Rep.* **2013**, *15*, 440–449. [CrossRef]
59. Caldito, E.G.; Ajiboye, O.; Flores, E.; Antia, C.; Demarais, P. The importance of skin exam in chronic pulmonary blastomycosis. *IDCases* **2020**, *20*, e00812. [CrossRef]
60. Reder, P.A.; Neel, H.B. Blastornycosis in Otolaryngology. *Laryngoscope* **1993**, *103 Pt 1*, 53–58. [CrossRef]
61. Saccente, M.; Woods, G.L. Clinical and Laboratory Update on Blastomycosis. *Clin. Microbiol. Rev.* **2010**, *23*, 367–381. [CrossRef]
62. Clinton, T.S.; Timko, A.L. Cutaneous Blastomycosis without Evidence of Pulmonary Involvement. *Mil. Med.* **2003**, *168*, 651–653. [CrossRef] [PubMed]
63. Jain, R.; Singh, K.; Lamzabi, I.; Harbhajanka, A.; Gattuso, P.; Reddy, V.B. Blastomycosis of Bone. *Am. J. Clin. Pathol.* **2014**, *142*, 609–616. [CrossRef] [PubMed]
64. Oppenheimer, M.; Embil, J.M.; Black, B.; Wiebe, L.; Limerick, B.; MacDonald, K.; Trepman, E. Blastomycosis of Bones and Joints. *South. Med. J.* **2007**, *100*, 570–578. [CrossRef] [PubMed]
65. Nokes, S.R.; Adametz, J.; Gardner, G.; Beaton, J.N. Radiological case of the month. Blastomycosis osteomyelitis with epidural and retropharyngeal abscess. *J. Ark. Med. Soc.* **1995**, *92*, 253–254.
66. Saccente, M.; Abernathy, R.S.; Pappas, P.G.; Shah, H.R.; Bradsher, R.W. Vertebral blastomycosis with paravertebral abscess: Report of eight cases and review of the literature. *Clin. Infect. Dis.* **1998**, *26*, 413–418. [CrossRef]
67. Saiz, P.; Gitelis, S.; Virkus, W.; Piasecki, P.; Bengana, C.; Templeton, A. Blastomycosis of Long Bones. *Clin. Orthop. Relat. Res.* **2004**, *421*, 255–259. [CrossRef]
68. Codifava, M.; Guerra, A.; Rossi, G.; Paolucci, P.; Iughetti, L. Unusual osseous presentation of blastomycosis in an immigrant child: A challenge for European pediatricians. *Ital. J. Pediatrics* **2012**, *38*, 69. [CrossRef]
69. Bergner, D.M.; Kraus, S.D.; Duck, G.B.; Lewis, R. Systemic Blastomycosis Presenting with Acute Prostatic Abscess. *J. Urol.* **1981**, *126*, 132–133. [CrossRef]
70. Sloan, M.J.; Manjee, K.G.; Kaplan, L.J.; Glaser, A.P. Prostatic abscess due to blastomycosis. *Urol. Case Rep.* **2022**, *42*, 102007. [CrossRef]
71. Eickenberg, H.-U.; Amin, M.; Lich, R. Blastomycosis of the Genitourinary Tract. *J. Urol.* **1975**, *113*, 650–652. [CrossRef]
72. Seo, R.; Oyasu, R.; Schaeffer, A. Blastomycosis of the epididymis and prostate. *Urology* **1997**, *50*, 980–982. [CrossRef]
73. Labastida, L.G.; Quintana, O.B.; Sánchez, I.P.R.; Delgado-Enciso, I.; Guzmán-Esquivel, J.; Macías, G.S.G.; Ancer-Rodríguez, J.; Garza-Guajardo, R. Prostatic Blastomycosis: An Infrequent Cause of Granulomatous Prostatitis. *Fungal Genom. Biol.* **2016**, *6*. [CrossRef]
74. Mouzin, E.L.; Seilke, M.A. Female Genital Blastomycosis: Case Report and Review. *Clin. Infect. Dis.* **1996**, *22*, 718–719. [CrossRef] [PubMed]
75. Murray, J.J.; Clark, C.A.; Lands, R.H.; Heim, C.R.; Burnett, L.S. Reactivation blastomycosis presenting as a tuboovarian abscess. *Obstet. Gynecol.* **1984**, *64*, 828. [PubMed]
76. Farber, E.R.; Leahy, M.S.; Meadows, T.R. Endometrial blastomycosis acquired by sexual contact. *Obstet. Gynecol.* **1968**, *32*, 195–199.
77. Bakleh, M.; Aksamit, A.J.; Tleyjeh, I.; Marshall, W.F. Successful Treatment of Cerebral Blastomycosis with Voriconazole. *Clin. Infect. Dis.* **2005**, *40*, e69–e71. [CrossRef]
78. Kravitz, G.R.; Davies, S.F.; Eckman, M.R.; Sarosi, G.A. Chronic blastomycotic meningitis. *Am. J. Med.* **1981**, *71*, 501–505. [CrossRef]
79. Friedman, J.A.; Wijdicks, E.F.; Fulgham, J.R.; Wright, A.J. Meningoencephalitis Due to Blastomyces dermatitidis: Case Report and Literature Review. *Mayo Clin. Proc.* **2000**, *75*, 403–408. [CrossRef]
80. Majdick, K.; Kaye, K.; Shorman, M.A. Central nervous system blastomycosis clinical characteristics and outcomes. *Med. Mycol.* **2020**, *59*, 87–92. [CrossRef]
81. Ryan, D.; Lippert, W.C.; Fenwick, A.J.; Bhatt, M.; Jones, C.R. Central Nervous System Blastomycosis With Multiple Brain Abscesses Presenting as Right Upper Extremity Weakness. *Neurohospitalist* **2019**, *9*, 230–234. [CrossRef]
82. Gonyea, E.F. The spectrum of primary blastomycotic meningitis: A review of central nervous system blastomycosis. *Ann. Neurol.* **1978**, *3*, 26–39. [CrossRef] [PubMed]
83. Benzel, E.C.; King, J.W.; Mirfakhraee, M.; West, B.C.; Misra, R.P.; Hadden, T.A. Blastomycotic meningitis. *Surg. Neurol.* **1986**, *26*, 192–196. [CrossRef]

84. Rippon, J.W.; Zvetina, J.R.; Reyes, C. Case report: Miliary blastomycosis with cerebral involvement. *Mycopathologia* **1977**, *60*, 121–125. [CrossRef] [PubMed]
85. Sarosi, G.A.; Eckman, M.R.; Davies, S.F.; Laskey, W.K. Canine Blastomycosis as a Harbinger of Human Disease. *Ann. Intern. Med.* **1979**, *91*, 733. [CrossRef] [PubMed]
86. Tekin, A.; Pinevich, Y.; Herasevich, V.; Pickering, B.W.; Vergidis, P.; Gajic, O.; O'Horo, J.C. Diagnostic delay in pulmonary blastomycosis: A case series reflecting a referral center experience. *Infection* **2022**, 1–9. [CrossRef] [PubMed]
87. Poplin, V.; Smith, C.; Milsap, D.; Zabel, L.; Bahr, N. Diagnosis of Pulmonary Infections Due to Endemic Fungi. *Diagnostics* **2021**, *11*, 856. [CrossRef]
88. Wu, S.J.; Valyi-Nagy, T.; Engelhard, H.H.; Do, M.A.; Janda, W.M. Secondary Intracerebral Blastomycosis with Giant Yeast Forms. *Mycopathologia* **2005**, *160*, 253–257. [CrossRef]
89. Alpern, J.D.; Bahr, N.C.; Vazquez-Benitez, G.; Boulware, D.R.; Sellman, J.S.; Sarosi, G.A. Diagnostic Delay and Antibiotic Overuse in Acute Pulmonary Blastomycosis. *Open Forum Infect. Dis.* **2016**, *3*, ofw078. [CrossRef]
90. Carlos, W.G.; Rose, A.S.; Wheat, L.J.; Norris, S.; Sarosi, G.A.; Knox, K.S.; Hage, C.A. Blastomycosis in Indiana. *Chest* **2010**, *138*, 1377–1382. [CrossRef]
91. Martynowicz, M.A.; Prakash, U.B.S. Pulmonary Blastomycosis. *Chest* **2002**, *121*, 768–773. [CrossRef]
92. Kochar, P.S.; O Lath, C.; Klein, A.P.; Ulmer, J.L. Multimodality imaging in cranial blastomycosis, a great mimicker: Case-based illustration with review of clinical and imaging findings. *Indian J. Radiol. Imaging* **2016**, *26*, 120–125. [CrossRef] [PubMed]
93. Schwartz, I.S.; Kauffman, C.A. Blastomycosis. *Semin. Respir. Crit. Care Med.* **2020**, *41*, 31–41. [CrossRef] [PubMed]
94. Connolly, P.; Hage, C.A.; Bariola, J.R.; Bensadoun, E.; Rodgers, M.; Bradsher, R.W.; Wheat, L.J. Blastomyces dermatitidis Antigen Detection by Quantitative Enzyme Immunoassay. *Clin. Vaccine Immunol.* **2012**, *19*, 53–56. [CrossRef] [PubMed]
95. Frost, H.M.; Novicki, T.J. Blastomyces Antigen Detection for Diagnosis and Management of Blastomycosis. *J. Clin. Microbiol.* **2015**, *53*, 3660–3662. [CrossRef]
96. Bariola, J.R.; Hage, C.A.; Durkin, M.; Bensadoun, E.; Gubbins, P.O.; Wheat, L.J.; Bradsher, R.W. Detection of Blastomyces dermatitidis antigen in patients with newly diagnosed blastomycosis. *Diagn. Microbiol. Infect. Dis.* **2011**, *69*, 187–191. [CrossRef]
97. Theel, E.S.; Rodino, K.G.; Granger, D. Detection of Blastomyces dermatitidis Antigen in Urine Using a Commercially Available Quantitative Enzyme Immunoassay. *J. Clin. Microbiol.* **2021**, *59*, e0144421. [CrossRef]
98. O'Dowd, T.R.; Mc Hugh, J.W.; Theel, E.S.; Wengenack, N.L.; O'Horo, J.C.; Enzler, M.J.; Vergidis, P. Diagnostic Methods and Risk Factors for Severe Disease and Mortality in Blastomycosis: A Retrospective Cohort Study. *J. Fungi* **2021**, *7*, 888. [CrossRef]
99. Hage, C.A.; Davis, T.E.; Fuller, D.; Egan, L.; Witt, J.R.; Wheat, L.J.; Knox, K.S. Diagnosis of Histoplasmosis by Antigen Detection in BAL Fluid. *Chest* **2010**, *137*, 623–628. [CrossRef]
100. Cari, E.V.; Leedy, N.; Ribes, J.A.; Soria, J.; Myint, T. Risk factors of severe blastomycosis and comparison of diagnosis and outcomes between immunocompetent and immunocompromised patients. *Mycoses* **2021**, *65*, 239–246. [CrossRef]
101. Durkin, M.; Witt, J.; LeMonte, A.; Wheat, B.; Connolly, P. Antigen Assay with the Potential To Aid in Diagnosis of Blastomycosis. *J. Clin. Microbiol.* **2004**, *42*, 4873–4875. [CrossRef]
102. Richer, S.M.; Smedema, M.L.; Durkin, M.M.; Brandhorst, T.T.; Hage, C.A.; Connolly, P.A.; Leland, D.S.; Davis, T.E.; Klein, B.S.; Wheat, L.J. Development of a Highly Sensitive and Specific Blastomycosis Antibody Enzyme Immunoassay Using Blastomyces dermatitidis Surface Protein BAD-1. *Clin. Vaccine Immunol.* **2013**, *21*, 143–146. [CrossRef] [PubMed]
103. Sidamonidze, K.; Peck, M.K.; Perez, M.; Baumgardner, D.; Smith, G.; Chaturvedi, V.; Chaturvedi, S. Real-Time PCR Assay for Identification of Blastomyces dermatitidis in Culture and in Tissue. *J. Clin. Microbiol.* **2012**, *50*, 1783–1786. [CrossRef]
104. Wang, N.; Luo, Z.; Deng, S.; Li, Q. A young male with chronic nonproductive cough diagnosed with blastomycosis in China: A case report. *BMC Pulm. Med.* **2020**, *20*, 189. [CrossRef]
105. Recht, L.D.; Philips, J.R.; Eckman, M.R.; A Sarosi, G. Self-limited blastomycosis: A report of thirteen cases. *Am. Rev. Respir. Dis.* **1979**, *120*, 1109–1112. [CrossRef] [PubMed]
106. Sarosi, G.A.; Davies, S.F.; Phillips, J.R. Self-limited blastomycosis: A report of 39 cases. *Semin. Respir. Infect.* **1986**, *1*, 40–44. [PubMed]
107. Smith, J.A.; Kauffman, C.A. Blastomycosis. *Proc. Am. Thorac. Soc.* **2010**, *7*, 173–180. [CrossRef] [PubMed]
108. Rauseo, A.M.; Mazi, P.; Lewis, P.; Burnett, B.; Mudge, S.; Spec, A. Bioavailability of Single-Dose SUBA-Itraconazole Compared to Conventional Itraconazole under Fasted and Fed Conditions. *Antimicrob. Agents Chemother.* **2021**, *65*, e0013421. [CrossRef]
109. Mayne Pharma. Mayne Pharma Received FDA Approval of TolsuraTM (SUBA®-Itraconazole Capsules) for the Treatment of Certain Fungal Infections [ess Release]. Cision PR Newswire. 2018. Available online: https://www.prnewswire.com/news-releases/mayne-pharma-receives-fda-approval-of-tolsura-suba-itraconazole-capsules-for-the-treatment-of-certain-fungal-infections-300763882.html (accessed on 1 August 2022).
110. Pappas, P.G.; Bradsher, R.W.; Kauffman, C.A.; Cloud, G.A.; Thomas, C.J.; Campbell, J.G.D.; Chapman, S.W.; Newman, C.; Dismukes, W.E. The National Institute of Allergy and Infectious Diseases Mycoses Study Group Treatment of Blastomycosis with Higher Doses of Fluconazole. *Clin. Infect. Dis.* **1997**, *25*, 200–205. [CrossRef]
111. National Institute of Allergy and Infectious Diseases Mycoses Study Group. Treatment of blastomycosis and histoplasmosis with ketoconazole. Results of a prospective randomized clinical trial. *Ann. Intern. Med.* **1985**, *103 Pt 1*, 861–872. [CrossRef]
112. Freifeld, A.; Proia, L.; Andes, D.; Baddour, L.M.; Blair, J.; Spellberg, B.; Arnold, S.; Lentnek, A.; Wheat, L.J. Voriconazole Use for Endemic Fungal Infections. *Antimicrob. Agents Chemother.* **2009**, *53*, 1648–1651. [CrossRef]

113. Falci, D.R.; Da Rosa, F.B.; Pasqualotto, A.C. Comparison of nephrotoxicity associated to different lipid formulations of amphotericin B: A real-life study. *Mycoses* **2015**, *58*, 104–112. [CrossRef] [PubMed]
114. Meyer, K.C.; McManus, E.J.; Maki, D.G. Overwhelming Pulmonary Blastomycosis Associated with the Adult Respiratory Distress Syndrome. *N. Engl. J. Med.* **1993**, *329*, 1231–1236. [CrossRef]
115. Lahm, T.; Neese, S.; Thornburg, A.T.; Ober, M.D.; Sarosi, G.A.; Hage, C.A. Corticosteroids for Blastomycosis-Induced ARDS. *Chest* **2008**, *133*, 1478–1480. [CrossRef]
116. Limper, A.H.; Knox, K.S.; Sarosi, G.A.; Ampel, N.M.; Bennett, J.E.; Catanzaro, A.; Davies, S.F.; Dismukes, W.E.; Hage, C.A.; Marr, K.A.; et al. An Official American Thoracic Society Statement: Treatment of Fungal Infections in Adult Pulmonary and Critical Care Patients. *Am. J. Respir. Crit. Care Med.* **2011**, *183*, 96–128. [CrossRef]
117. Felton, T.; Troke, P.F.; Hope, W.W.; Simonsen, K.A.; Anderson-Berry, A.L.; Delair, S.F.; Dele Davies, H. Tissue Penetration of Antifungal Agents. *Clin. Microbiol. Rev.* **2014**, *27*, 68–88. [CrossRef] [PubMed]
118. Wheat, L.J.; Freifeld, A.G.; Kleiman, M.B.; Baddley, J.W.; McKinsey, D.S.; Loyd, J.; Kauffman, C.A. Clinical Practice Guidelines for the Management of Patients with Histoplasmosis: 2007 Update by the Infectious Diseases Society of America. *Clin. Infect. Dis.* **2007**, *45*, 807–825. [CrossRef]
119. Johnson, M.D.; Perfect, J.R. Fungal Infections of the Bones and Joints. *Curr. Infect. Dis. Rep.* **2001**, *3*, 450–460. [CrossRef]
120. Groll, A.H.; Giri, N.; Petraitis, V.; Petraitiene, R.; Candelario, M.; Bacher, J.S.; Piscitelli, S.C.; Walsh, T.J. Comparative Efficacy and Distribution of Lipid Formulations of Amphotericin B in Experimental *Candida albicans* Infection of the Central Nervous System. *J. Infect. Dis.* **2000**, *182*, 274–282. [CrossRef]
121. Ward, B.A.; Parent, A.D.; Raila, F. Indications for the surgical management of central nervous system blastomycosis. *Surg. Neurol.* **1995**, *43*, 379–388. [CrossRef]
122. Pappas, P.G.; Pottage, J.C.; Powderly, W.G.; Fraser, V.J.; Stratton, C.W.; McKenzie, S.; Tapper, M.L.; Chmel, H.; Bonebrake, F.C.; Blum, R.; et al. Blastomycosis in Patients with the Acquired Immunodeficiency Syndrome. *Ann. Intern. Med.* **1992**, *116*, 847–853. [CrossRef]
123. Pappas, P.G.; Threlkeld, M.G.; Bedsole, G.D.; Cleveland, K.O.; Gelfand, M.S.; Dismukes, W.E. Blastomycosis in Immunocompromised Patients. *Medicine* **1993**, *72*, 311–325. [CrossRef] [PubMed]
124. McBride, J.A.; Sterkel, A.K.; Matkovic, E.; Broman, A.T.; Gibbons-Burgener, S.N.; Gauthier, G.M. Clinical Manifestations and Outcomes in Immunocompetent and Immunocompromised Patients With Blastomycosis. *Clin. Infect. Dis.* **2020**, *72*, 1594–1602. [CrossRef] [PubMed]
125. Lemos, L.B.; Soofi, M.; Amir, E. Blastomycosis and pregnancy. *Ann. Diagn. Pathol.* **2002**, *6*, 211–215. [CrossRef]
126. Scolarici, M.J.; King, C.; Sterkel, A.; Smith, J.; Gauthier, G.; Saddler, C. The Role of Isavuconazonium Sulphate for the Treatment of Blastomycosis: A Case Series and Antifungal Susceptibility. *Open Forum. Infect. Dis.* **2022**, *9*, ofac220. [CrossRef]
127. Mehta, T.I.; Kurman, J.; Dolan, S.; Gill, H.; Thapa, B. Blastomycosis in Solid Organ Transplant Recipients—A Retrospective Series from Southeastern Wisconsin. *Transpl. Infect. Dis.* **2021**, *123*, e13671. [CrossRef]

Journal of
Fungi

MDPI

Review

Current Progress on Epidemiology, Diagnosis, and Treatment of Sporotrichosis and Their Future Trends

Anderson Messias Rodrigues [1,2,*], Sarah Santos Gonçalves [3], Jamile Ambrósio de Carvalho [1,2], Luana P. Borba-Santos [4], Sonia Rozental [4] and Zoilo Pires de Camargo [1,2]

1 Laboratory of Emerging Fungal Pathogens, Department of Microbiology, Immunology, and Parasitology, Discipline of Cellular Biology, Federal University of São Paulo (UNIFESP), Sao Paulo 04023062, Brazil; jamileambrosio@hotmail.com (J.A.d.C.); zpcamargo1@gmail.com (Z.P.d.C.)
2 Department of Medicine, Discipline of Infectious Diseases, Federal University of São Paulo (UNIFESP), Sao Paulo 04023062, Brazil
3 Infectious Diseases Postgraduate Program, Center for Research in Medical Mycology, Federal University of Espírito Santo (UFES), Vitoria 29043900, Brazil; sarahunifesp@yahoo.com.br
4 Cell Biology and Parasitology Program, Institute of Biophysics Carlos Chagas Filho, Federal University of Rio de Janeiro (UFRJ), Rio de Janeiro 21941902, Brazil; luana.p.borba@gmail.com (L.P.B.-S.); rozental@biof.ufrj.br (S.R.)
* Correspondence: amrodrigues@unifesp.br; Tel.: +55-1155764551 (ext. 1540)

Abstract: Sporotrichosis, a human and animal disease caused by *Sporothrix* species, is the most important implantation mycosis worldwide. *Sporothrix* taxonomy has improved in recent years, allowing important advances in diagnosis, epidemiology, and treatment. Molecular epidemiology reveals that *S. brasiliensis* remains highly prevalent during the cat-transmitted sporotrichosis outbreaks in South America and that the spread of *S. brasiliensis* occurs through founder effects. *Sporothrix globosa* and *S. schenckii* are cosmopolitan on the move, causing major sapronoses in Asia and the Americas, respectively. In this emerging scenario, one-health approaches are required to develop a creative, effective, and sustainable response to tackle the spread of sporotrichosis. In the 21st century, it has become vital to speciate *Sporothrix*, and PCR is the main pillar of molecular diagnosis, aiming at the detection of the pathogen DNA from clinical samples through multiplex assays, whose sensitivity reaches remarkably three copies of the target. The treatment of sporotrichosis can be challenging, especially after the emergence of resistance to azoles and polyenes. Alternative drugs arising from discoveries or repositioning have entered the radar of basic research over the last decade and point to several molecules with antifungal potential, especially the hydrazone derivatives with great in vitro and in vivo activities. There are many promising developments for the near future, and in this review, we discuss how these trends can be applied to the *Sporothrix*-sporotrichosis system to mitigate the advance of an emerging and re-emerging disease.

Keywords: *Sporothrix brasiliensis*; *Sporothrix schenckii*; *Sporothrix globosa*; sporotrichosis; implantation mycosis; subcutaneous mycosis; epidemiology; treatment; antifungal; diagnosis

Citation: Rodrigues, A.M.; Gonçalves, S.S.; de Carvalho, J.A.; Borba-Santos, L.P.; Rozental, S.; Camargo, Z.P.d. Current Progress on Epidemiology, Diagnosis, and Treatment of Sporotrichosis and Their Future Trends. *J. Fungi* **2022**, *8*, 776. https://doi.org/10.3390/jof8080776

Academic Editor: Alessandro C. Pasqualotto

Received: 29 June 2022
Accepted: 16 July 2022
Published: 26 July 2022

Publisher's Note: MDPI stays neutral with regard to jurisdictional claims in published maps and institutional affiliations.

1. A Brief Introduction to the System *Sporothrix*-Sporotrichosis

Sporotrichosis is a subcutaneous mycosis caused by the dimorphic fungus *Sporothrix schenckii* and related species, which are found worldwide in vegetation, decaying organic matter, *Sphagnum* moss, and soil [1]. Sporotrichosis is transmitted through traumatic inoculation of *Sporothrix* propagules into skin tissue [2]. The classical transmission route refers to sapronosis (i.e., environment to warm-blooded vertebrate host). Therefore, it is an occupational mycosis usually associated with trauma during outdoor work in gardeners, farmers, extractivist, and florists, among others. The alternative route of infection is related to horizontal animal transmission, mainly affecting domestic cats and armadillos [3,4]. In the cat-transmitted sporotrichosis, these animals spread the disease through scratches and bites or direct contact with their secretions to other cats, causing epizootics, or directly to humans (zoonosis) [5].

Most cases of human sporotrichosis manifest in the skin and subcutaneous tissues. The disease may vary according to the immune status of the infected host, with the lymphocutaneous form being the most common manifestation (~80% of cases) [6]. The fungus spread to bones and viscera is uncommon and occurs more frequently in immunosuppressed patients, especially in AIDS [7,8]. Pulmonary sporotrichosis, resulting from the inhalation of fungal propagules (conidia or yeasts), is uncommon [9].

Cats are the most susceptible hosts to contamination by *Sporothrix* and commonly develop the most severe forms of the disease, which can progress to death [10]. Multiple ulcerative lesions are usually observed in the cephalic region, mainly in the nose and paw region, due to feline behavior that involves scratching and biting during fights [10–12]. Sporotrichosis is higher among adult male cats, without owners, and those not neutered [10]. Different from what occurs in lesions in humans, a high number of yeasts can be observed in felines [13].

Spontaneous cure of human and animal sporotrichosis is rare, and treatment with antifungals is indispensable for most patients. Although localized sporotrichosis is readily treated, managing osteoarticular sporotrichosis, disseminated visceral forms, and feline sporotrichosis is laborious [2,14,15].

For over a century, *S. schenckii sensu lato* was described as the sole agent of human and animal sporotrichosis [16,17]. However, advances in molecular taxonomy revealed that it is not a monotypic taxon [18]. *Sporothrix* comprises approximately 53 species [15], including *S. brasiliensis*, *S. schenckii*, *S. globosa*, and *S. luriei*, forming a clade of clinical interest as they are frequently recovered from cases of sporotrichosis. The other *Sporothrix* species are embedded in the environmental clade and show little or no virulence to the warm-blooded vertebrate host [19]. Strictly environmental *Sporothrix* species are often associated with soil, insects, and plants. However, we highlight the members of the *S. pallida* complex (*S. chilensis*, *S. gemella*, *S. humicola*, *S. mexicana*, *S. pallida*, *S. palmiculminata*, *S. protea-sedis*, and *S. stylites*) which include soil-inhabitants fungi with mild-pathogenic potential for humans and animals (Figure 1) [20].

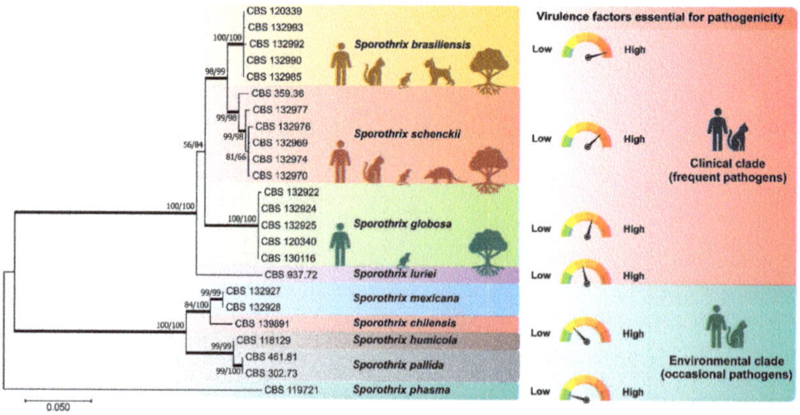

Figure 1. Phylogenetic analysis of the main members of medical relevance in the genus *Sporothrix* using sequences from the partial calmodulin-encoding gene (exons 3–5) and the ITS region (ITS1/2+5.8s). In the clinical clade, *S. brasiliensis* is highly virulent for the warm-blooded vertebrate host, followed by *S. schenckii*, *S. globosa*, and *S. luriei*. In the environmental clade, *S. chilensis*, *S. humicola*, *S. mexicana*, and *S. pallida* are occasional pathogens with mild-pathogenic potential to mammals. *Sporothrix phasma*, a species with no virulence to mammals, was used as an outgroup in the phylogenetic analysis. Numbers close to the branches represent bootstraps values (ML/NJ).

The differential pathogenicity in *Sporothrix* may be related to the efficiency in the temperature-induced morphological transition. Thermal dimorphism is an important morphological adaptation for infection, shared with other human pathogens, phylogenetically distant in the Onygenales and Eurotiales [21]. *Sporothrix* species nested in the clinical clade are 'professional' thermodimorphic fungi responding more efficiently to thermal stimuli. In addition, *S. brasiliensis* express important virulence attributes such as thermotolerance, adhesins, and melanin [22], being the most virulent species in murine models such as BALB/c [23,24], C57BL/6 [25], and OF-1 mice [26]. Such exacerbated virulence in animals is also observed in the human host, and *S. brasiliensis* is associated with atypical [27–29] and more severe forms of the disease, including disseminated skin infection in immunocompetent hosts and systemic disease [2,29–31].

2. Trends in the Epidemiology of *Sporothrix* Species

Sporotrichosis is a cosmopolitan mycosis whose etiological agents are constantly on the move. Species of clinical interest are not evenly distributed worldwide, and many are associated with different transmission routes [32]. No official statistics show the burden of human and animal sporotrichosis globally, but only case series denounce the problem. Therefore, it is a fact that sporotrichosis has classically been a mysterious disease from an epidemiological point of view.

Notoriously, the history of sporotrichosis shows repeated manifestations in the form of outbreaks and epidemics. The most famous epidemic occurred in the mid-1940s in South Africa, where more than 3000 native Bantu miners were infected with *Sporothrix* growing in the soil and the supporting timbers of the Witwatersrand gold mines [33–35]. Recently, to a lesser extent, a new case series was described in South Africa, showing that the fungus can persist in nature for decades. In South Africa, sapronotic transmission of the disease predominates, where *S. schenckii s. str.* is the main agent identified by molecular methods [36–41]. Animal sporotrichosis is rare in Africa [42], and environmental isolation shows the presence of members of the *S. pallida* complex [36]. Data on the occurrence of sporotrichosis on the African continent are scarce, being reported mainly in Madagascar [43], Zimbabwe [44], Nigeria [45], and Sudan [46]. In these areas, there is no correlation between the incidence of sporotrichosis and the HIV/AIDS epidemic. Despite a discreet series, the reports from the 1940s were fundamental for understanding the sapronotic route of the disease, clarifying, for the first time, the ecoepidemiological aspects of sporotrichosis [33–35].

On the Asian continent, epidemiological data arise mainly from Japan, China, India, and Malaysia, where there is a predominance of cases of human sporotrichosis due to *S. globosa*. Human sporotrichosis is endemic in India, occurring with high prevalence in the northern sub-Himalayan region, from Himachal Pradesh in the northwest to Assam and West Bengal in the east [47–51]. Historically, sporotrichosis was very common in Japan between the 1940s and 1980s, with significant remission since then [52–55]. In this country, molecular epidemiology was significantly influenced by RFLPs and PCR-RFLPs analysis of mtDNA, which revealed two main clades, groups A and B [54–58]. Currently, the reinterpretation of epidemiological data in light of taxonomic changes in *Sporothrix* confirms that *S. globosa* (group B) was the main agent of human sporotrichosis in Japan in the 1980s, followed by *S. schenckii s. str.* (group A) [59,60].

The incidence of human sporotrichosis in China is among the highest globally [61–65]. Cases are concentrated in the northeast region of China, in an area with a temperate continental monsoon climate, including Jilin, Liaoning, and Heilongjiang provinces [64–67]. Interestingly, there is a higher incidence of cases during the winter, possibly associated with contamination of the home environment with wood, twigs, and sticks used as an important energy matrix for cooking and heating [65,68]. Therefore, similar to epidemics in Africa, in Asia, sporotrichosis is a sapronosis whose main transmission route is the traumatic inoculation of plant material. However, unlike what happens in Africa, the etiological agent is *S. globosa* [3]. The exception to the rule is Malaysia, where *S. schenckii s. str.* can cause epizootics in domestic cats, increasing zoonotic transmission levels [69–71].

Human sporotrichosis was common on the European continent at the beginning of the last century. The cases emerged mainly in France and were richly reported in the literature [72–74]. However, the disease has decreased its incidence considerably since then, with rare case reports mainly from the United Kingdom, Spain, and Italy [18,75–79]. Notwithstanding, with the introduction of *S. brasiliensis* in England [80], our attention should be focused on the evolution of the number of cases in the coming years.

In Australia, sporadic reports of human sporotrichosis do not exceed a few hundred infections. Australian cases are generally associated with *S. schenckii s. str.* and *S. globosa* following a sapronotic route, mainly in Queensland, New South Wales, and Western Australia [81–83]. In these areas, outbreaks of sporotrichosis attributed to environmental sources such as hay are not uncommon. Feline sporotrichosis is rare in Australia, as described in the mid-1980s [84]. New cases have now been associated with *S. pallida* in cats, an even rarer association [85].

Sporotrichosis is relatively common in the Americas. In the USA, where the disease was first described in 1898 [16], *S. schenckii s. str.* causes illness in professionals linked to agricultural activities such as rose gardeners and farmers. The largest reported outbreak in the USA occurred in 1988 and affected 84 patients in 15 states exposed to the fungus in mosses of the genus *Sphagnum*, used in gardening procedures [86,87]. In the USA, between 2000 and 2013, 1471 hospitalizations were reported in patients with opportunistic conditions such as HIV/AIDS, immune-mediated inflammatory diseases, and chronic obstructive pulmonary disease [88].

In Latin America, sporotrichosis emerges as the most common implantation mycosis [89,90], with areas of high endemicity in Brazil [91], Colombia [92], Peru [93,94], and Venezuela [95]. However, we noticed significant differences that reflect on the species transmitted and on the route of transmission of the disease. For example, the sapronotic route of sporotrichosis is common throughout Latin America, where *S. schenckii s. str.* and *S. globosa* are spread through contact with fungal propagules present in the environment. On the other hand, the zoonotic route of sporotrichosis is more common in Brazilian territory, where cat-transmitted sporotrichosis is the main type of infection for humans, dogs, and other cats, and *S. brasiliensis* is the major agent in these cases [5].

Case reports of human sporotrichosis occur in 25 of the 26 Brazilian states [3,91,96–100]. However, due to the emergence of sporotrichosis in cats, there is a marked temporal variation concerning the succession of species involved in transmissions (Figure 2). Before the 1990s, the classical sapronotic transmission of human sporotrichosis prevailed, similar in Latin American countries [91]. After the 1990s, with the entry of the domestic cat into the sporotrichosis transmission chain, it is possible to detect a considerable increase in epizootic manifestations in felines and zoonotic transmission to humans. This scenario has the metropolitan region of Rio de Janeiro as its epicenter, and between the 1990s and 2000s we observed a gradual spread of the epidemic to other states in the South and Southeast regions. Recently, in full expansion, we described the emergence of *S. brasiliensis* in the northeast region of the country [91,101], mainly in the states of Pernambuco [102], Paraíba [103], and the Rio Grande do Norte [31,104,105]. Interestingly, in these areas of feline sporotrichosis, *S. schenckii s. str.* is no longer the main transmitted species, and *S. brasiliensis* becomes the main agent during the cat-transmitted sporotrichosis events (Figure 2).

Judging from a public health point of view, the major drawback in the above scenario is the absence of a national notification system to report disease cases. Since 2011, the notification of sporotrichosis has been mandatory in the State of Rio de Janeiro, but not in other Brazilian states, with rare specific exceptions at regional and municipal levels [13,100,106–108].

In general terms, the geographic fluctuation of sporotrichosis agents is fascinating. *Sporothrix globosa* is the predominant molecular type in Asia. *S. schenckii s. str.* is highly prevalent in Australia, South Africa, western South and Central America, and North

America. *Sporothrix brasiliensis* is highly prevalent in Brazil. Among all medically relevant *Sporothrix* species, *S. brasiliensis* has the greatest potential for geographic dispersal. In areas where it occurs, *S. brasiliensis* easily outperforms other clinically relevant species due to feline transmission [101]. To date, Argentina and Paraguay have reported the occurrence of *S. brasiliensis* in humans and cats outside Brazil [109–111], and there are suspected cases in Bolivia, Colombia, and Panama [112,113]. Recently, a zoonotic case was reported in the UK after a veterinarian treated a cat with sporotrichosis imported from Brazil [80]. In the absence of official epidemiological data, case reports show the importance of the disease, its local and regional escalation, and the urgent need to establish sanitary barriers to mitigate the advance of *S. brasiliensis* and the cat-transmitted sporotrichosis.

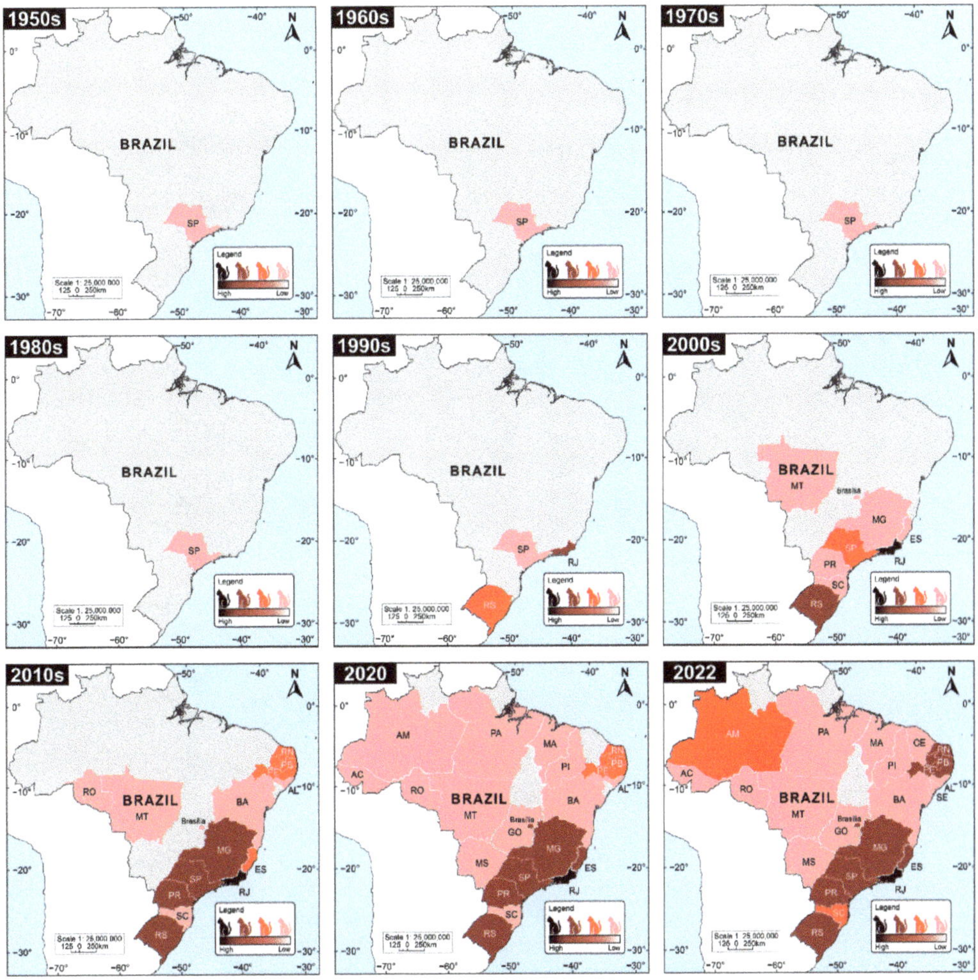

Figure 2. Temporal evolution of feline sporotrichosis cases in Brazil between 1950 and 2022. The current scenario of sporotrichosis shows signs of frank expansion. The map was drawn based on case reports available on the literature [5,10,12,15,31,32,68,91,98–101,105,108,114–153].

In feline sporotrichosis, it is generally accepted that a single diseased cat introduced to a new location can trigger an outbreak that will quickly evolve into an epidemic. A new dissemination area may include locations as close as a neighborhood, a new city, or even more distant areas such as other states. Introducing diseased animals to new areas has occurred repeatedly within the natural history of cat-transmitted sporotrichosis. The metropolitan region of Rio de Janeiro is described as the possible center of origin, from where the disease initially spread to other border states in the southeastern region (e.g., São Paulo, Espírito Santo, and Minas Gerais) and later to the southern region. (e.g., Paraná). The most recent migration event occurred towards the Brazilian northeast in mid-2015 (Figure 3).

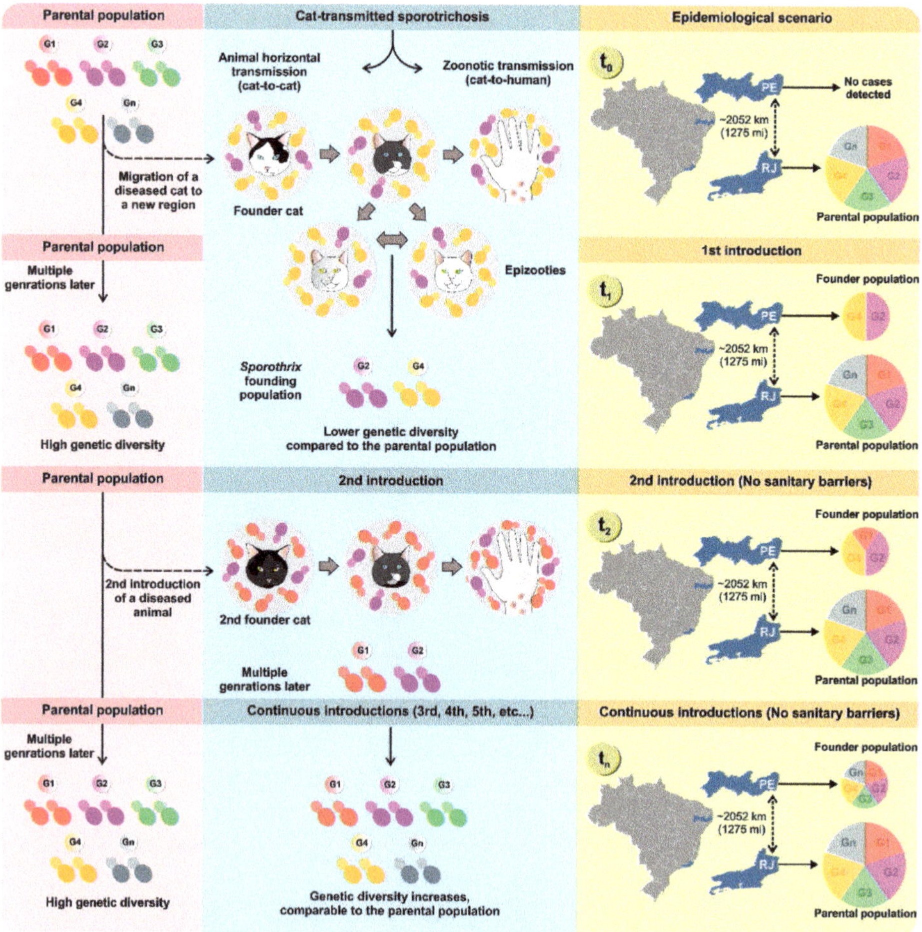

Figure 3. Founder effect events explain the expansion dynamics of cat-transmitted sporotrichosis. The molecular studies developed by de Carvalho et al. [101] offer new bases for proposing public policies to mitigate sporotrichosis. The *S. brasiliensis* genotypes (e.g., G1, G2, G3, G4, G5, Gn) infect cats living in the metropolitan region of Rio de Janeiro and are considered the parental population. Eventually, a sick cat infected with a single or a group of genotypes (e.g., G2 and G4) is taken to a new area (e.g., Pernambuco), where it will establish a founder population, transmitting *S. brasiliensis* to other cats (epizootics) or humans (zoonoses). A study of genetic diversity at time one (t1) will reveal that the founder population has less genetic diversity when compared to the parental population. However, the absence of sanitary barriers and the continuous exchange of diseased animals taken by their tutors from the parental-to-founder population will gradually (t2, t3, t4, tn) reconstitute the genetic diversity in the founding population and accelerate the pace of diversification.

De Carvalho and colleagues reported that cat-transmitted sporotrichosis progresses through founder effects (Figure 3) [101]. In population genetics, a founder effect refers to the reduction in genetic variability that occurs when a small group of individuals not genetically representative of the parental population migrates to a new area and establishes a new population. Over time, the resulting new subpopulation will have genotypes and phenotypic characteristics similar to the founding individual, which may differ greatly from the parent population. Therefore, a founder effect may explain, for example, the low genetic diversity found during the initial outbreaks of cat-transmitted sporotrichosis. However, the absence of sanitary barriers and the constant introduction of sick animals (parental population → founder population) to new areas can reconstitute genetic diversity in the founder population, leading to comparable genetic diversity.

Cats adapt to a wide range of environments, and in general, the roaming area of domestic cats (0.02–10 ha) overlaps with the human residential area, where they can more easily secure food [154–156]. Therefore, the only viable hypothesis to justify the detection of a genotype from the parental population of Rio de Janeiro in areas as remote as the state of Pernambuco in the northeast region (>2000 km) is the introduction of sick cats via humans who migrate with their pets, since sporotrichosis is not a disease of direct person-to-person transmission or even a zooanthroponosis. This hypothesis may also explain the introduction of sick cats in other South American countries, such as Argentina and Paraguay [109–111], or even the European continent, as recently reported in England [80]. Establishing sanitary barriers to contain the migration of sick felines is a fundamental measure for controlling the expansion of *S. brasiliensis* (Figure 3).

The recent outbreaks in Brazil due to cat-transmitted sporotrichosis and the widespread expansion in South America are important reminders of how human and non-human health are essentially connected. Animals are the source of 70% of emerging and re-emerging infectious disease threats to human health and more than half of all recognized human pathogens [157–160]. We observed that the recent entry of the domestic cat into the transmission chain of sporotrichosis associated with the emergence of *S. brasiliensis*, a more virulent species adapted to animal transmission, produced a significant revolution in the classical epidemiological pattern [101], confirming that such threats are dynamic [161].

Although the absence of official and reliable data makes it difficult to measure the problem, cat-transmitted sporotrichosis is responsible for a significant burden of sporotrichosis in Brazil [15]. Geoepidemiological analyses of zoonotic sporotrichosis cases in Rio de Janeiro, Brazil, reveal that the social determinants of the disease are linked to social vulnerability. The disease mainly affects women (25 to 59 years old), especially in socioeconomically disadvantaged neighborhoods of Rio de Janeiro, expressed by low per capita income and deficient supply of treated water to households [162]. This shows that sporotrichosis can be aggravated in scenarios of greater social vulnerability [160,163], a regretful development that tends to be repeated in other areas of the country [133,164].

Therefore, a public health problem that encompasses human, animal, and environmental health issues requires solutions based on one-health approaches (Figure 4). One-health approaches consider the interactions among different spheres of global health to develop a creative, effective, and sustainable response. Therefore, interdisciplinary research is mandatory, as is interventionist practice at local, national, and international levels, involving public managers, physicians, veterinarians, biologists, public and animal health authorities, environmental health agents, and microbiologists, among other allies.

The current Brazilian environmental scenario results from climate change, intense deforestation, and biodiversity loss [165]. The Intergovernmental Panel on Climate Change's sixth assessment report reveals that a 1.5 °C rise in global temperature would result in a 100–200% increase in the population affected by floods in Colombia, Brazil, and Argentina, 300% in Ecuador, and 400% in Peru (medium confidence) [166]. Higher temperatures, heavy rainfall, and flooding are associated with an increase in emerging zoonotic diseases [166,167]. Medically relevant *Sporothrix* can be detected in the soil of endemic areas where it remains for years [168,169], and it is interesting to hypothesize that water from floods and inundations, a phenomenon increasingly common in Brazil, may also promote the diffusion of *Sporothrix* propagules in the soil.

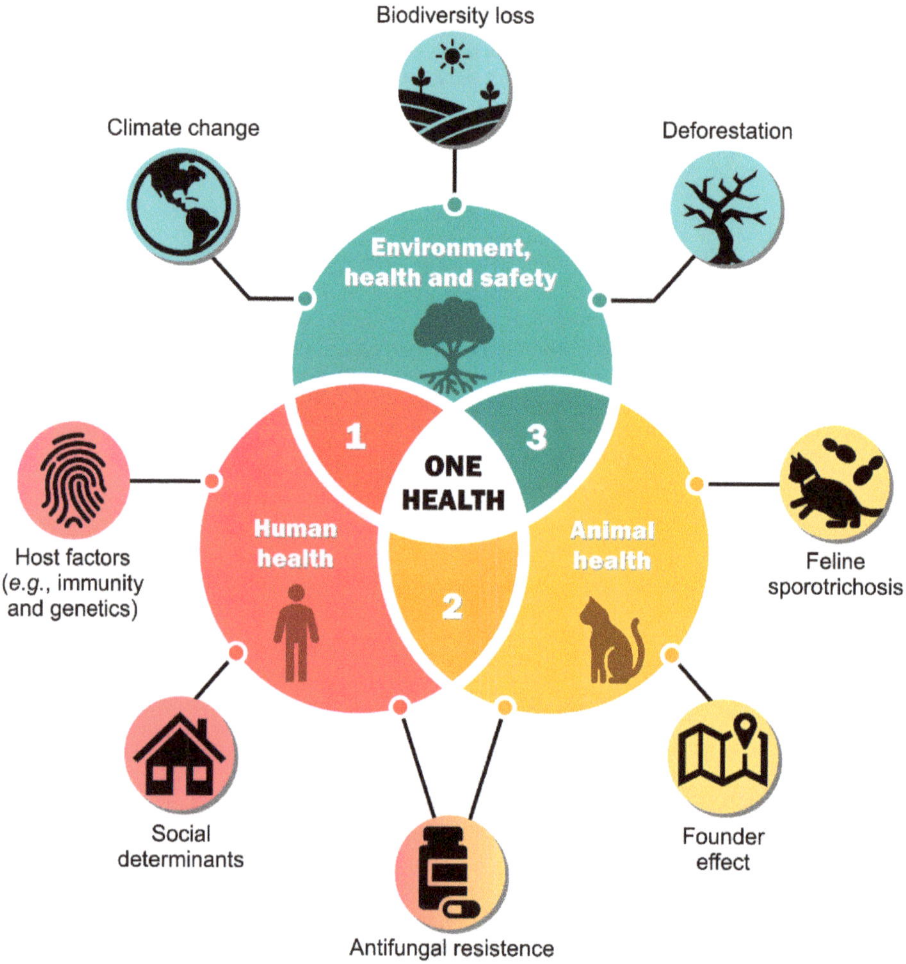

Figure 4. A one-health approach to mitigating the spread of cat-transmitted sporotrichosis considers human health (1), animal health (2), and environmental health and safety (3).

In this chaotic scenario, we incorporate the fact that in endemic areas, cats with sporotrichosis are often buried directly in the soil, producing latent foci of the pathogen. Soil functions as the main reservoir of fungal propagules and certainly does not act as a passive reservoir. The advance of deforestation in the Amazon, Cerrado, and Atlantic Forest is worrying, leading to biodiversity loss in these biomes. It is known that soil microbial composition can be altered due to deforestation, reflecting co-occurrence patterns among microorganism taxa, leading to ecological imbalance [170,171]. For example, soil amoebas (e.g., *Acanthamoeba castellanii*) rapidly change the composition of the bacterial community in the soil [172], and it is well known that many of these protozoa interact with *Sporothrix*, predating the microorganism in the soil [173,174]. Therefore, it is expected that environmental stresses (e.g., higher temperatures, humidity, pH, etc.) that affect *A. castellanii* in the soil [175], leading to alterations in biodiversity, may reflect population imbalances in the fluctuation of *Sporothrix* species (Figure 4).

These scenarios demonstrate that one-health solutions are complex to implement in their totality, yet they are crucial to combat the spread of emerging *Sporothrix* species (Figure 4) [176].

3. Trends in the Diagnosis of Sporotrichosis

Sporotrichosis can be diagnosed through a correlation of clinical, epidemiological, and laboratory data [32]. The early and accurate laboratory diagnosis of sporotrichosis is of substantial importance since the clinical aspect of cutaneous lesions can mimic other dermatologic manifestations, such as mycobacteriosis, actinomycosis, American tegumentary leishmaniasis, blastomycosis, cryptococcosis, paracoccidioidomycosis, among others [2]. In addition, ulcerative lesions can mimic pyoderma gangrenosum [177]. Clinical suspicion and the patient's epidemiological context are key to assembling this puzzle and thus promptly establishing the diagnosis.

The diagnosis of sporotrichosis is based on the isolation and identification of *Sporothrix* in culture, cytopathology, histopathology, sporotrichosis skin test, serology, immunohistochemistry, and molecular techniques [10,15,78,116,178]. Moreover, complementary laboratory tests such as blood count and biochemical profile should be requested in systemic forms. Anemia, neutrophilic leukocytosis, gammopathies, and hypoalbuminemia are commonly observed [2,7].

3.1. Mycological Test

Currently, the reference method for sporotrichosis diagnosis remains the isolation and identification of microorganisms in culture media and characterization of the agent by morphological parameters [179–182]. Although it has been widely used in clinical routine, the method sensitivity is not 100%, which can generate false-negative results due to contamination of the samples with bacteria and anemophilous fungi and inadequate transport of the material [181–183]. While considered a low-cost diagnosis, it is noteworthy that it is not suitable for diagnosing atypical and extracutaneous forms of the disease [184].

Depending on the clinical form and laboratory approaches, several biological samples can be investigated. The most frequent specimens are tissue fragments, serosanguineous exudates, purulent secretion, scraping of hyperkeratotic crusts, aspirate of lymph nodes, and organ fragments obtained during necropsy [5,185]. It must be emphasized that some care must be observed not to attenuate the technique sensitivity, such as material transport temperature, time storage, and appropriate clinical sample processing [186].

To ensure the fungus viability, some criteria should be pursued, including (i) seed the material as soon as possible in culture media; (ii) transport swabs on Stuart media; (iii) preserve tissue fragment biopsies in sterile saline solution (mycological tests) or formalin solution (histological examinations). Furthermore, the sooner the biological material is seeded in culture media, the greater the chances of recovery of the microorganism. Otherwise, amid eventualities, the clinical sample should be kept at 4 °C for a time less than 8–10 h [130].

The average growth time of *Sporothrix* spp. in mycelial form (25 °C) is 3–5 days to two weeks [187]. The isolates obtained can be accurately identified from positive cultures, and antifungal susceptibility testing in vitro and other assays can be performed [2]. Macroscopically, in media such as malt extract agar (MEA) or potato dextrose agar (PDA), the colonies start to grow as hyaline filamentous fungi and then turn brown to black after a few days, mainly in the colony's center [118,188,189]. Sometimes the mycelium can grow entirely white, designated as an albino [118,188,189]. The *Sporothrix* colonies' diameter is smaller, ranging from 19 to 41 mm, compared to other filamentous fungi, such as *Aspergillus fumigatus* (65–70 mm) [118,190]. Microscopically, the mycelial form is observed as septate, thin, branched, and hyaline hyphae, with conidiogenous cells arising from undifferentiated hyphae, forming conidia thick-walled (hyaline or brown), small (2–3 × 3–6µm), with a different arrangement, such as sympodial form, appearing in small groups of denticles in a slight apical dilation of the conidiophore or as sessile [118,188,189].

Sporothrix is a thermodimorphic fungus. To ensure its identification, it is recommended to stimulate the morphological transition. The fungus should be seeded in enriched media such as blood glucose-cysteine agar or Brain Heart Infusion agar (BHI) and subsequent incubation at 35–37 °C to obtain the yeast form [177,187]. The growth time is the same as the mycelial form. In some cases, the isolates grow slowly, so the fungi should be incubated for up to 30 days for the outcome [181,191,192]. In yeast form, the colonies are tan or cream-colored and smooth. Micromorphologically, it is possible to observe spindle-shaped and oval cells measuring 2.5–5 μm in diameter, like cigar-shaped buds on a narrow base [15,118,181,192].

Although the phenotypic characterization is not distinctly effective to speciate *Sporothrix*, it allows the presumptive identification of some species belonging to the clinical clade. Some characteristics, such as the color and shape of conidia, suggest some clinical species, such as *S. brasiliensis* and *S. schenckii s. str.* [78,193]. The latter predominates more elongated conidia, and some isolates showed triangular pigmented conidia, thereof being characteristic of the species. Concerning *S. brasiliensis*, the conidia are mostly more globose dematiaceous conidia, yet they may have the presence or absence of melanin [2,78,118]. Micromorphological characteristics may also hint at identifying *S. chilensis*, *S. mexicana*, and *S. pallida*, commonly environmental species, with mid-pathogenic potential for humans [20,78]. All these aspects are subtle, and their variations can lead to errors in identifying the species [75,98,118,194].

Marimon et al. [78] and Rudramurthy et al. [51] demonstrated that physiological characteristics such as growth rate, thermotolerance, and sugar assimilation might be helpful in the differentiation of morphologically similar species embedded in the clinical clade. Besides, some studies imply that medically relevant *Sporothrix* can be distinguished from environmental, according to the analysis of growth rates and thermotolerance [195]. Although not entirely elucidated, clinical strains probably have specific characteristics acquired during their evolution that undoubtedly contributed to their pathogenicity [195].

Toward accurately speciating *Sporothrix*, the polyphasic approach is most advisable for assembled morphological, physiological, molecular, and ecological characteristics [19,98]. Whereas culture remains a reference in the sporotrichosis diagnosis, the delayed results may impact the severity and compromise the disease treatment, decreasing the probability of cure in cats and humans and improving the transmission risk.

3.2. *Direct Microscopic, Cytopathological, and Histopathological Examinations*

Biological material obtained from human skin lesions and tissue fragments has a low fungal burden, and the yeast size (2–6 μm) hinders their visualization in the direct microscopic examination (DME) of fresh material. Thus, DME of samples treated with potassium hydroxide solution (KOH, 10–30%) or with fast staining techniques should not be recommended [185,196,197]. However, the DME shows better positivity in immunosuppressed patients [2,198]. The DME specificity and sensitivity of the tissue samples are still unknown, as most investigators consider this tool inefficacious [177].

According to Orofino-Costa et al. [2], purulent secretion imprints or biopsies stained with Giemsa increase the test's sensitivity in humans. In extracutaneous forms, the DME sensitivity is even lower; occasionally, it is possible to observe the fungal structures in cigar or shuttles forms [199].

Cytopathological examination stained with periodic acid-Schiff (PAS) or Gomori-methenamine silver (GMS), the aspiration puncture of the lesions, especially in the extracutaneous and disseminated forms, eventually allows the observation of granuloma of epithelioid cells, asteroid bodies, and yeast cells [177]. In 20% of cases, asteroid bodies may be observed in the center of the granuloma [196,200,201]. Asteroid bodies are globular or oval yeast cells surrounded by radiated eosinophilic material (Splendore-Hoeppli reaction), including antibodies (IgM and IgG), with the role of defense against phagocytes [32,202].

Findings of asteroid bodies in the stained cytopathological examination are very unpredictable since it depends on the staining method and displays minor reproducibility. It is not a sporotrichosis pathognomonic structure, as it may be observed in other granulomatous and infectious diseases [177,185,196,203]. Gram, Giemsa, PAS, and GMS may be successfully used in disseminated manifestations [200,201].

Conversely, Gram, quick Panoptic, Wright, Giemsa, or Rosenfeld cytopathological staining techniques are more sensitive in animals, particularly felines (Figure 5A) [116]. The cytopathological examination from exudates and skin lesions shows a high fungal burden, making it possible to observe *Sporothrix* yeast cells that range from being rounded, oval, or cigar-shaped, surrounded by a transparent, capsule-like halo, as seen in *Cryptococcus* spp. and *Histoplasma* spp. [10,32,179]. These structures may be disposed of inwardly in macrophages, neutrophils, and multinucleated or free giant cells [130,131]. Concerning feline sporotrichosis, asteroid bodies are uncommon [204].

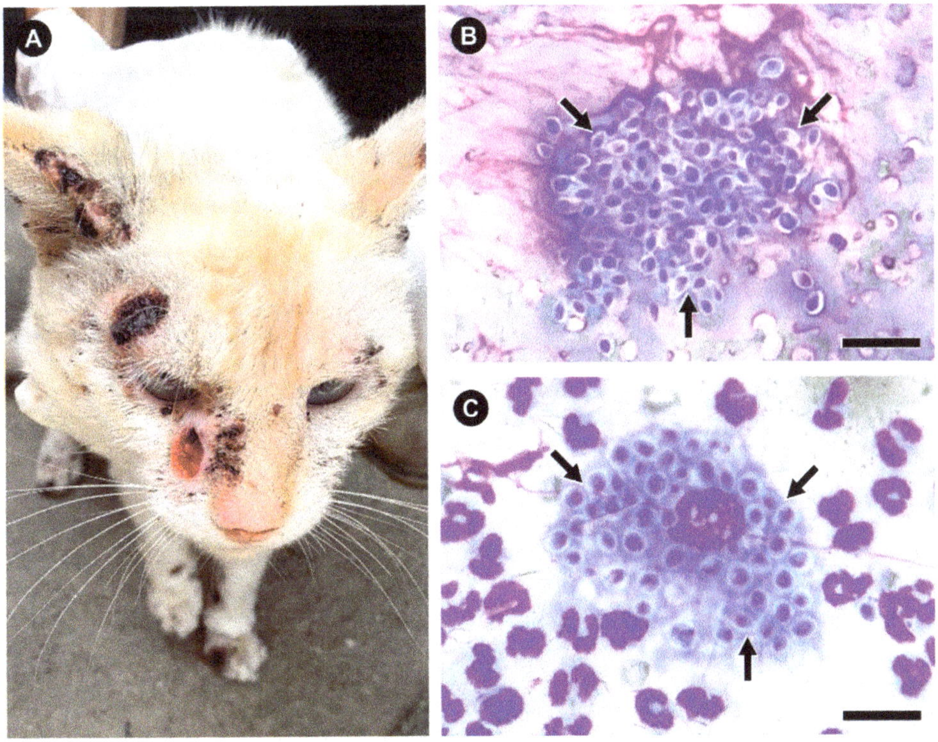

Figure 5. The diagnosis of feline sporotrichosis can employ simple, fast, and inexpensive methods such as the quick Panoptic method. (**A**) Clinical aspect of feline sporotrichosis with ulcerated lesions in the cephalic region of a cat from the state of Espírito Santo, Brazil. (**B**,**C**) Feline macrophages infected with numerous *S. brasiliensis* yeasts cells (arrows), stained using the quick Panoptic method. Bar = 15 μm.

The quick Panoptic method, a Romanowsky-type staining technique like Diff-Quik, has become relatively common in veterinary clinics due to its practicality, low cost, and great return (Figure 5B,C). A staining kit profits around 1000 slides [116,205]. This diagnosis has a sensitivity of 52.6% to 95% in cats compared to culture, the reference method [116,120,206]. However, for non-ulcerated or low exudative lesions, treatment with antifungals in high doses seems to interfere negatively with this method's sensitivity [116,120]. In the last years, the feline sporotrichosis laboratory diagnosis starts with cytology by imprinting the lesions in glass slides and isolating the fungus in culture [207]. In this scenario, an old-fashioned method such as cell block cytology achieves an impressive 97.5% sensitivity when diagnosing feline sporotrichosis during outbreaks and epidemics [208].

As in cytopathology, the histopathological findings in human samples may be nonspecific, only signaling the human sporotrichosis diagnosis. The paucity of fungal structures is also similar [183,204,209]. The histopathological pattern is associated with granulomatous and pyogenic reactions, which may lodge with epidermal hyperplasia (with or without ulceration), papillomatous acanthosis, hyperkeratosis, intraepidermal microabscess, and fungal elements such as yeast cells and asteroid bodies [196,209]. In approximately 50% of cases, the yeasts may be visible in tissue smears stained with PAS and GMS [196]. Briefly, the granuloma caused by *Sporothrix* may show three distinct zones: (i) center with abscesses or necrosis (central zone); (ii) area with granulomatous inflammation constituted by giant cells (tuberculoid zone); and (iii) lymphocytes and plasm cells, with granulation tissue and fibrosis (syphiloid zone) [209]. The inflammatory infiltrates are best observed by hematoxylin-eosin (HE) staining [177,196]. In patients with AIDS, the histopathological findings are uncommon; the inflammatory response is lower, and there is no presence of asteroid bodies. However, the amount of yeast is profuse [210].

3.3. Serology

The first serological tests for sporotrichosis diagnosis were performed in 1910 using tube agglutination (TA) and complement fixation (FC) tests [211]. The tests using precipitins were described in 1947, employing a polysaccharide antigen [212]. In 1973, Blumer et al. [213], besides testing these methods, also evaluated immunodiffusion (ID), slide agglutination test (SLA), and indirect fluorescent antibody (IFA). At that time, the SLA and ID methods were the most specific. SLA had the highest sensitivity (94%), proving to be an easy-to-perform technique that generates quick results and is highly recommended for clinical routine.

Sporotrichin, an antigenic complex consisting of a peptide-rhamnomannan, is obtained from a crude extract of the mycelial phase of *S. schenckii sensu lato*. This antigenic fraction was developed for intradermal skin reactions to measure the degree of immunity or receptivity of the individual with suspected disease, determining first contact with the fungus without developing the disease [212,214]. Sporotrichin has a long history in surveillance studies depicting areas of high endemicity [6,215].

The absence of a standardized or commercial antigen impacts the serological diagnosis of sporotrichosis in Europe and the USA [6,216]. Nevertheless, sporotrichin effectiveness in highly endemic regions ranges from 89 to 96%, making it an interesting additional test [216–220]. Bonifaz et al. [221] reported that false-positive results might be attributed to patients in constant contact with the agent, such as those who live in endemic areas or cases in which the patient preserves immunological memory. However, false-negative points correspond to patients with different immunosuppression degrees.

Many attempts have been made to adopt serological tests such as sporotrichosis diagnostic methods, given that they are fast, highly accurate, and not invasive [222]. The fungal cell wall antigens and anti-cell wall antibodies became the main target of studies searching to develop serological tools that are more sensitive and specific [222–224].

It is well known that *S. schenckii* displays a mixture antigen complex, including peptide-rhamnomannan, a cell wall glycoconjugate (CWPR) of the yeast phase of the fungus. This structure can be fractionated by affinity chromatography on Sepharose 4B with concanavalin A (Con A), generating two fractions: one binding to Con A (SsCBF) and the other non-binding to Con A (SsNBF). The fraction bound to Con A is relevant for serological diagnosis [225,226]. Besides, techniques such as immunoblot, fluorescent antibodies, counterimmunoelectrophoresis (CIE), double immunodiffusion (DID), and enzyme-linked immunosorbent assays (ELISA) have been quoted as auspicious [227,228].

A study by Penha and Lopes-Bezerra [213], using the ELISA method, evaluated 35 patients with the sporotrichosis cutaneous form, and SsCBF showed 100% specificity with the investigated sera. Subsequent studies have also demonstrated high levels of sensitivity and specificity for the various clinical presentations of human sporotrichosis, including extracutaneous, lymphocutaneous, fixed, and disseminated forms [222,229].

Bernardes-Engemann et al. [184] validated the ELISA test applying purified antigenic fraction SsCBF for the human sporotrichosis diagnosis by detecting the anti-SsCBF immunoglobulin G (IgG) antibody fraction. Thus, 177 serum samples from different clinical forms were evaluated. The investigators observed high specificity (82%) and sensitivity (89%) with a reproducibility of 98%, including for emerging species such as *S. brasiliensis*, and may be made disposable for routines of the health services.

One year after that, Alvarado et al. [227] evaluated the potential of the crude antigen obtained from the mycelial form of *S. schenckii s. str.* in serum from sporotrichosis patients by applying different methods, including DID, CIE, and ELISA. The assays were validated using serum from disease patients such as paracoccidioidomycosis, histoplasmosis, leishmaniasis, tuberculosis, lupus, and serum from healthy patients. Investigators achieved 100% sensitivity and >98% specificity for all tests.

Fernandes et al. [225] evaluated ELISA's test using SsCBF and an exoantigen for the domestic cats' serological assay with suspected sporotrichosis. The authors reported 90% sensitivity and 96% specificity for SsCBF. Excellent results were also reached by Rodrigues et al. [122], testing different antigen types obtained from crude extracts of *S. schenckii s. str.* and *S. brasiliensis* yeast cells. The results suggest that the ELISA technique with distinct antigens may be applied in diagnosing feline sporotrichosis. Furthermore, all the antigens studied reacted similarly, with no significant difference in titer. Therefore, the results generated by antigens of the different etiological agents should probably not interfere, likely because antigenic epitopes are shared and well conserved between *S. schenckii* and *S. brasiliensis* [24,230].

Baptista et al. [217] recently validated felines serology; however, it has been available only in private clinics and laboratories. The authors modified the SsCBF-ELISA test for human serological diagnosis and the quantification of IgG antibodies for all clinical forms of feline sporotrichosis.

Recombinant *Sporothrix* antigens have also been studied as diagnostic markers [224], and in 2019, Martinez-Alvarez and colleagues [231] evaluated an ELISA test for the human sporotrichosis detection from a recombinant glycoprotein obtained from the cell wall of *S. schenckii*, Gp70. Nonetheless, the findings were not as promising for the diagnosis compared to the SsCBF antigen [231].

The hyperepidemic worsening of feline and human sporotrichosis has reached proportions in different regions and countries. Public policies should be established to contain the disease, and new diagnostic methods should be developed to prevent the spread of sporotrichosis. Therefore, serological methods can qualitatively and quantitatively evaluate the condition, generating fast results with high levels of specificity and sensitivity. It would already be a big step, as it would directly impact the diagnosis of the disease and the diagnostic screening and therapeutic monitoring of human and animal sporotrichosis, including those who develop atypical and severe forms [232].

3.4. Molecular Diagnosis

For the last decades, the classical identification of the species belonging to the *Sporothrix* genus was based only on phenotypic methods [22,87,233–235]. Although these methods supply reduced cost, they are laborious, time-consuming between collection and final diagnosis, and do not allow to speciate *Sporothrix* since they have overlapping morphological and physiological features [15,95,98,181,192,194]. It is essential to highlight that the phenotypic plasticity in *Sporothrix* may exist even intraspecifically [20,78,193].

With the advance and improvement of molecular tools, it is possible to identify, reclassify, and recognize new species of fungi, allowing a greater understanding of the biology of these microorganisms [236,237]. Regarding the molecular identification of *Sporothrix*, some critical points should be considered; among them, we highlight the type of sample (e.g., soil, biopsy, isolated culture, exudate, vegetable), the cost of technique, the time that the process is executed, sensitivity/specificity, and DNA quality [238,239]. The advances in molecular diagnosis of sporotrichosis were recently reviewed by de Carvalho et al. [240].

The polymerase chain reaction (PCR) and its variants are widely used in human and feline sporotrichosis diagnosis [241,242]. It is possible to carry out DNA amplification using several molecular markers and different methods depending on the proposed objective from clinical specimens and fungi obtained from growth in culture. Kano et al. [242] pioneered the non-culture-dependent PCR technique. From human tissue, they identified *S. schenckii sensu lato* using chitin synthase I (*CHS1*) as a target gene to amplify DNA. In 2003, Kano and collaborators [243] applied this same technique in biopsy samples obtained from six patients with human sporotrichosis, confirmed by histopathology and mycological findings. The generated DNA fragments showed 99% similarity with *S. schenckii* reference strain.

After that, other molecular tools with the potential to detect the *Sporothrix* DNA directly from environmental samples (soils, plants, decaying wood, tree bark) and clinical specimens obtained from humans and animals (biopsy, skin lesion, aspirate of abscess, exudate swabs, and pus) were proposed. The most frequently used techniques have been species-specific PCR (Figure 6A), nested PCR, restriction fragment length polymorphism (RFLP), and qPCR (quantitative real-time PCR). Among the different targets, the most common are 18S rRNA [241,244], Large Subunit (LSU), internal transcribed spacer (ITS) [245], β-tubulin (*BT2*) [238], calmodulin (*CAL*) [96,246–248], and mitochondrial DNA genes [59,60]. Nevertheless, it is worth mentioning that few techniques allow for detecting *Sporothrix* DNA from clinical samples and identifying it down to a species level.

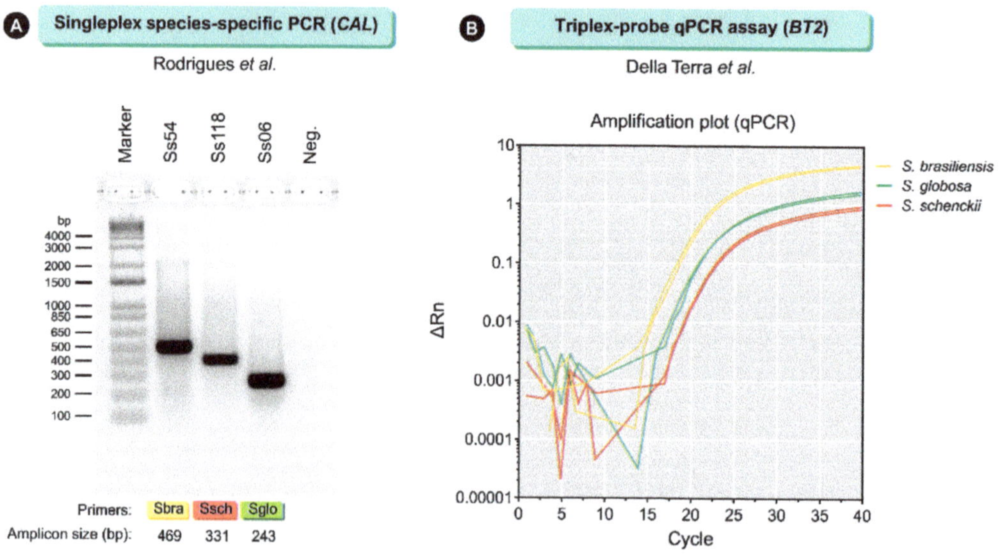

Figure 6. The molecular diagnosis of sporotrichosis relies on PCR. (**A**) Species-specific PCR is a fast and inexpensive method that combines conventional PCR and agarose gel electrophoresis to detect *Sporothrix* DNA, mainly in vitro culture samples [96]. Eventually, species-specific PCR is used for detection from clinical samples [249]. Although it has great specificity, the method has low sensitivity (up to 10–100 fg of DNA). (**B**) For the rapid and accurate diagnosis of human and feline sporotrichosis, a multiplex qPCR assay was developed for the simultaneous detection and speciation of *S. brasiliensis*, *S. schenckii*, and *S. globosa* from cultured DNA and clinical samples (up to 3 copies of the target) [238]. Ss54 = *S. brasiliensis*; Ss118 = *S. schenckii*; Ss06 = *S. globosa*.

Della-Terra et al. [238] developed and evaluated a multiplex qPCR assay for sporotrichosis diagnosis exploiting polymorphisms found in the β-tubulin gene (Figure 6B). Samples of cat lesions and environmental samples spiked with *S. brasiliensis*, *S. schenckii*, and *S. globosa* propagules were used to reveal the feasibility of the method to detect *Sporothrix* DNA. High specificity (100%) and sensitivity (98.6%) were reported [238]. Moreover, the technique did not show cross-reaction with other pathogenic fungi, human, feline, or murine DNA, allowing the identification of all major *Sporothrix* species in a single tube. The key advantages of the multiplex qPCR assay include decreasing amount of reagents, reduction of the process steps, greater sensitivity when compared to culture [114], species-specific PCR [96,249], rolling circle amplification (RCA) [247], and other qPCR assays [245,250].

Regarding identifying and recognizing cryptic species belonging to *Sporothrix*, the reference method is DNA sequencing (Sanger) followed by phylogenetic analysis. In addition to amplification and partial sequencing of the ITS region (ITS1/2+5.8s), protein-coding loci such as *BT2* [18], translation elongation factor (EF-1α) [20], and *CAL* [18,78,193] should be included [189]. Nowadays, the region between exons 3 and 5 of the *CAL* gene is the primary marker for identifying clinical species for this genus [19,20,62,79,91,95,98,99,109,114,251,252].

Techniques like *CAL*-RFLP [246], RCA [247], and species-specific PCR [96] can also identify *Sporothrix* down to species level. The major difference among methods is that RCA and species-specific PCR can detect *Sporothrix* DNA, whereas *CAL*-RFLP was designed for strains isolated in culture.

DNA fingerprint methods such as amplified fragment length polymorphisms (AFLPs) have demonstrated broad applicability in speciating pathogenic fungi and describing genetic diversity. AFLP is widely used in evolutionary, population, epidemiological, and conservation studies of different taxa [253]. AFLP's main advantage is the simultaneous assessment of several loci, randomly distributed throughout the genome, without prior knowledge of the DNA sequence. This makes the AFLP singularly helpful for species with no genomic knowledge, and a powerful tool to be used to explore *Sporothrix* genetic diversity, answer questions related to the structure of the population, transmission routes, intra and interspecific variability, as well as modes of recombination and reproduction, among many other biological issues [254–256].

In the last decades, mass profiles of ribosomal proteins generated by Matrix-Assisted Laser Desorption/Ionization Time-of-Flight Mass Spectrometry (MALDI-ToF MS) have been a powerful tool in yeast identification in routine clinical laboratories [257,258]. In recent years, MALDI-ToF MS databases have been built and expanded to identify filamentous fungi; for the technique to be reliable, rapid, and economical, databases must be accurate [259,260]. The method was standardized to speciate *Sporothrix* isolates growing in vitro during the yeast phase, which allows the recognition of *S. brasiliensis*, *S. schenckii*, *S. globosa*, *S. luriei*, and members of the *S. pallida* complex. Furthermore, speciation by MALDI-ToF or DNA sequencing methods are in full agreement [259]. However, the greatest limitation of MALDI-TOF relies on the need for prior culture. Only a few reports in the literature use this methodology to speciate *Sporothrix*. Matos et al. [261] constructed an in-house database enriched with spectra generated from reference *Sporothrix* strains, enabling the identification of an isolate from *S. brasiliensis* obtained from a patient with subconjunctival infiltrative injury in a right eye.

Despite the relevance of molecular diagnostic tools, unfortunately, these techniques are not widely available in health services, being restricted to research laboratories and reference centers, thus negatively impacting the diagnosis of the disease.

4. Trends in the Treatment of Sporotrichosis

The treatment choice for sporotrichosis depends on the disease's clinical form, the host's immunological status, and the species *Sporothrix* involved. The most virulent species (*S. brasiliensis*, *S schenckii*, and *S. globosa*) exhibit different susceptibility profiles to antifungals; thus, the response to therapy can be variable [262].

Sporotrichosis was first described in 1898 when specific antifungal drugs were unavailable, and potassium iodide (KI) was used for several infectious and non-infectious diseases. KI saturated solution has been used for treating sporotrichosis since 1903 [2]. KI has immunomodulatory activity; it can suppress the production of toxic intermediates from oxygen by the polymorphonuclear leukocytes and, therefore, exert its anti-inflammatory effect. KI's ability to directly destroy microorganisms is still a matter of speculation.

Notwithstanding, reports in the literature suggest that when *Sporothrix* yeasts are exposed to increasing drug concentrations, cell lysis occurs through the release of lysosomal enzymes [263]. Recent studies show that KI inhibits biofilm development in *Sporothrix* [264]. The main adverse events of KI are metallic taste and nausea, followed by an acneiform eruption. Nowadays, its use in humans has been replaced by itraconazole, but due to its low cost, it is still used to treat cutaneous human forms and felines, in association or not with itraconazole [14,265,266].

Itraconazole started to be used in sporotrichosis treatment in the late 1980s, during the advent of the triazoles generation. Itraconazole is the drug of choice due to its effectiveness, safety, and posologic convenience for lymphocutaneous and cutaneous sporotrichosis. In Brazil, it is also used in animal treatment [15,189,266]. Itraconazole is a fungistatic drug that inhibits the synthesis of ergosterol, the main sterol from the fungus cell membrane (Figure 7) [267]. Depending on the disease severity and the host's immunological status, the therapeutic dose may range from 100 to 400 mg/day [188]. Although it has good efficacy, treatment with itraconazole can cause several side effects and interact with more than 200 other drugs, inducing adverse events or lacking effectiveness [268]. The main adverse effects reported are headaches and gastrointestinal disorders. It is also hepatotoxic, teratogenic, and embryotoxic, and may not be used in patients with liver diseases or pregnant women.

Another inhibitor of ergosterol synthesis used for sporotrichosis treatment is terbinafine (Figure 7). This allylamine treats cutaneous form in humans when itraconazole or KI is not tolerated or cannot be used [269]. It is available in 125 and 250 mg tablets, facilitating pediatric administration. The recommended dose is 250 mg/day, but it may increase to 500 mg/day for adults [2].

In severe life-threatening cases, amphotericin B (deoxycholate or, preferably, liposomal, because such formulation has fewer adverse effects) is recommended until clinical improvement has been achieved, when it should be replaced with itraconazole. Amphotericin B is a polyene antifungal drug developed in the 1950s. There are four models proposed for the polyene mode of antifungal action: (i) the pore-forming model, (ii) the surface adsorption model, (iii) the sterol sponge model, and (iv) the oxidative damage model. In every suggested model, the binding of the polyene with ergosterol is essential to its antifungal effect [270] (Figure 7). The total cumulative dose recommended for amphotericin B ranges from 1 to 3 g. Although effective and acts with fungicidal properties, treatment with amphotericin B is not recommended for cutaneous and lymphocutaneous sporotrichosis because of its high toxicity and the inconvenience of intravenous administration [7,8].

Although the sporotrichosis treatment is mainly based on the prescription of the antifungals described above, other commercial drugs can also inhibit the *Sporothrix* growth in vitro. Table 1 shows the activities of different commercial antifungals according to the minimum inhibitory concentration (MIC) values obtained in vitro by the broth microdilution methods described by the Clinical and Laboratory Standards Institute [271,272]. Most of these antifungals are not used in treating sporotrichosis and are only used in the topical treatment of other fungal infections. Amphotericin B, itraconazole, and terbinafine exhibit higher in vitro activity against *Sporothrix* cells, with MIC described in the literature as lower than 1 μg/mL (Table 1).

Little is known about in vivo activity of other azoles for sporotrichosis treatment (Table 1), except for voriconazole and posaconazole. Voriconazole exhibits modest efficacy, while posaconazole is effective against disseminated sporotrichosis murine models [273,274]. Although they have been used for other fungal infections, they are not currently used for sporotrichosis treatment.

Non-pharmacological measures are also used in treating sporotrichosis with good results, such as cryosurgery and thermotherapy. Cryosurgery is indicated when a slow itraconazole response is observed for the resolution of chronic lesions and when adverse effects lead to interruption of the antifungals [275]. Local hyperthermia and cryosurgery are safe and efficacy options for treating pregnant women with cutaneous sporotrichosis [275,276].

Photodynamic therapy could also be applied as a non-pharmacological treatment for sporotrichosis, considering the promising in vitro and in vivo studies; however, further clinical observation is still necessary [277].

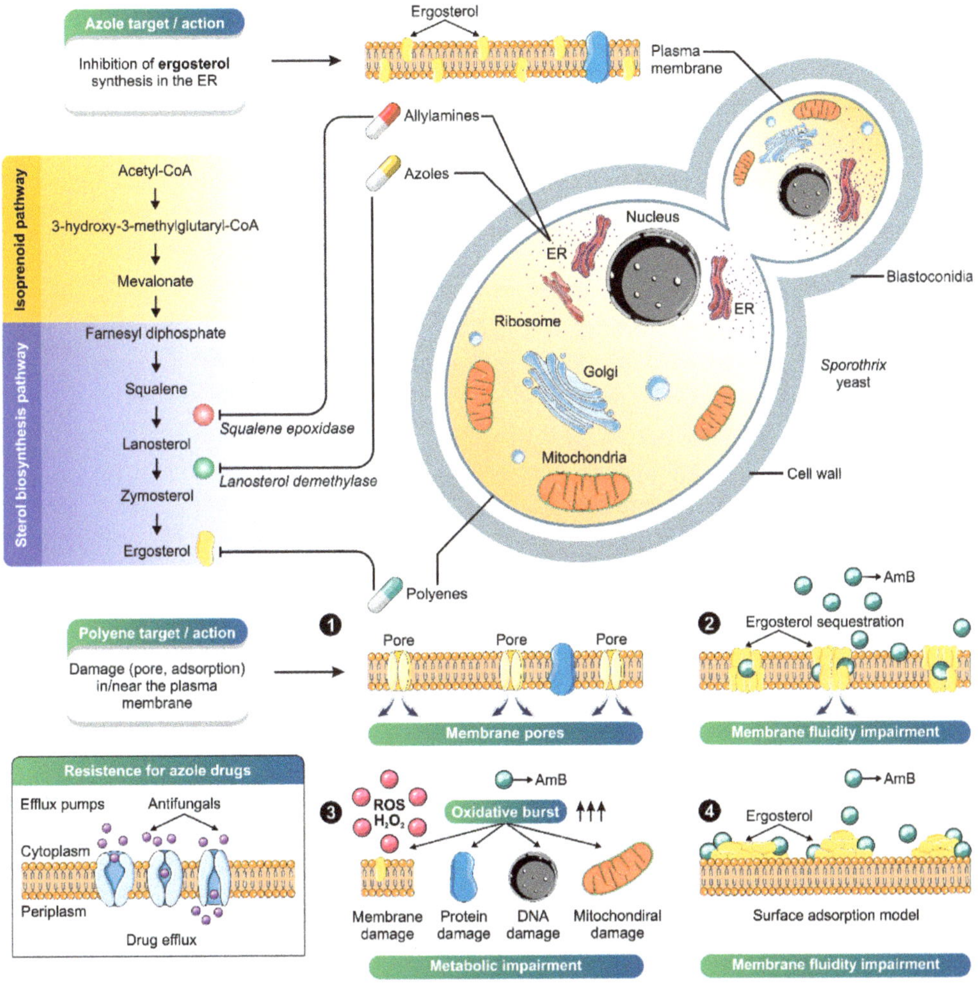

Figure 7. The main antifungal agents used in treating sporotrichosis and their cellular targets are depicted. Azoles (e.g., itraconazole) and allylamines (e.g., terbinafine) are fungistatic drugs that slow fungal growth; the azoles by inhibiting cytochrome P-450-dependent synthesis of ergosterol (purple chart: sterol biosynthesis pathway) and the allylamines by competitive inhibition of squalene epoxidase, blocking the conversion of squalene to lanosterol. Amphotericin B is a fungicidal drug whose main target is the ergosterol molecule, producing pores in the plasma membrane (1), leading to the leakage of cytoplasmic material, sequestering (2), absorbing or extracting ergosterol from the membrane (4). Oxidative damage is reported as an alternative mode of action of AmB (3). The main adaptation that can lead to resistance in *Sporothrix* species is the increased expression of efflux pumps, especially to azoles, although other mechanisms may be involved. The illustration was partially based on Servier Medical Art elements and licensed under a Creative Commons Attribution 3.0 Unported License. ER: endoplasmic reticulum; ROS: reactive oxygen species.

In recent years, there has been an increase in therapeutic failures in treating sporotrichosis and reports of isolates with low sensitivity to itraconazole [97,267,278–284]. The decrease of itraconazole effectiveness against *S. brasiliensis* observed in clinical isolates from Brazil over the last years could be related to resistance mechanisms developed by this species, such as overexpression of efflux pumps (Figure 4). The increased expression of efflux pumps in the fungal membrane reduces the antifungal activity due to the extrusion of the drug, decreasing its effect. The overexpression of efflux pumps corresponds to the main mechanism of acquired resistance to azoles in fungi of medical relevance [285]. However, other resistance mechanisms could be related to *S. brasiliensis* decreasing susceptibility to itraconazole, such as melanin production and overexpression or mutation of the target enzyme [281]. On the other hand, studies investigating new molecules with anti-*Sporothrix* activity have increased.

Table 1. In vitro antifungal activity against pathogenic *Sporothrix* species.

	In Vitro Antifungal Activity [a]			
	High (MIC $\leq$ 1 μg/mL)	**Moderate (1 < MIC $\leq$ 4 μg/mL)**	**Low (MIC > 4 μg/mL)**	**Reference**
Polyenes				
Amphotericin B	■			[262]
Azoles				
Albaconazole			■	[286]
Clotrimazole		■		[287]
Eberconazole		■		[286]
Fluconazole			■	[286]
Itraconazole	■			[262]
Isavuconazole		■		[288]
Ketoconazole	■			[262]
Miconazole		■		[286]
Posaconazole	■			[262]
Ravuconazole		■		[286]
Voriconazole			■	[262]
Allylamines				
Terbinafine	■			[262]
Naftifine		■		[289]
Echinocandins				
Anidulafungin		■		[286]
Caspofungin			■	[286]
Micafungin			■	[286]
Pirimidine				
Flucytosine			■	[286]

[a] In vitro antifungal activity is determined according to MIC values obtained by the microdilution technique [271,272].

We performed a retroactive literature search for studies published between 2012 and July 2022 using the databases Cortellis Drug Discovery Intelligence (https://www.cortellis.com/drugdiscovery/ (accessed on 14 July 2022) [268] and PubMed (https://pubmed.ncbi.nlm.nih.gov/ (accessed on 14 July 2022), with "*Sporothrix*" as the Keywords. We considered promising compounds with MIC against *Sporothrix* species less or equal to 4 μg/mL or 1 μM, as determined by Clinical and Laboratory Standards Institute protocols [271,272]. Based on these criteria, we found seventeen new molecules or repositionable drugs over the last ten years (Table 2). Most studies listed in Table 2 are Brazilian research papers highlighting the importance of this pathogen in Brazil. Only two of the seventeen papers are not from Brazilian groups.

Table 2. New compounds, natural products, and repositionable drugs exhibited promising activity against *Sporothrix* spp. in the last ten years.

Group	Compound	Minimum Inhibitory Concentration [a]	Antifungal Effect or Mechanism of Action	Reference
Hydrazone derivatives	22-hydrazone-imidazolin-2-yl-chol-5-ene-3β-ol	0.01–0.5 μg/mL	Inhibition of ergosterol biosynthesis	[291]
	4-bromo-N′-(3,5-dibromo-2-hydroxybenzylidene)-benzohydrazide	0.12–1 μg/mL	Inhibition of vesicular transport and cell cycle progression	[290]
Oxadiazole	N-[5-(4-chlorophenyl)-1,3,4-oxadiazol-2-yl]-3-(trifluoromethyl)benzamide	0.25–0.5 μM	Cell membrane disruption and neutral lipid accumulation	[292]
Alkylphospholipid analogc	TCAN26	0.25–2 μg/mL	Cell membrane disruption	[293]
Pentathiepin	23	0.5–1 μg/mL	Unknown	[294]
Benzisothiazolone	1.9	0.5 μg/mL	Apoptosis induction	[295]
Metal complex	Zn(itraconazole)2Cl_2	0.08 μM	Unknown	[296]
	Zn(ketoconazole)2(Ac)2·H_2O	0.125 μM	Unknown	[297]
	[Cu(PPh3)2(ketoconazole)2]NO_3	0.006 μM	Unknown	[298]
Naphthoquinone derivative	2,5-dichloro-3,6-bis(4-methylpiperazin-1-yl)cyclohexa-2,5-diene-1,4-dione	1.56 μg/mL	Unknown	[299]
Natural products	farnesol	0.003–0.222 μg/mL	Unknown	[300]
Repositionable drugs	Miltefosine	1–2 μg/mL 0.25–2 μg/mL	Cell membrane disruption	[278] [301]
	Iodoquinol	0.5–1 μM	Cell membrane disruption	[302]
	Buparvaquone [b]	0.005–0.16 μg/mL	Mitochondrial dysfunction	[303]
	Ibuprofen	0.03–0.5 μg/mL [c]	Cell membrane disruption and ROS accumulation	[292]
	Pentamidine	0.06–0.25 μg/mL	DNA intercalation	[304]

[a] MIC range or MIC mean [271,272].; [b] veterinary use; [c] combined with itraconazole, amphotericin B, or terbinafine.

Furthermore, twelve of these works were published in the last five years, reflecting the increase in studies using *Sporothrix* as a model due to its importance as a human pathogen. Besides, most compounds induce disruption of the cell membrane in *Sporothrix* cells. The MIC values show that farnesol and buparvaquone are the most promising compounds (Table 2). However, considering in vitro and in vivo approaches, the hydrazone derivative 4-bromo-N′-(3,5-dibromo-2-hydroxybenzylidene)-benzohydrazide (reported as D13 in the original study) exhibited the most interesting results, with good in vitro and in vivo activities [290].

5. Conclusions

The current scenario shows the emergence and re-emergence of sporotrichosis as a cosmopolitan mycosis whose etiological agents are in constant movement and associated with different transmission routes. Therefore, public policies should vary according to the epidemiological scenario, preferably using one-health strategies. Policies aimed at the human–environment interface are mandatory for areas where the sapronotic route prevails, driven by *S. globosa* and *S. schenckii s. str.* For areas where cat-transmitted sporotrichosis is emerging, policies aimed at the human–animal–environment interface are necessary (e.g., responsible animal ownership; limiting the number of cats per house; neutering campaigns; limiting cat access to the streets; cremation of dead cats). The invasive capacity of *S. brasiliensis* is impressive, becoming the predominant species shortly after its introduction into a new area. Such dynamics of the expansion of cat-transmitted sporotrichosis associated with *S. brasiliensis* occur through successive founder effect events. Therefore, the imposition of sanitary barriers preventing the free movement of sick animals is vital to tackle the geographic expansion of *S. brasiliensis* beyond the borders of Brazil.

In a rapidly expanding epidemic scenario, diagnostic tools need to keep up with the pace of the problem. Therefore, fast and accurate methods are important tools for laboratory diagnosis. In feline sporotrichosis, the quick Panoptic method is an efficient alternative for diagnosis given the high fungal load in the lesions, in addition to the robustness and low price. However, speciation will only be possible through the application of molecular tools. The pillar of the molecular diagnosis of sporotrichosis relies on PCR, with conventional assays that combine in vitro amplification and agarose gel electrophoresis (e.g., species-specific PCR) to multiplex reactions capable of real-time detecting mixed infections in a single tube (e.g., qPCR). Rapid diagnosis allows specific treatment, which positively impacts patients' clinical outcomes and reduces transmission rates.

The choice of treatment for sporotrichosis depends on the triad, (i) the clinical form of the disease, (ii) the immune status of the host, and (iii) the *Sporothrix* species involved. The main challenge for the coming years is the sudden emergence of isolates resistant to itraconazole, the first choice to treat sporotrichosis. The good news is that a search in the Cortellis Drug Discovery Intelligence database revealed 17 new molecules or repositionable drugs in the last ten years, highlighting the hydrazone derivatives as a promising alternative based on great in vitro and in vivo activities.

All the scenarios contemplated above are interdependent and must be considered to mitigate the advance of the major mycosis of implantation worldwide.

Author Contributions: Conceptualization, A.M.R.; A.M.R., J.A.d.C. and Z.P.d.C. contributed to writing the 'Trends in the Epidemiology of *Sporothrix* Species' section. S.S.G. contributed to writing the 'Trends in the Diagnosis of Sporotrichosis' section. L.P.B.-S. and S.R. contributed to writing the 'Trends in the Treatment of Sporotrichosis' section. A.M.R. prepared all figures and was responsible for editing the manuscript. All authors have read and agreed to the published version of the manuscript.

Funding: This research was funded by Fundação de Amparo à Pesquisa do Estado de São Paulo (2017/27265-5) http://www.fapesp.br/, Fundação de Amparo à Pesquisa do Estado do Rio de Janeiro (E-26/202.513/2019 and E-26/210.873/2019) http://www.faperj.br/, Conselho Nacional de Desenvolvimento Científico e Tecnológico (CNPq 429594/2018-6) http://www.cnpq.br/, and Coordenação de Aperfeiçoamento de Pessoal de Nível Superior (CAPES 88887.177846/2018-00 and 88882.316588/2019-01) https://www.capes.gov.br. AMR and SR are CNPq Research Productivity Fellow (CNPq 304902/2020-9 and 306358/2021-2, respectively).

Institutional Review Board Statement: Not applicable.

Informed Consent Statement: Not applicable.

Data Availability Statement: The data presented in this study are available within the article.

Conflicts of Interest: The authors declare no conflict of interest. The funders had no role in the design of the study; in the collection, analyses, or interpretation of data; in the writing of the manuscript, or in the decision to publish the results.

References

1. Ramírez-Soto, M.; Aguilar-Ancori, E.; Tirado-Sánchez, A.; Bonifaz, A. Ecological determinants of sporotrichosis etiological agents. *J. Fungi* **2018**, *4*, 95. [CrossRef] [PubMed]
2. Orofino-Costa, R.C.; Macedo, P.M.; Rodrigues, A.M.; Bernardes-Engemann, A.R. Sporotrichosis: An update on epidemiology, etiopathogenesis, laboratory and clinical therapeutics. *An. Bras. De Dermatol.* **2017**, *92*, 606–620. [CrossRef] [PubMed]
3. Rodrigues, A.M.; de Hoog, G.S.; de Camargo, Z.P. *Sporothrix* species causing outbreaks in animals and humans driven by animal-animal transmission. *PLoS Pathog.* **2016**, *12*, e1005638. [CrossRef]
4. Rodrigues, A.M.; Bagagli, E.; de Camargo, Z.P.; Bosco, S.M.G. *Sporothrix schenckii sensu stricto* isolated from soil in an armadillo's burrow. *Mycopathologia* **2014**, *177*, 199–206. [CrossRef] [PubMed]
5. Gremião, I.D.; Miranda, L.H.; Reis, E.G.; Rodrigues, A.M.; Pereira, S.A. Zoonotic epidemic of sporotrichosis: Cat to human transmission. *PLoS Pathog.* **2017**, *13*, e1006077. [CrossRef]

6. Bonifaz, A.; Vázquez-González, D. Diagnosis and treatment of lymphocutaneous sporotrichosis: What are the options? *Curr. Fungal Infect. Rep.* **2013**, *7*, 252–259. [CrossRef]
7. Kauffman, C.A.; Bustamante, B.; Chapman, S.W.; Pappas, P.G. Clinical practice guidelines for the management of sporotrichosis: 2007 update by the Infectious Diseases Society of America. *Clin. Infect. Dis.* **2007**, *45*, 1255–1265. [CrossRef]
8. Kauffman, C.A.; Hajjeh, R.; Chapman, S.W.; Group, M.S. Practice guidelines for the management of patients with sporotrichosis. *Clin. Infect. Dis.* **2000**, *30*, 684–687. [CrossRef]
9. Aung, A.K.; Teh, B.M.; McGrath, C.; Thompson, P.J. Pulmonary sporotrichosis: Case series and systematic analysis of literature on clinico-radiological patterns and management outcomes. *Med. Mycol.* **2013**, *51*, 534–544. [CrossRef]
10. Gremião, I.D.; Menezes, R.C.; Schubach, T.M.; Figueiredo, A.B.; Cavalcanti, M.C.; Pereira, S.A. Feline sporotrichosis: Epidemiological and clinical aspects. *Med. Mycol.* **2015**, *53*, 15–21. [CrossRef]
11. Pereira, S.A.; Gremião, I.D.F.; Menezes, R.C. Sporotrichosis in Animals: Zoonotic Transmission. In *Sporotrichosis: New Developments and Future Prospects*; Zeppone Carlos, I., Ed.; Springer International Publishing: Cham, Switzerland, 2015; pp. 83–102. [CrossRef]
12. Pereira, S.A.; Gremião, I.D.; Kitada, A.A.; Boechat, J.S.; Viana, P.G.; Schubach, T.M. The epidemiological scenario of feline sporotrichosis in Rio de Janeiro, State of Rio de Janeiro, Brazil. *Rev. Da Soc. Bras. De Med. Trop.* **2014**, *47*, 392–393. [CrossRef] [PubMed]
13. Schubach, A.; Schubach, T.M.; Barros, M.B.; Wanke, B. Cat-transmitted sporotrichosis, Rio de Janeiro, Brazil. *Emerg. Infect. Dis.* **2005**, *11*, 1952–1954. [CrossRef] [PubMed]
14. Gremião, I.D.F.; Martins da Silva da Rocha, E.; Montenegro, H.; Carneiro, A.J.B.; Xavier, M.O.; de Farias, M.R.; Monti, F.; Mansho, W.; de Macedo Assunção Pereira, R.H.; Pereira, S.A.; et al. Guideline for the management of feline sporotrichosis caused by *Sporothrix brasiliensis* and literature revision. *Braz. J. Microbiol. Publ. Braz. Soc. Microbiol.* **2021**, *52*, 107–124. [CrossRef]
15. Rodrigues, A.M.; Della Terra, P.P.; Gremiao, I.D.; Pereira, S.A.; Orofino-Costa, R.; de Camargo, Z.P. The threat of emerging and re-emerging pathogenic *Sporothrix* species. *Mycopathologia* **2020**, *185*, 813–842. [CrossRef] [PubMed]
16. Schenck, B.R. On refractory subcutaneous abscesses caused by a fungus possibly related to the *Sporotricha*. *Bull. Johns Hopkins Hosp.* **1898**, *9*, 286–290.
17. Hektoen, L.; Perkins, C.F. Refractory subcutaneous abscesses caused by *Sporothrix schenckii*: A new pathogenic fungus. *J. Exp. Med.* **1900**, *5*, 77–89. [CrossRef] [PubMed]
18. Marimon, R.; Gené, J.; Cano, J.; Trilles, L.; Dos Santos Lazéra, M.; Guarro, J. Molecular phylogeny of *Sporothrix schenckii*. *J. Clin. Microbiol.* **2006**, *44*, 3251–3256. [CrossRef]
19. de Beer, Z.W.; Duong, T.A.; Wingfield, M.J. The divorce of *Sporothrix* and *Ophiostoma*: Solution to a problematic relationship. *Stud. Mycol.* **2016**, *83*, 165–191. [CrossRef]
20. Rodrigues, A.M.; Cruz Choappa, R.; Fernandes, G.F.; De Hoog, G.S.; Camargo, Z.P. *Sporothrix chilensis* sp. nov. (Ascomycota: Ophiostomatales), a soil-borne agent of human sporotrichosis with mild-pathogenic potential to mammals. *Fungal Biol.* **2016**, *120*, 246–264. [CrossRef]
21. Sil, A.; Andrianopoulos, A. Thermally dimorphic human fungal pathogens—Polyphyletic pathogens with a convergent pathogenicity trait. *Cold Spring Harb. Perspect. Med.* **2014**, *5*, a019794. [CrossRef]
22. Fernandes, G.F.; dos Santos, P.O.; Amaral, C.C.; Sasaki, A.A.; Godoy-Martinez, P.; Camargo, Z.P.d. Characteristics of 151 Brazilian *Sporothrix schenckii* isolates from 5 different geographic regions of Brazil: A forgotten and re-emergent pathogen. *Open Mycol. J.* **2009**, *3*, 48–58. [CrossRef]
23. Fernandes, G.F.; dos Santos, P.O.; Rodrigues, A.M.; Sasaki, A.A.; Burger, E.; de Camargo, Z.P. Characterization of virulence profile, protein secretion and immunogenicity of different *Sporothrix schenckii sensu stricto* isolates compared with *S. globosa* and *S. brasiliensis* species. *Virulence* **2013**, *4*, 241–249. [CrossRef] [PubMed]
24. Della Terra, P.P.; Rodrigues, A.M.; Fernandes, G.F.; Nishikaku, A.S.; Burger, E.; de Camargo, Z.P. Exploring virulence and immunogenicity in the emerging pathogen *Sporothrix brasiliensis*. *PLoS Negl. Trop. Dis.* **2017**, *11*, e0005903. [CrossRef]
25. Almeida-Paes, R.; de Oliveira, L.C.; Oliveira, M.M.E.; Gutierrez-Galhardo, M.C.; Nosanchuk, J.D.; Zancopé Oliveira, R.M. Phenotypic characteristics associated with virulence of clinical isolates from the *Sporothrix* complex. *BioMed Res. Int.* **2015**, *2015*, 1–10. [CrossRef]
26. Arrillaga-Moncrieff, I.; Capilla, J.; Mayayo, E.; Marimon, R.; Mariné, M.; Gené, J.; Cano, J.; Guarro, J. Different virulence levels of the species of *Sporothrix* in a murine model. *Clin. Microbiol. Infect.* **2009**, *15*, 651–655. [CrossRef] [PubMed]
27. de Macedo, P.M.; Sztajnbok, D.C.; Camargo, Z.P.; Rodrigues, A.M.; Lopes-Bezerra, L.M.; Bernardes-Engemann, A.R.; Orofino-Costa, R. Dacryocystitis due to *Sporothrix brasiliensis*: A case report of a successful clinical and serological outcome with low-dose potassium iodide treatment and oculoplastic surgery. *Br. J. Dermatol.* **2015**, *172*, 1116–1119. [CrossRef]
28. Silva-Vergara, M.L.; de Camargo, Z.P.; Silva, P.F.; Abdalla, M.R.; Sgarbieri, R.N.; Rodrigues, A.M.; dos Santos, K.C.; Barata, C.H.; Ferreira-Paim, K. Disseminated *Sporothrix brasiliensis* infection with endocardial and ocular involvement in an HIV-infected patient. *Am. J. Trop. Med. Hyg.* **2012**, *86*, 477–480. [CrossRef]
29. Almeida-Paes, R.; de Oliveira, M.M.; Freitas, D.F.; do Valle, A.C.; Zancope-Oliveira, R.M.; Gutierrez-Galhardo, M.C. Sporotrichosis in Rio de Janeiro, Brazil: *Sporothrix brasiliensis* is associated with atypical clinical presentations. *PLoS Negl. Trop. Dis.* **2014**, *8*, e3094. [CrossRef] [PubMed]

30. Nepomuceno Araujo, M.; Nihei, C.H.; Rodrigues, A.M.; Higashino, H.; Ponzio, V.; Campos Pignatari, A.C.; Barcellos, M.A.; Braga, O.; Duayer, I.F. Case report: Invasive sinusitis due to *Sporothrix brasiliensis* in a renal transplant recipient. *Am. J. Trop. Med. Hyg.* **2021**, *105*, 1218–1221. [CrossRef]
31. do Monte Alves, M.; Pipolo Milan, E.; da Silva-Rocha, W.P.; Soares de Sena da Costa, A.; Araujo Maciel, B.; Cavalcante Vale, P.H.; de Albuquerque, P.R.; Lopes Lima, S.; Salles de Azevedo Melo, A.; Messias Rodrigues, A.; et al. Fatal pulmonary sporotrichosis caused by *Sporothrix brasiliensis* in Northeast Brazil. *PLoS Negl. Trop. Dis.* **2020**, *14*, e0008141. [CrossRef]
32. Barros, M.B.; de Almeida Paes, R.; Schubach, A.O. *Sporothrix schenckii* and sporotrichosis. *Clin. Microbiol. Rev.* **2011**, *24*, 633–654. [CrossRef]
33. Dangerfield, L.F.; Gear, J. Sporotrichosis among miners on the Witwatersrand gold mines. *S.A. Med. J.* **1941**, *15*, 128–131.
34. Pijper, A.; Pullinger, B.D. An outbreak of sporotrichosis among South African native miners. *Lancet* **1927**, *210*, 914–916. [CrossRef]
35. Brown, R.; Bowen, J.W.; Weintroub, D.; Cluver, E.H.; Buchanan, G.; Simson, F.W.; Lurie, H.I.; Du Toit, C.J.; Golberg, L.; Goedvolk, C.J.; et al. Sporotrichosis infection on mines of the Witwatersrand. In Proceedings of the Transvaal Mine Medical Officer's Association, Cape Town, South Africa, 1 January 1947; pp. 1–67.
36. Govender, N.P.; Maphanga, T.G.; Zulu, T.G.; Patel, J.; Walaza, S.; Jacobs, C.; Ebonwu, J.I.; Ntuli, S.; Naicker, S.D.; Thomas, J. An outbreak of lymphocutaneous sporotrichosis among mine-workers in South Africa. *PLoS Negl. Trop. Dis.* **2015**, *9*, e0004096. [CrossRef]
37. Vismer, H.F.; Hull, P.R. Prevalence, epidemiology and geographical distribution of *Sporothrix schenckii* infections in Gauteng, South Africa. *Mycopathologia* **1997**, *137*, 137–143. [CrossRef] [PubMed]
38. Nenoff, P.; Reinel, D.; Kruger, C.; Grob, H.; Mugisha, P.; Suss, A.; Mayser, P. Tropical and travel-related dermatomycoses: Part 2: Cutaneous infections due to yeasts, moulds, and dimorphic fungi. *Der Hautarzt Z. Fur. Dermatol. Venerol. Und Verwandte Geb.* **2015**, *66*, 522–532. [CrossRef]
39. Callens, S.F.; Kitetele, F.; Lukun, P.; Lelo, P.; Van Rie, A.; Behets, F.; Colebunders, R. Pulmonary Sporothrix schenckii infection in a HIV positive child. *J. Trop. Pediatrics* **2006**, *52*, 144–146. [CrossRef] [PubMed]
40. Ponnighaus, M.; Grosser, S.; Baum, H.P.; Mischke, D.; Kowalzick, L. Sporotrichosis as the cause of a leg ulcer. *Der Hautarzt; Z. Fur Dermatol. Venerol. Und Verwandte Geb.* **2003**, *54*, 64–66. [CrossRef]
41. Findlay, G. Sporotrichosis research in the Transvaal-how it began 60 years ago. *S. Afr. Med. J. Suid-Afrik. Tydskr. Vir Geneeskd.* **1985**, *68*, 117–118.
42. Dalis, J.S.; Kazeem, H.M.; Kwaga, J.K.; Kwanashie, C.N. Severe generalized skin lesions due to mixed infection with *Sporothrix schenkii* and *Dermatophilus congolensis* in a bull from Jos, Nigeria. *Vet. Microbiol.* **2014**, *172*, 475–478. [CrossRef]
43. Rasamoelina, T.; Maubon, D.; Raharolahy, O.; Razanakoto, H.; Rakotozandrindrainya, N.; Rakotomalala, F.A.; Bailly, S.; Sendrasoa, F.; Ranaivo, I.; Andrianarison, M. Sporotrichosis in the Highlands of Madagascar, 2013–2017. *Emerg. Infect. Dis.* **2019**, *25*, 1893–1902. [CrossRef] [PubMed]
44. Ross, M.D.; Gelfand, M. Deep fungal infections in Rhodesia—A 10-year survey of histological material. *Part I Cent. Afr. J. Med.* **1978**, *24*, 208–212. [PubMed]
45. Jacyk, W.K.; Lawande, R.V.; Tulpule, S.S. Deep mycoses in West Africa: A report of 13 cases and review of the Nigerian literature. *J. Natl. Med. Assoc.* **1981**, *73*, 251–256. [PubMed]
46. Gumaa, S.A. Sporotrichosis in Sudan. *Trans. R. Soc. Trop. Med. Hyg.* **1978**, *72*, 637–640. [CrossRef]
47. Chakrabarti, A.; Bonifaz, A.; Gutierrez-Galhardo, M.C.; Mochizuki, T.; Li, S. Global epidemiology of sporotrichosis. *Med. Mycol.* **2015**, *53*, 3–14. [CrossRef]
48. Sharma, N.L.; Mehta, K.I.; Mahajan, V.K.; Kanga, A.K.; Sharma, V.C.; Tegta, G.R. Cutaneous sporotrichosis of face: Polymorphism and reactivation after intralesional triamcinolone. *Indian J. Dermatol. Venereol. Leprol.* **2007**, *73*, 188–190. [CrossRef]
49. Mehta, K.I.S.; Sharma, N.L.; Kanga, A.K.; Mahajan, V.K.; Ranjan, N. Isolation of *Sporothrix schenckii* from the environmental sources of cutaneous sporotrichosis patients in Himachal Pradesh, India: Results of a pilot study. *Mycoses* **2007**, *50*, 496–501. [CrossRef]
50. Verma, S.; Verma, G.K.; Singh, G.; Kanga, A.; Shanker, V.; Singh, D.; Gupta, P.; Mokta, K.; Sharma, V. Sporotrichosis in Sub-Himalayan India. *PLoS Negl. Trop. Dis.* **2012**, *6*, e1673. [CrossRef]
51. Rudramurthy, S.M.; Shankarnarayan, S.A.; Hemashetter, B.M.; Verma, S.; Chauhan, S.; Nath, R.; Savio, J.; Capoor, M.; Kaur, H.; Ghosh, A.K.; et al. Phenotypic and molecular characterisation of *Sporothrix globosa* of diverse origin from India. *Braz. J. Microbiol. Publ. Braz. Soc. Microbiol.* **2021**, *52*, 91–100. [CrossRef]
52. Fukushiro, R. Epidemiology and ecology of sporotrichosis in Japan. *Zent. Bakteriol. Mikrobiol. Hyg. A* **1984**, *257*, 228–233. [CrossRef]
53. Ishizaki, H.; Kawasaki, M. Molecular epidemiology of *Sporothrix schenckii*. *Nihon Ishinkin Gakkai Zasshi* **2000**, *41*, 245–249. [CrossRef] [PubMed]
54. Takeda, Y.; Kawasaki, M.; Ishizaki, H. Phylogeny and molecular epidemiology of *Sporothrix schenckii* in Japan. *Mycopathologia* **1991**, *116*, 9–14. [CrossRef] [PubMed]
55. Suzuki, K.; Kawasaki, M.; Ishizaki, H. Analysis of restriction profiles of mitochondrial DNA from Sporothrix schenckii and related fungi. *Mycopathologia* **1988**, *103*, 147–151. [CrossRef] [PubMed]
56. Ishizaki, H.; Kawasaki, M.; Aoki, M.; Wu, S.; Lin, J.; Kim, J.A.; Won, Y.H.; Calvo, C.R. Mitochondrial DNA analysis of *Sporothrix schenckii* from China, Korea and Spain. *Nihon Ishinkin Gakkai Zasshi* **2004**, *45*, 23–25. [CrossRef] [PubMed]

57. Ishizaki, H.; Kawasaki, M.; Aoki, M.; Vismer, H.; Muir, D. Mitochondrial DNA analysis of *Sporothrix schenckii* in South Africa and Australia. *Med. Mycol.* **2000**, *38*, 433–436. [CrossRef]
58. Ishizaki, H.; Kawasaki, M.; Aoki, M.; Matsumoto, T.; Padhye, A.A.; Mendoza, M.; Negroni, R. Mitochondrial DNA analysis of Sporothrix schenckii in North and South America. *Mycopathologia* **1998**, *142*, 115–118. [CrossRef]
59. Mochizuki, H.; Anzawa, K.; Mochizuki, T. Genotyping of intraspecies polymorphisms of *Sporothrix globosa* using partial sequence of mitochondrial DNA. *J. Derm.* **2022**, *49*, 263–271. [CrossRef]
60. Kawasaki, M.; Anzawa, K.; Mochizuki, T.; Ishizaki, H. New strain typing method with *Sporothrix schenckii* using mitochondrial DNA and polymerase chain reaction restriction fragment length polymorphism (PCR–RFLP) technique. *J. Derm.* **2012**, *39*, 362–365. [CrossRef] [PubMed]
61. Moussa, T.A.; Kadasa, N.M.; Al Zahrani, H.S.; Ahmed, S.A.; Feng, P.; Gerrits van den Ende, A.H.; Zhang, Y.; Kano, R.; Li, F.; Li, S.; et al. Origin and distribution of *Sporothrix globosa* causing sapronoses in Asia. *J. Med. Microbiol.* **2017**, *66*, 560–569. [CrossRef]
62. Liu, T.-T.; Zhang, K.; Zhou, X. Molecular identification of *Sporothrix* clinical isolates in China. *J. Zhejiang Univ. Sci. B* **2014**, *15*, 100–108. [CrossRef]
63. Yu, X.; Wan, Z.; Zhang, Z.; Li, F.; Li, R.; Liu, X. Phenotypic and molecular identification of *Sporothrix* isolates of clinical origin in Northeast China. *Mycopathologia* **2013**, *176*, 67–74. [CrossRef] [PubMed]
64. Lv, S.; Hu, X.; Liu, Z.; Lin, Y.; Wu, H.; Li, F. Clinical epidemiology of sporotrichosis in Jilin province, China (1990-2019): A series of 4969 cases. *Infect. Drug Resist.* **2022**, *15*, 1753–1765. [CrossRef]
65. Song, Y.; Li, S.S.; Zhong, S.X.; Liu, Y.Y.; Yao, L.; Huo, S.S. Report of 457 sporotrichosis cases from Jilin province, northeast China, a serious endemic region. *J. Eur. Acad. Derm. Venereol.* **2013**, *27*, 313–318. [CrossRef] [PubMed]
66. Yao, L.; Song, Y.; Cui, Y.; Zhou, J.F.; Zhong, S.X.; Zhao, D.Y.; Li, S.S. Pediatric sporotrichosis in Jilin province of China (2010-2016): A retrospective study of 704 cases. *J. Pediatric Infect. Dis. Soc.* **2019**, *9*, 342–348. [CrossRef]
67. Yao, L.; Song, Y.; Zhou, J.F.; Cui, Y.; Li, S.S. Epidemiological and clinical comparisons of pediatric and adult sporotrichosis in Jilin Province, China. *Mycoses* **2020**, *63*, 308–313. [CrossRef] [PubMed]
68. Zhang, Y.; Hagen, F.; Stielow, B.; Rodrigues, A.M.; Samerpitak, K.; Zhou, X.; Feng, P.; Yang, L.; Chen, M.; Deng, S.; et al. Phylogeography and evolutionary patterns in *Sporothrix* spanning more than 14,000 human and animal case reports. *Persoonia* **2015**, *35*, 1–20. [CrossRef]
69. Zamri-Saad, M.; Salmiyah, T.S.; Jasni, S.; Cheng, B.Y.; Basri, K. Feline sporotrichosis: An increasingly important zoonotic disease in Malaysia. *Vet. Rec.* **1990**, *127*, 480.
70. Tang, M.M.; Tang, J.J.; Gill, P.; Chang, C.C.; Baba, R. Cutaneous sporotrichosis: A six-year review of 19 cases in a tertiary referral center in Malaysia. *Int. J. Derm.* **2012**, *51*, 702–708. [CrossRef]
71. Kano, R.; Okubo, M.; Siew, H.H.; Kamata, H.; Hasegawa, A. Molecular typing of *Sporothrix schenckii* isolates from cats in Malaysia. *Mycoses* **2015**, *58*, 220–224. [CrossRef]
72. Beurmann, L.; Ramond, L. Abcès sous-cutanés multiples d'origine mycosique. *Ann. Derm. Syph.* **1903**, *4*, 678–685.
73. Matruchot, L. Sur un nouveau groupe de champignons pathogènes, agents des sporotrichoses. *Comptes Rendus De L'Académie De Sci.* **1910**, *150*, 543–545.
74. Beurmann, L.; Gougerot, H. *Les Sporotrichose*; Librairie Felix Alcan: Paris, France, 1912.
75. Romeo, O.; Scordino, F.; Criseo, G. New insight into molecular phylogeny and epidemiology of *Sporothrix schenckii* species complex based on calmodulin-encoding gene analysis of Italian isolates. *Mycopathologia* **2011**, *172*, 179–186. [CrossRef] [PubMed]
76. Criseo, G.; Romeo, O. Ribosomal DNA sequencing and phylogenetic analysis of environmental *Sporothrix schenckii* strains: Comparison with clinical isolates. *Mycopathologia* **2010**, *169*, 351–358. [CrossRef] [PubMed]
77. Criseo, G.; Malara, G.; Romeo, O.; Puglisi Guerra, A. Lymphocutaneous sporotrichosis in an immunocompetent patient: A case report from extreme southern Italy. *Mycopathologia* **2008**, *166*, 159–162. [CrossRef]
78. Marimon, R.; Cano, J.; Gené, J.; Sutton, D.A.; Kawasaki, M.; Guarro, J. *Sporothrix brasiliensis, S. globosa,* and *S. mexicana,* three new *Sporothrix* species of clinical interest. *J. Clin. Microbiol.* **2007**, *45*, 3198–3206. [CrossRef]
79. Madrid, H.; Cano, J.; Gene, J.; Bonifaz, A.; Toriello, C.; Guarro, J. *Sporothrix globosa,* a pathogenic fungus with widespread geographical distribution. *Rev. Iberoam. De Micol.* **2009**, *26*, 218–222. [CrossRef]
80. Rachman, R.; Ligaj, M.; Chinthapalli, S.; Serafino Wani, R. Zoonotic acquisition of cutaneous *Sporothrix brasiliensis* infection in the UK. *BMJ Case Rep.* **2022**, *15*, e248418. [CrossRef]
81. McGuinness, S.L.; Boyd, R.; Kidd, S.; McLeod, C.; Krause, V.L.; Ralph, A.P. Epidemiological investigation of an outbreak of cutaneous sporotrichosis, Northern Territory, Australia. *BMC Infect. Dis.* **2016**, *16*, 16. [CrossRef]
82. New, D.; Beukers, A.G.; Kidd, S.E.; Merritt, A.J.; Weeks, K.; van Hal, S.J.; Arthur, I. Identification of multiple species and subpopulations among Australian clinical *Sporothrix* isolates using whole genome sequencing. *Med. Mycol.* **2019**, *57*, 905–908. [CrossRef]
83. Schauer, A.; Campbell, A.J.; Arthur, I.; Blyth, C.C.; Bowen, A.C. Spotting sporotrichosis skin infection: The first Australian paediatric case series. *J. Paediatr. Child Health* **2020**, *56*, 408–410. [CrossRef]
84. Mackay, B.; Menrath, V.; Ridley, M.; Kelly, W. Sporotrichosis in a cat. *Aust. Vet Pr.* **1986**, *16*, 3–5.
85. Thomson, J.; Trott, D.J.; Malik, R.; Galgut, B.; McAllister, M.M.; Nimmo, J.; Renton, D.; Kidd, S.E. An atypical cause of sporotrichosis in a cat. *Med. Mycol. Case Rep.* **2019**, *23*, 72–76. [CrossRef] [PubMed]

86. Dixon, D.M.; Salkin, I.F.; Duncan, R.A.; Hurd, N.J.; Haines, J.H.; Kemna, M.E.; Coles, F.B. Isolation and characterization of *Sporothrix schenckii* from clinical and environmental sources associated with the largest U.S. epidemic of sporotrichosis. *J. Clin. Microbiol.* **1991**, *29*, 1106–1113. [CrossRef] [PubMed]
87. Coles, F.B.; Schuchat, A.; Hibbs, J.R.; Kondracki, S.F.; Salkin, I.F.; Dixon, D.M.; Chang, H.G.; Duncan, R.A.; Hurd, N.J.; Morse, D.L. A multistate outbreak of sporotrichosis associated with *Sphagnum* moss. *Am. J. Epidemiol.* **1992**, *136*, 475–487. [CrossRef] [PubMed]
88. Gold, J.A.W.; Derado, G.; Mody, R.K.; Benedict, K. Sporotrichosis-associated hospitalizations, United States, 2000–2013. *Emerg. Infect. Dis.* **2016**, *22*, 1817–1820. [CrossRef] [PubMed]
89. Queiroz-Telles, F.; Fahal, A.H.; Falci, D.R.; Caceres, D.H.; Chiller, T.; Pasqualotto, A.C. Neglected endemic mycoses. *Lancet Infect. Dis.* **2017**, *17*, e367–e377. [CrossRef]
90. Hernández-Castro, R.; Pinto-Almazán, R.; Arenas, R.; Sánchez-Cárdenas, C.D.; Espinosa-Hernández, V.M.; Sierra-Maeda, K.Y.; Conde-Cuevas, E.; Juárez-Durán, E.R.; Xicohtencatl-Cortes, J.; Carrillo-Casas, E.M.; et al. Epidemiology of clinical sporotrichosis in the Americas in the last ten years. *J. Fungi* **2022**, *8*, 588. [CrossRef]
91. Rodrigues, A.M.; de Hoog, G.S.; Zhang, Y.; Camargo, Z.P. Emerging sporotrichosis is driven by clonal and recombinant *Sporothrix* species. *Emerg. Microbes Infect.* **2014**, *3*, e32. [CrossRef]
92. Florez-Munoz, S.V.; Alzate, J.F.; Mesa-Arango, A.C. Molecular identification and antifungal susceptibility of clinical isolates of *Sporothrix schenckii* complex in Medellin, Colombia. *Mycopathologia* **2019**, *184*, 53–63. [CrossRef]
93. Ramirez Soto, M.C. Sporotrichosis: The Story of an Endemic Region in Peru over 28 Years (1985 to 2012). *PLoS ONE* **2015**, *10*, e0127924. [CrossRef]
94. Pappas, P.G.; Tellez, I.; Deep, A.E.; Nolasco, D.; Holgado, W.; Bustamante, B. Sporotrichosis in Peru: Description of an area of hyperendemicity. *Clin. Infect Dis.* **2000**, *30*, 65–70. [CrossRef] [PubMed]
95. Camacho, E.; León-Navarro, I.; Rodríguez-Brito, S.; Mendoza, M.; Niño-Vega, G.A. Molecular epidemiology of human sporotrichosis in Venezuela reveals high frequency of *Sporothrix globosa*. *BMC Infect Dis.* **2015**, *15*, 94. [CrossRef] [PubMed]
96. Rodrigues, A.M.; de Hoog, G.S.; de Camargo, Z.P. Molecular diagnosis of pathogenic *Sporothrix* species. *PLoS Negl. Trop. Dis.* **2015**, *9*, e0004190. [CrossRef] [PubMed]
97. Rodrigues, A.M.; de Hoog, G.S.; de Cassia Pires, D.; Brilhante, R.S.N.; da Costa Sidrim, J.J.; Gadelha, M.F.; Colombo, A.L.; de Camargo, Z.P. Genetic diversity and antifungal susceptibility profiles in causative agents of sporotrichosis. *BMC Infect. Dis.* **2014**, *14*, 219. [CrossRef]
98. Rodrigues, A.M.; de Hoog, S.; de Camargo, Z.P. Emergence of pathogenicity in the *Sporothrix schenckii* complex. *Med. Mycol.* **2013**, *51*, 405–412. [CrossRef]
99. Rodrigues, A.M.; de Melo Teixeira, M.; de Hoog, G.S.; Schubach, T.M.P.; Pereira, S.A.; Fernandes, G.F.; Bezerra, L.M.L.; Felipe, M.S.; de Camargo, Z.P. Phylogenetic analysis reveals a high prevalence of *Sporothrix brasiliensis* in feline sporotrichosis outbreaks. *PLoS Negl. Trop. Dis.* **2013**, *7*, e2281. [CrossRef]
100. Falcão, E.M.M.; de Lima Filho, J.B.; Campos, D.P.; Valle, A.C.F.d.; Bastos, F.I.; Gutierrez-Galhardo, M.C.; Freitas, D.F.S. [Hospitalizations and deaths related to sporotrichosis in Brazil (1992–2015)]. *Cad. De Saude Publica* **2019**, *35*, e00109218. [CrossRef]
101. de Carvalho, J.A.; Beale, M.A.; Hagen, F.; Fisher, M.C.; Kano, R.; Bonifaz, A.; Toriello, C.; Negroni, R.; Rego, R.S.M.; Gremiao, I.D.F.; et al. Trends in the molecular epidemiology and population genetics of emerging *Sporothrix* species. *Stud. Mycol.* **2021**, *100*, 100129. [CrossRef]
102. Araujo, A.K.L.; de Santana Leal, C.A. Feline sporotrichosis in the municipality of Bezerros, Agreste Pernambucano: Case report. *Pubvet* **2016**, *10*, 795–872. [CrossRef]
103. Nunes, G.D.L.; dos Santos Carneiro, R.; Filgueira, K.D.; Filgueira, F.G.F.; Fernandes, T.H.T. Feline sporotrichosis in Itaporanga municipality, Paraíba state, Brazil: Case report. *Arq. De Ciências Veterinárias E Zool. Da UNIPAR* **2011**, *14*.
104. Filgueira, K.D. Sporotrichosis in the canine species: A case report on city of Mossoro, RN. *Cienc. Anim. Bras.* **2009**, *10*, 673–677.
105. de Oliveira Bento, A.; de Sena Costa, A.S.; Lima, S.L.; do Monte Alves, M.; de Azevedo Melo, A.S.; Rodrigues, A.M.; da Silva-Rocha, W.P.; Milan, E.P.; Chaves, G.M. The spread of cat-transmitted sporotrichosis due to *Sporothrix brasiliensis* in Brazil towards the Northeast region. *PLoS Negl. Trop. Dis.* **2021**, *15*, e0009693. [CrossRef] [PubMed]
106. Schubach, T.M.; Schubach, A.; Okamoto, T.; Barros, M.B.; Figueiredo, F.B.; Cuzzi, T.; Fialho-Monteiro, P.C.; Reis, R.S.; Perez, M.A.; Wanke, B. Evaluation of an epidemic of sporotrichosis in cats: 347 cases (1998–2001). *J. Am. Vet. Med. Assoc.* **2004**, *224*, 1623–1629. [CrossRef]
107. Schubach, A.O.; Schubach, T.M.; Barros, M.B. Epidemic cat-transmitted sporotrichosis. *N. Engl. J. Med.* **2005**, *353*, 1185–1186. [CrossRef] [PubMed]
108. Falcao, E.M.M.; Pires, M.C.S.; Andrade, H.B.; Goncalves, M.L.C.; Almeida-Paes, R.; do Valle, A.C.F.; Bastos, F.I.; Gutierrez-Galhardo, M.C.; Freitas, D.F.S. Zoonotic sporotrichosis with greater severity in Rio de Janeiro, Brazil: 118 hospitalizations and 11 deaths in the last 2 decades in a reference institution. *Med. Mycol.* **2019**, *58*, 141–143. [CrossRef]
109. Córdoba, S.; Isla, G.; Szusz, W.; Vivot, W.; Hevia, A.; Davel, G.; Canteros, C.E. Molecular identification and susceptibility profile of *Sporothrix schenckii sensu lato* isolated in Argentina. *Mycoses* **2018**, *61*, 441–448. [CrossRef] [PubMed]
110. Etchecopaz, A.N.; Lanza, N.; Toscanini, M.A.; Devoto, T.B.; Pola, S.J.; Daneri, G.L.; Iovannitti, C.A.; Cuestas, M.L. Sporotrichosis caused by *Sporothrix brasiliensis* in Argentina: Case report, molecular identification and in vitro susceptibility pattern to antifungal drugs. *J. De Mycol. Med.* **2019**, *30*, 100908. [CrossRef]

111. García Duarte, J.M.; Wattiez Acosta, V.R.; Fornerón Viera, P.M.L.; Aldama Caballero, A.; Gorostiaga Matiauda, G.A.; Rivelli de Oddone, V.B.; Pereira Brunelli, J.G. Sporotrichosis transmitted by domestic cat. A family case report. *Rev. Del Nac.* **2017**, *9*, 67–76. [CrossRef]
112. Rios, M.E.; Suarez, J.M.D.; Moreno, J.; Vallee, J.; Moreno, J.P. Zoonotic sporotrichosis related to cat contact: First case report from Panama in Central America. *Cureus* **2018**, *10*, e2906. [CrossRef]
113. PAHO. *Sporothrix brasiliensis*, an Emerging Fungal Pathogen, Notable for its Zoonotic Transmission and Epidemic Potential for Human and Animal Health in the Americas. Available online: https://www.someve.com.ar/images/noticias/2019/S-brasiliensis_lasAmericas_30082019_ES.pdf (accessed on 30 August 2019).
114. Montenegro, H.; Rodrigues, A.M.; Galvão Dias, M.A.; da Silva, E.A.; Bernardi, F.; Camargo, Z.P. Feline sporotrichosis due to *Sporothrix brasiliensis*: An emerging animal infection in São Paulo, Brazil. *BMC Vet Res.* **2014**, *10*, 269. [CrossRef]
115. Zhou, X.; Rodrigues, A.M.; Feng, P.; Hoog, G.S. Global ITS diversity in the *Sporothrix schenckii* complex. *Fungal Divers.* **2014**, *66*, 153–165. [CrossRef]
116. Rediguieri, B.C.; da Cruz Bahiense, I.; de Carvalho, J.A.; Leite, G.R.; Falqueto, A.; Rodrigues, A.M.; Gonçalves, S.S. Clinical, epidemiological, and epizootic features of *Sporothrix brasiliensis* in Espírito Santo, Brazil. *Ecohealth* **2022**, *19*, 124–134. [CrossRef] [PubMed]
117. Maschio-Lima, T.; Marques, M.D.R.; Lemes, T.H.; Brizzotti-Mazuchi, N.S.; Caetano, M.H.; de Almeida, B.G.; Bianco, L.M.; Monteiro, R.C.; Rodrigues, A.M.; de Camargo, Z.P.; et al. Clinical and epidemiological aspects of feline sporotrichosis caused by *Sporothrix brasiliensis* and in vitro antifungal susceptibility. *Vet. Res. Commun.* **2021**, *45*, 171–179. [CrossRef]
118. da Cruz Bahiense Rocha, I.; Terra, P.P.D.; Cardoso de Oliveira, R.; Lubianca Zanotti, R.; Falqueto, A.; de Camargo, Z.P.; Rodrigues, A.M.; Goncalves, S.S. Molecular-based assessment of diversity and population structure of *Sporothrix* spp. clinical isolates from Espirito Santo-Brazil. *Mycoses* **2021**, *64*, 420–427. [CrossRef] [PubMed]
119. Macêdo-Sales, P.A.; Souza, L.O.P.; Della-Terra, P.P.; Lozoya-Pérez, N.E.; Machado, R.L.D.; Rocha, E.M.d.S.d.; Lopes-Bezerra, L.M.; Guimarães, A.J.; Rodrigues, A.M.; Mora-Montes, H.M.; et al. Coinfection of domestic felines by distinct *Sporothrix brasiliensis* in the Brazilian sporotrichosis hyperendemic area. *Fungal Genet. Biol.* **2020**, *140*, 103397. [CrossRef] [PubMed]
120. Macêdo-Sales, P.A.; Souto, S.R.L.S.; Destefani, C.A.; Lucena, R.P.; Machado, R.L.D.; Pinto, M.R.; Rodrigues, A.M.; Lopes-Bezerra, L.M.; Rocha, E.M.S.; Baptista, A.R.S. Domestic feline contribution in the transmission of *Sporothrix* in Rio de Janeiro State, Brazil: A comparison between infected and non-infected populations. *BMC Vet. Res.* **2018**, *14*, 19. [CrossRef]
121. Teixeira, M.d.M.; Rodrigues, A.M.; Tsui, C.K.M.; de Almeida, L.G.P.; Van Diepeningen, A.D.; Gerrits van den Ende, B.; Fernandes, G.F.; Kano, R.; Hamelin, R.C.; Lopes-Bezerra, L.M.; et al. Asexual propagation of a virulent clone complex in human and feline outbreak of sporotrichosis. *Eukaryot Cell* **2015**, *14*, 158–169. [CrossRef]
122. Rodrigues, A.M.; Fernandes, G.F.; Araujo, L.M.; Della Terra, P.P.; Dos Santos, P.O.; Pereira, S.A.; Schubach, T.M.; Burger, E.; Lopes-Bezerra, L.M.; de Camargo, Z.P. Proteomics-based characterization of the humoral immune response in sporotrichosis: Toward discovery of potential diagnostic and vaccine antigens. *PLoS Negl. Trop. Dis.* **2015**, *9*, e0004016. [CrossRef]
123. Freitas, D.C.; Migliano, M.F.; Zani Neto, L. Sporotrichosis. Observation of spontaneous case in domestic cat (*Felis catus*). *Rev. Fac. Med. Vet. Univ. Sao Paulo* **1956**, *5*, 601–604. [CrossRef]
124. Freitas, D.C.; Moreno, G.; Saliba, A.M.; Botino, J.Á.; Mós, E.M. Sporotrichosis in dogs and cats. *Rev. Fac. Med. Vet Univ. Sao Paulo* **1965**, *7*, 381–387. [CrossRef]
125. Almeida, F.; Sampaio, S.A.P.; Lacaz, C.S.; Fernandes, J.C. Statistical data on sporotrichosis. Analysis of 344 cases. *An. Bras. De Dermatol.* **1955**, *30*, 9–12.
126. Lacaz, C.S.; Porto, E.; Martins, J.E.C.; Heins-Vaccari, E.M.; de Melo, N.T. (Eds.) *Tratado de Micologia Médica*, 9th ed.; Sarvier: São Paulo, Brazil, 2002; p. 1104.
127. Marques, S.A.; Franco, S.R.; de Camargo, R.M.; Dias, L.D.; Haddad Junior, V.; Fabris, V.E. Sporotrichosis of the domestic cat (*Felis catus*): Human transmission. *Rev. Do Inst. De Med. Trop. De Sao Paulo* **1993**, *35*, 327–330.
128. Baroni, F.; Campos, S.; Direito, G. Sporotrichosis in a cat (a case description). *Rev. Bras. Med. Vet.* **1998**, *20*, 25–27.
129. Barros, M.B.d.L.; Schubach, T.P.; Coll, J.O.; Gremião, I.D.; Wanke, B.; Schubach, A. Sporotrichosis: Development and challenges of an epidemic. *Rev. Panam. Salud Publica* **2010**, *27*, 455–460. [CrossRef] [PubMed]
130. Gremião, I.D.F.; Oliveira, M.M.E.; Monteiro de Miranda, L.H.; Saraiva Freitas, D.F.; Pereira, S.A. Geographic expansion of sporotrichosis, Brazil. *Emerg. Infect. Dis.* **2020**, *26*, 621–624. [CrossRef]
131. Larsson, C.E.; Goncalves Mde, A.; Araujo, V.C.; Dagli, M.L.; Correa, B.; Fava Neto, C. Feline sporotrichosis: Clinical and zoonotic aspects. *Rev. Do Inst. De Med. Trop. De Sao Paulo* **1989**, *31*, 351–358.
132. Borges, T.S.; Rossi, C.N.; Fedullo, J.D.; Taborda, C.P.; Larsson, C.E. Isolation of *Sporothrix schenckii* from the claws of domestic cats (indoor and outdoor) and in captivity in Sao Paulo (Brazil). *Mycopathologia* **2013**, *176*, 129–137. [CrossRef]
133. Silva, C.E.; Valeriano, C.A.T.; Ferraz, C.E.; Neves, R.P.; Oliveira, M.M.E.; Silva, J.C.A.L.; Magalhães, V.; Lima-Neto, R.G. Epidemiological features and geographical expansion of sporotrichosis in the state of Pernambuco, northeastern Brazil. *Future Microbiol.* **2021**, *16*, 1371–1379. [CrossRef]
134. Larsson, C.E. Sporotrichosis. *Braz. J. Vet. Res. Anim. Sci.* **2011**, *48*, 250–259. [CrossRef]
135. Almeida, F.P. *Micologia Médica: Estudo das Micoses Humanas e de Seus Cogumelos*; Companhia Melhoramentos de São Paulo: São Paulo, Brazil, 1939; p. 710.

136. Fernandes, C.G.N.; Moura, S.T.d.; Dantas, A.F.M.; Blatt, M.C.S. Feline sporotrichosis: Clinical epidemiological aspects: Case reports (Cuiabá, Mato Grosso, Brasil). *MEDVEP. Rev. Cient. Med. Vet.* **2004**, *2*, 39–43.
137. SES-MS. Sporotrichosis Technical Note—Nº 1/2021-GTZ/CEVE/DGVS/SES. Available online: https://www.vs.saude.ms.gov.br/wp-content/uploads/2021/08/Nota-tecnica-esporotricose.pdf (accessed on 17 June 2022).
138. CRMV-SE. Technical Note No. 01/2021. Available online: https://www.crmvse.org.br/wp-content/uploads/2021/06/Nota-Tecnica-01-2021.pdf (accessed on 23 June 2022).
139. SES-AM. Joint Technical Note—Nº 006/2021—GEVEP/DEVAE/DAP/DRA/SUBGS. Available online: https://semsa.manaus.am.gov.br/wp-content/uploads/2021/05/Nota-Tecnica_Fluxo_APS_e_Vigilancia_e_Ficha_de_Notificacao_da_Esporotricose_assinadas.pdf (accessed on 17 June 2022).
140. COVISA. Technical Note 09: DVE/DVZ/COVISA/2020—Surveillance and Clinical Management of Human Sporotrichosis in the Municipality of São Paulo. Available online: https://docs.bvsalud.org/biblioref/2020/08/1102196/nota-tecnica-09-dve-zoo-2020_esporotricose_v6-alterada-a-pedid_CBJA7E3.pdf (accessed on 17 June 2022).
141. Bison, I. Feline Sporotrichosis: Literature Review. Bachelor's Thesis, Federal University of Rondônia, Rolim de Moura, Rondônia, 2019.
142. Boechat, J.S.; Oliveira, M.M.E.; Almeida-Paes, R.; Gremiao, I.D.F.; Machado, A.C.S.; Oliveira, R.V.C.; Figueiredo, A.B.F.; Rabello, V.B.S.; Silva, K.B.L.; Zancope-Oliveira, R.M.; et al. Feline sporotrichosis: Associations between clinical-epidemiological profiles and phenotypic-genotypic characteristics of the etiological agents in the Rio de Janeiro epizootic area. *Mem. Do Inst. Oswaldo Cruz* **2018**, *113*, 185–196. [CrossRef] [PubMed]
143. de Souza, E.W.; Borba, C.M.; Pereira, S.A.; Gremiao, I.D.F.; Langohr, I.M.; Oliveira, M.M.E.; de Oliveira, R.V.C.; da Cunha, C.R.; Zancope-Oliveira, R.M.; de Miranda, L.H.M.; et al. Clinical features, fungal load, coinfections, histological skin changes, and itraconazole treatment response of cats with sporotrichosis caused by *Sporothrix brasiliensis*. *Sci. Rep.* **2018**, *8*, 9074. [CrossRef] [PubMed]
144. de Araujo, M.L.; Rodrigues, A.M.; Fernandes, G.F.; de Camargo, Z.P.; de Hoog, G.S. Human sporotrichosis beyond the epidemic front reveals classical transmission types in Espírito Santo, Brazil. *Mycoses* **2015**, *58*, 485–490. [CrossRef] [PubMed]
145. Colodel, M.M.; Jark, P.C.; Ramos, C.J.R.; Martins, V.M.V.; Schneider, A.F.; Pilati, C. Feline cutaneous sporotrichosis in the state of Santa Catarina: Case reports. *Vet. Foco* **2009**, *7*, 18–27.
146. Cordeiro, F.N.; Bruno, C.B.; Paula, C.D.; Motta Jde, O. Familial occurrence of zoonotic sporotrichosis. *An. Bras. De Dermatol.* **2011**, *86*, S121–S124. [CrossRef] [PubMed]
147. Silva, G.M.; Howes, J.C.F.; Leal, C.A.S.; Mesquita, E.P.; Pedrosa, C.M.; Oliveira, A.A.F.; Silva, L.B.G.; Mota, R.A. Feline sporotrichosis outbreak in metropolitan Recife. *Pesqui. Vet. Bras.* **2018**, *38*, 1767–1771. [CrossRef]
148. Marques-Melo, E.H.; Lessa, D.F.d.S.; Garrido, L.H.A.; Nunes, A.C.B.T.; Chaves, K.P.; Porto, W.J.N.; Notomi, M. Domestic feline as sporotricosis transmitting agent for human: Report of the first case in the state of Alagoas. *Rev. Baiana De Saúde Pública* **2014**, *38*, 490–498.
149. Madrid, I.M.; Mattei, A.; Martins, A.; Nobre, M.; Meireles, M. Feline sporotrichosis in the southern region of Rio Grande do Sul, Brazil: Clinical, zoonotic and therapeutic aspects. *Zoonoses Public Health* **2010**, *57*, 151–154. [CrossRef]
150. Sanchotene, K.O.; Madrid, I.M.; Klafke, G.B.; Bergamashi, M.; Terra, P.P.D.; Rodrigues, A.M.; de Camargo, Z.P.; Xavier, M.O. *Sporothrix brasiliensis* outbreaks and the rapid emergence of feline sporotrichosis. *Mycoses* **2015**, *58*, 652–658. [CrossRef]
151. Madrid, I.M.; Mattei, A.S.; Fernandes, C.G.; Oliveira Nobre, M.; Meireles, M.C.A. Epidemiological findings and laboratory evaluation of sporotrichosis: A description of 103 cases in cats and dogs in Southern Brazil. *Mycopathologia* **2012**, *173*, 265–273. [CrossRef]
152. Eudes Filho, J.; Santos, I.B.D.; Reis, C.M.S.; Patané, J.S.L.; Paredes, V.; Bernardes, J.; Poggiani, S.; Castro, T.C.B.; Gomez, O.M.; Pereira, S.A.; et al. A novel *Sporothrix brasiliensis* genomic variant in Midwestern Brazil: Evidence for an older and wider sporotrichosis epidemic. *Emerg. Microbes Infect.* **2020**, *9*, 2515–2525. [CrossRef] [PubMed]
153. Bernardes-Engemann, A.R.; Almeida, M.A.; Bison, I.; Rabello, V.B.S.; Ramos, M.L.M.; Pereira, S.A.; Almeida-Paes, R.; de Lima Brasil, A.W.; Zancopé-Oliveira, R.M. Anti-*Sporothrix* Antibody Detection in Domestic Cats as an Indicator of a Possible New Occurrence Area for Sporotrichosis in North Brazil. *Mycopathologia* **2022**, 1–10. [CrossRef] [PubMed]
154. Schmidt, P.M.; Lopez, R.R.; Collier, B.A. Survival, fecundity, and movements of free-roaming cats. *J. Wildl. Manag.* **2007**, *71*, 915–919. [CrossRef]
155. Jongman, E.C. Adaptation of domestic cats to confinement. *J. Vet. Behav.* **2007**, *2*, 193–196. [CrossRef]
156. Horn, J.A.; Mateus-Pinilla, N.; Warner, R.E.; Heske, E.J. Home range, habitat use, and activity patterns of free-roaming domestic cats. *J. Wildl. Manag.* **2011**, *75*, 1177–1185. [CrossRef]
157. Keesing, F.; Belden, L.K.; Daszak, P.; Dobson, A.; Harvell, C.D.; Holt, R.D.; Hudson, P.; Jolles, A.; Jones, K.E.; Mitchell, C.E.; et al. Impacts of biodiversity on the emergence and transmission of infectious diseases. *Nature* **2010**, *468*, 647–652. [CrossRef]
158. Jones, K.E.; Patel, N.G.; Levy, M.A.; Storeygard, A.; Balk, D.; Gittleman, J.L.; Daszak, P. Global trends in emerging infectious diseases. *Nature* **2008**, *451*, 990–993. [CrossRef]
159. Shaheen, M.N.F. The concept of one health applied to the problem of zoonotic diseases. *Rev. Med. Virol.* **2022**, *32*, e2326. [CrossRef] [PubMed]
160. Winck, G.R.; Raimundo, R.L.G.; Fernandes-Ferreira, H.; Bueno, M.G.; D'Andrea, P.S.; Rocha, F.L.; Cruz, G.L.T.; Vilar, E.M.; Brandão, M.; Cordeiro, J.L.P.; et al. Socioecological vulnerability and the risk of zoonotic disease emergence in Brazil. *Sci. Adv.* **2022**, *8*, eabo5774. [CrossRef]

161. Bloom, D.E.; Cadarette, D. Infectious disease threats in the twenty-first century: Strengthening the global response. *Front. Immunol.* **2019**, *10*, 549. [CrossRef]
162. Alzuguir, C.L.C.; Pereira, S.A.; Magalhães, M.A.F.M.; Almeida-Paes, R.; Freitas, D.F.S.; Oliveira, L.F.A.; Pimentel, M.I.F. Geo-epidemiology and socioeconomic aspects of human sporotrichosis in the municipality of Duque de Caxias, Rio de Janeiro, Brazil, between 2007 and 2016. *Trans. R. Soc. Trop. Med. Hyg.* **2020**, *114*, 99–106. [CrossRef] [PubMed]
163. Silva, M.B.; Costa, M.M.; Torres, C.C.; Galhardo, M.C.; Valle, A.C.; Magalhaes Mde, A.; Sabroza, P.C.; Oliveira, R.M. [Urban sporotrichosis: A neglected epidemic in Rio de Janeiro, Brazil]. *Cad. De Saude Publica* **2012**, *28*, 1867–1880. [CrossRef] [PubMed]
164. Veasey, J.V.; Carvalho, G.d.S.M.; Ruiz, L.R.B.; Neves Neto, M.F.; Zaitz, C. Epidemiological and geographical distribution profile of urban sporotrichosis in the city of São Paulo. *An. Bras. De Dermatol.* **2022**, *97*, 228–230. [CrossRef] [PubMed]
165. Ellwanger, J.H.; Kulmann-Leal, B.; Kaminski, V.L.; Valverde-Villegas, J.M.; Veiga, A.; Spilki, F.R.; Fearnside, P.M.; Caesar, L.; Giatti, L.L.; Wallau, G.L.; et al. Beyond diversity loss and climate change: Impacts of Amazon deforestation on infectious diseases and public health. *An. Da Acad. Bras. De Cienc.* **2020**, *92*, e20191375. [CrossRef]
166. Pörtner, H.-O.; Roberts, D.C.; Adams, H.; Adelekan, I.; Adler, C.; Adrian, R.; Aldunce, P.; Ali, E.; Ara Begum, R.; Bednar-Friedl, B.; et al. *IPCC, 2022: Climate Change 2022: Impacts, Adaptation, and Vulnerability: Contribution of Working Group II to the Sixth Assessment Report of the Intergovernmental Panel on Climate Change*; Pörtner, H.-O., Roberts, D.C., Tignor, M., Poloczanska, E.S., Mintenbeck, K., Alegría, A., Craig, M., Langsdorf, S., Löschke, S., Möller, V., et al., Eds.; Cambridge University Press: Cambridge, UK, 2022.
167. Jones Bryony, A.; Grace, D.; Kock, R.; Alonso, S.; Rushton, J.; Said Mohammed, Y.; McKeever, D.; Mutua, F.; Young, J.; McDermott, J.; et al. Zoonosis emergence linked to agricultural intensification and environmental change. *Proc. Natl. Acad. Sci. USA* **2013**, *110*, 8399–8404. [CrossRef]
168. Rabello, V.B.S.; Almeida-Silva, F.; Scramignon-Costa, B.d.S.; Motta, B.d.S.; de Macedo, P.M.; Teixeira, M.d.M.; Almeida-Paes, R.; Irinyi, L.; Meyer, W.; Zancopé-Oliveira, R.M. Environmental isolation of *Sporothrix brasiliensis* in an area with recurrent feline sporotrichosis cases. *Front. Cell. Infect. Microbiol.* **2022**, *12*, 894297. [CrossRef]
169. Almeida-Silva, F.; Rabello, V.B.; Scramignon-Costa, B.D.; Zancopé-Oliveira, R.M.; de Macedo, P.M.; Almeida-Paes, R. Beyond domestic cats: Environmental detection of *Sporothrix brasiliensis* DNA in a hyperendemic area of sporotrichosis in Rio de Janeiro state, Brazil. *J. Fungi* **2022**, *8*, 604. [CrossRef]
170. Navarrete, A.A.; Tsai, S.M.; Mendes, L.W.; Faust, K.; de Hollander, M.; Cassman, N.A.; Raes, J.; van Veen, J.A.; Kuramae, E.E. Soil microbiome responses to the short-term effects of Amazonian deforestation. *Mol. Ecol.* **2015**, *24*, 2433–2448. [CrossRef]
171. Khan, M.A.W.; Bohannan, B.J.M.; Nüsslein, K.; Tiedje, J.M.; Tringe, S.G.; Parlade, E.; Barberán, A.; Rodrigues, J.L.M. Deforestation impacts network co-occurrence patterns of microbial communities in Amazon soils. *FEMS Microbiol. Ecol.* **2019**, *95*, fiy230. [CrossRef]
172. Rosenberg, K.; Bertaux, J.; Krome, K.; Hartmann, A.; Scheu, S.; Bonkowski, M. Soil amoebae rapidly change bacterial community composition in the rhizosphere of *Arabidopsis thaliana*. *ISME J.* **2009**, *3*, 675–684. [CrossRef]
173. Steenbergen, J.N.; Nosanchuk, J.D.; Malliaris, S.D.; Casadevall, A. Interaction of *Blastomyces dermatitidis*, *Sporothrix schenckii*, and *Histoplasma capsulatum* with *Acanthamoeba castellanii*. *Infect. Immun.* **2004**, *72*, 3478–3488. [CrossRef] [PubMed]
174. Lemos Tavares, P.; Carvalho Ribeiro, A.; Kercher Berte, F.; da Silva Hellwig, A.H.; Machado Pagani, D.; Tavares de Souza, C.C.; Brittes Rott, M.; Scroferneker, M.L. The interaction between *Sporothrix schenckii sensu stricto* and *Sporothrix brasiliensis* with *Acanthamoeba castellanii*. *Mycoses* **2020**, *63*, 302–307. [CrossRef] [PubMed]
175. Aqeel, Y.; Siddiqui, R.; Iftikhar, H.; Khan, N.A. The effect of different environmental conditions on the encystation of *Acanthamoeba castellanii* belonging to the T4 genotype. *Exp. Parasitol.* **2013**, *135*, 30–35. [CrossRef] [PubMed]
176. Rossow, J.A.; Queiroz-Telles, F.; Caceres, D.H.; Beer, K.D.; Jackson, B.R.; Pereira, J.G.; Ferreira Gremião, I.D.; Pereira, S.A. A one health approach to combatting *Sporothrix brasiliensis*: Narrative review of an emerging zoonotic fungal pathogen in South America. *J. Fungi* **2020**, *6*, 247. [CrossRef] [PubMed]
177. Mahajan, V.K. Sporotrichosis: An overview and therapeutic options. *Dermatol. Res. Pract.* **2014**, *2014*, 272376. [CrossRef] [PubMed]
178. Lopes-Bezerra, L.M.; Mora-Montes, H.M.; Zhang, Y.; Nino-Vega, G.; Rodrigues, A.M.; de Camargo, Z.P.; de Hoog, S. Sporotrichosis between 1898 and 2017: The evolution of knowledge on a changeable disease and on emerging etiological agents. *Med. Mycol.* **2018**, *56*, 126–143. [CrossRef]
179. Kwon-Chung, J.K.; Bennett, J.E. *Medical Mycology*; Lea & Febiger: Philadelphia, PA, USA, 1992.
180. Rippon, J.W. *Medical Mycology—The Pathogenic Fungi and the Pathogenic Actinomycetes*, 3rd ed.; W. B. Saunders Company: Philadelphia, PA, USA, 1988.
181. Rodrigues, A.M.; Orofino-Costa, R.; de Camargo, Z.P. *Sporothrix* spp. In *Pocket Guide to Mycological Diagnosis*, 1st ed.; Cordeiro Rde, A., Ed.; CRC Press: Boca Raton, FL, USA, 2019; pp. 99–113. [CrossRef]
182. Rodrigues, A.M.; de Hoog, G.S.; de Camargo, Z.P. Feline Sporotrichosis. In *Emerging and Epizootic Fungal Infections in Animals*; Seyedmousavi, S., de Hoog, G.S., Guillot, J., Verweij, P.E., Eds.; Springer International Publishing: Cham, Switzerland, 2018; Volume 1, pp. 199–231.
183. Miranda, L.H.; Quintella, L.P.; Menezes, R.C.; dos Santos, I.B.; Oliveira, R.V.; Figueiredo, F.B.; Lopes-Bezerra, L.M.; Schubach, T.M. Evaluation of immunohistochemistry for the diagnosis of sporotrichosis in dogs. *Vet. J.* **2011**, *190*, 408–411. [CrossRef]

184. Bernardes-Engemann, A.R.; de Lima Barros, M.; Zeitune, T.; Russi, D.C.; Orofino-Costa, R.; Lopes-Bezerra, L.M. Validation of a serodiagnostic test for sporotrichosis: A follow-up study of patients related to the Rio de Janeiro zoonotic outbreak. *Med. Mycol.* **2015**, *53*, 28–33. [CrossRef]
185. Gezuele, E.; Da Rosa, D. Importance of the sporotrichosis asteroid body for the rapid diagnosis of sporotrichosis. *Rev. Iberoam. De Micol.* **2005**, *22*, 147–150. [CrossRef]
186. Schwarz, J. The diagnosis of deep mycoses by morphologic methods. *Hum. Pathol.* **1982**, *13*, 519–533. [CrossRef]
187. Morris-Jones, R. Sporotrichosis. *Clin. Exp. Dermatol.* **2002**, *27*, 427–431. [CrossRef] [PubMed]
188. de Lima Barros, M.B.; Schubach, A.O.; de Vasconcellos Carvalhaes de Oliveira, R.; Martins, E.B.; Teixeira, J.L.; Wanke, B. Treatment of cutaneous sporotrichosis with Itraconazole—Study of 645 patients. *Clin. Infect. Dis.* **2011**, *52*, e200–e206. [CrossRef] [PubMed]
189. Rodrigues, A.M.; Hagen, F.; de Camargo, Z.P. A spotlight on *Sporothrix* and sporotrichosis. *Mycopathologia* **2022**, 1–5, *in press*. [CrossRef]
190. Gonçalves, S.S.; Cano, J.F.; Stchigel, A.M.; Melo, A.S.; Godoy-Martinez, P.C.; Correa, B.; Guarro, J. Molecular phylogeny and phenotypic variability of clinical and environmental strains of *Aspergillus flavus*. *Fungal Biol.* **2012**, *116*, 1146–1155. [CrossRef] [PubMed]
191. Silva, J.N.; Miranda, L.H.M.; Menezes, R.C.; Gremiao, I.D.F.; Oliveira, R.V.C.; Vieira, S.M.M.; Conceicao-Silva, F.; Ferreiro, L.; Pereira, S.A. Comparison of the sensitivity of three methods for the early diagnosis of sporotrichosis in cats. *J. Comp. Pathol.* **2018**, *160*, 72–78. [CrossRef]
192. Rodrigues, A.M.; Fernandes, G.F.; de Camargo, Z.P. Sporotrichosis. In *Emerging and Re-Emerging Infectious Diseases of Livestock*; Bayry, J., Ed.; Springer: Berlin/Heidelberg, Germany, 2017; pp. 391–421. [CrossRef]
193. Marimon, R.; Gené, J.; Cano, J.; Guarro, J. *Sporothrix luriei*: A rare fungus from clinical origin. *Med. Mycol.* **2008**, *46*, 621–625. [CrossRef]
194. Oliveira, M.M.; Almeida-Paes, R.; Muniz, M.M.; Gutierrez-Galhardo, M.C.; Zancope-Oliveira, R.M. Phenotypic and molecular identification of *Sporothrix* isolates from an epidemic area of sporotrichosis in Brazil. *Mycopathologia* **2011**, *172*, 257–267. [CrossRef]
195. Zhao, M.-d.; Zhou, X.; Liu, T.-t.; Yang, Z.-b. Morphological and physiological comparison of taxa comprising the Sporothrix schenckii complex. *J. Zhejiang Univ. Sci. B* **2015**, *16*, 940–947. [CrossRef]
196. Vásquez-del-Mercado, E.; Arenas, R.; Padilla-Desgarenes, C. Sporotrichosis. *Clin. Derm.* **2012**, *30*, 437–443. [CrossRef]
197. Richardson, M.D.; Warnock, D.W. *Fungal Infection: Diagnosis and Management*, 4th ed.; Wiley-Blackwell: Oxford, UK, 2012.
198. Oyarce, J.A.; García, C.; Alave, J.; Bustamante, B. Epidemiological clinical and laboratory characterization of sporotrichosis in patients of a tertiary care hospital in Lima, Peru, from 1991 to 2014. *Rev. Chil. Infectol.* **2016**, *33*, 315–321. [CrossRef]
199. Arenas, R.; Sanchez-Cardenas, C.D.; Ramirez-Hobak, L.; Ruiz Arriaga, L.F.; Vega Memije, M.E. Sporotrichosis: From KOH to molecular biology. *J. Fungi* **2018**, *4*, 62. [CrossRef]
200. Quintella, L.P.; Passos, S.R.; do Vale, A.C.; Galhardo, M.C.; Barros, M.B.; Cuzzi, T.; Reis Rdos, S.; de Carvalho, M.H.; Zappa, M.B.; Schubach Ade, O. Histopathology of cutaneous sporotrichosis in Rio de Janeiro: A series of 119 consecutive cases. *J. Cutan. Pathol.* **2011**, *38*, 25–32. [CrossRef]
201. Quintella, L.P.; Passos, S.R.; de Miranda, L.H.; Cuzzi, T.; Barros, M.B.; Francesconi-do-Vale, A.C.; Galhardo, M.C.; Madeira Mde, F.; Figueiredo de Carvalho, M.H.; Schubach Ade, O. Proposal of a histopathological predictive rule for the differential diagnosis between American tegumentary leishmaniasis and sporotrichosis skin lesions. *Br. J. Dermatol.* **2012**, *167*, 837–846. [CrossRef]
202. da Rosa, W.D.; Gezuele, E.; Calegari, L.; Goni, F. Asteroid body in sporotrichosis. Yeast viability and biological significance within the host immune response. *Med. Mycol.* **2008**, *46*, 443–448. [CrossRef] [PubMed]
203. Civila, E.S.; Bonasse, J.; Conti-Diaz, I.A.; Vignale, R.A. Importance of the direct fresh examination in the diagnosis of cutaneous sporotrichosis. *Int. J. Derm.* **2004**, *43*, 808–810. [CrossRef] [PubMed]
204. Miranda, L.H.; Conceicao-Silva, F.; Quintella, L.P.; Kuraiem, B.P.; Pereira, S.A.; Schubach, T.M. Feline sporotrichosis: Histopathological profile of cutaneous lesions and their correlation with clinical presentation. *Comp. Immunol. Microbiol. Infect. Dis.* **2013**, *36*, 425–432. [CrossRef] [PubMed]
205. Welsh, R.D. Sporotrichosis. *J. Am. Vet. Med. Assoc.* **2003**, *223*, 1123–1126. [CrossRef]
206. Pereira, S.A.; Menezes, R.C.; Gremião, I.D.F.; Silva, J.N.; de O. Honse, C.; Figueiredo, C.; da Silva, D.T.; Kitada, A.A.B.; dos Reis, É.G.; Schubach, T.M.P. Sensitivity of cytopathological examination in the diagnosis of feline sporotrichosis. *J. Feline Med. Surg.* **2011**, *13*, 220–223. [CrossRef]
207. Seyedmousavi, S.; Bosco, S.d.M.G.; de Hoog, S.; Ebel, F.; Elad, D.; Gomes, R.R.; Jacobsen, I.D.; Jensen, H.E.; Martel, A.; Mignon, B.; et al. Fungal infections in animals: A patchwork of different situations. *Med. Mycol.* **2018**, *56*, 165–187. [CrossRef]
208. Gonsales, F.F.; Fernandes, N.; Mansho, W.; Montenegro, H.; Guerra, J.M.; de Araujo, L.J.T.; da Silva, S.M.P.; Benites, N.R. Feline *Sporothrix* spp. detection using cell blocks from brushings and fine-needle aspirates: Performance and comparisons with culture and histopathology. *Vet. Clin. Pathol.* **2019**, *48*, 143–147. [CrossRef]
209. Zhang, Y.Q.; Xu, X.G.; Zhang, M.; Jiang, P.; Zhou, X.Y.; Li, Z.Z.; Zhang, M.F. Sporotrichosis: Clinical and histopathological manifestations. *Am. J. Dermatopathol.* **2011**, *33*, 296–302. [CrossRef] [PubMed]
210. Hussein, M.R. Mucocutaneous Splendore-Hoeppli phenomenon. *J. Cutan. Pathol.* **2008**, *35*, 979–988. [CrossRef] [PubMed]
211. Widal, F.; Aerami, P.; Joltrain, E.; Brissaud, E.T.; Weill, A. Serodiagnostic mycosique, applications au diagnostic de la sporotrichose et de l'actinomycose. Les coagulations et cofixations mycosiques. *Ann. Inst. Pasteur.* **1910**, *24*, 1–33.

212. Ochoa, A.G.; Figueroa, E. Polisacaridos del *Sporotrichum schenckii*. Datos immunologicos: Intradermoreaccion en el diagnostico de la esporotrichosis. *Rev. Inst. Salubr. Enferm. Trop.* **1947**, *8*, 143–153.
213. Blumer, S.O.; Kaufman, L.; Kaplan, W.; McLaughlin, D.W.; Kraft, D.E. Comparative evaluation of five serological methods for the diagnosis of sporotrichosis. *Appl. Microbiol.* **1973**, *26*, 4–8. [CrossRef]
214. Velasco, O.; Ochoa, A.G. Sporotrichosis in patients with previous positive sporotrichin reaction. *Rev. De Investig. En Salud Publica* **1971**, *31*, 53–55.
215. Ghosh, A.; Chakrabarti, A.; Sharma, V.K.; Singh, K.; Singh, A. Sporotrichosis in Himachal Pradesh (north India). *Trans. R. Soc. Trop. Med. Hyg.* **1999**, *93*, 41–45. [CrossRef]
216. Bonifaz, A.; Vázquez-González, D. Sporotrichosis: An update. *G Ital Derm. Venereol* **2010**, *145*, 659–673.
217. Kusuhara, M. Sporotrichosis and dematiaceous fungal skin infections. *Nihon Ishinkin Gakkai Zasshi* **2009**, *50*, 213–217. [CrossRef]
218. Itoh, M.; Okamoto, S.; Kariya, H. Survey of 200 cases of sporotrichosis. *Dermatologica* **1986**, *172*, 209–213. [CrossRef]
219. Kusuhara, M.; Hachisuka, H.; Sasai, Y. Statistical survey of 150 cases with sporotrichosis. *Mycopathologia* **1988**, *102*, 129–133. [CrossRef] [PubMed]
220. Posada, H.R. Prueba cutanea con esporotricina. *Mycopathologia* **1968**, *36*, 42–54. [CrossRef]
221. Bonifaz, A.; Toriello, C.; Araiza, J.; Ramírez-Soto, M.C.; Tirado-Sánchez, A. Sporotrichin skin test for the diagnosis of sporotrichosis. *J. Fungi* **2018**, *4*, 55. [CrossRef]
222. Bernardes-Engemann, A.R.; Orofino Costa, R.C.; Miguens, B.P.; Penha, C.V.L.; Neves, E.; Pereira, B.A.S.; Dias, C.M.P.; Mattos, M.; Gutierrez, M.C.; Schubach, A.; et al. Development of an enzyme-linked immunosorbent assay for the serodiagnosis of several clinical forms of sporotrichosis. *Med. Mycol.* **2005**, *43*, 487–493. [CrossRef] [PubMed]
223. López-Ribot, J.L.; Casanova, M.; Murgui, A.; Martínez, J.P. Antibody response to *Candida albicans* cell wall antigens. *FEMS Immunol. Med. Microbiol.* **2004**, *41*, 187–196. [CrossRef] [PubMed]
224. Alba-Fierro, C.A.; Pérez-Torres, A.; López-Romero, E.; Cuéllar-Cruz, M.; Ruiz-Baca, E. Cell wall proteins of *Sporothrix schenckii* as immunoprotective agents. *Rev. Iberoam. De Micol.* **2014**, *31*, 86–89. [CrossRef]
225. Fernandes, G.F.; Lopes-Bezerra, L.M.; Bernardes-Engemann, A.R.; Schubach, T.M.; Dias, M.A.; Pereira, S.A.; de Camargo, Z.P. Serodiagnosis of sporotrichosis infection in cats by enzyme-linked immunosorbent assay using a specific antigen, SsCBF, and crude exoantigens. *Vet. Microbiol.* **2011**, *147*, 445–449. [CrossRef]
226. Penha, C.V.; Bezerra, L.M. Concanavalin A-binding cell wall antigens of *Sporothrix schenckii*: A serological study. *Med. Mycol.* **2000**, *38*, 1–7. [CrossRef]
227. Alvarado, P.; Ostos, A.; Franquiz, N.; Roschman-González, A.; Zambrano, E.A.; Mendoza, M. Serological diagnosis of sporotrichosis using an antigen of *Sporothrix schenckii sensu stricto* mycelium. *Investig. Clin.* **2015**, *56*, 111–122.
228. Fernandes, G.F.; Amaral, C.C.D.; Sasaki, A.; Godoy, P.M.; De Camargo, Z.P. Heterogeneity of proteins expressed by Brazilian *Sporothrix schenckii* isolates. *Med. Mycol.* **2009**, *47*, 855–861. [CrossRef]
229. Bernardes-Engemann, A.R.; Loureiro y Penha, C.V.; Benvenuto, F.; Braga, J.U.; Barros, M.L.; Orofino-Costa, R.; Lopes-Bezerra, L.M. A comparative serological study of the SsCBF antigenic fraction isolated from three *Sporothrix schenckii* strains. *Med. Mycol.* **2009**, *47*, 874–878. [CrossRef]
230. Rodrigues, A.M.; Kubitschek-Barreira, P.H.; Fernandes, G.F.; de Almeida, S.R.; Lopes-Bezerra, L.M.; de Camargo, Z.P. Immunoproteomic analysis reveals a convergent humoral response signature in the *Sporothrix schenckii* complex. *J. Proteom.* **2015**, *115*, 8–22. [CrossRef] [PubMed]
231. Martínez-Álvarez, J.A.; García-Carnero, L.C.; Kubitschek-Barreira, P.H.; Lozoya-Pérez, N.E.; Belmonte-Vázquez, J.L.; de Almeida, J.R.; Gómez-Infante, A.D.J.; Curty, N.; Villagómez-Castro, J.C.; Peña-Cabrera, E.; et al. Analysis of some immunogenic properties of the recombinant *Sporothrix schenckii* Gp70 expressed in *Escherichia coli*. *Future Microbiol.* **2019**, *14*, 397–410. [CrossRef]
232. Baptista, V.S.; Mothé, G.B.; Santos, G.M.P.; Melivilu, C.S.I.; Santos, T.O.; Virginio, E.D.; de Macêdo-Sales, P.A.; Pinto, M.R.; Machado, R.L.D.; Rocha, E.M.S.; et al. Promising application of the SsCBF ELISA test to monitor the therapeutic response of feline sporotrichosis caused by *Sporothrix brasiliensis* from Brazilian epidemics. *Braz. J. Microbiol.* **2021**, *52*, 145–153. [CrossRef] [PubMed]
233. Elias Costa, M.R.; Da Silva Lacaz, C.; Kawasaki, M.; De Camargo, Z.P. Conventional versus molecular diagnostic tests. *Med. Mycol.* **2000**, *38* (Suppl. 1), 139–145. [CrossRef]
234. Ghosh, A.; Maity, P.K.; Hemashettar, B.M.; Sharma, V.K.; Chakrabarti, A. Physiological characters of *Sporothrix schenckii* isolates. *Mycoses* **2002**, *45*, 449–554.
235. Mesa-Arango, A.C.; del Rocío Reyes-Montes, M.; Pérez-Mejía, A.; Navarro-Barranco, H.; Souza, V.; Zúñiga, G.; Toriello, C. Phenotyping and genotyping of *Sporothrix schenckii* isolates according to geographic origin and clinical form of sporotrichosis. *J. Clin. Microbiol.* **2002**, *40*, 3004–3011. [CrossRef]
236. Lücking, R.; Aime, M.C.; Robbertse, B.; Miller, A.N.; Aoki, T.; Ariyawansa, H.A.; Cardinali, G.; Crous, P.W.; Druzhinina, I.S.; Geiser, D.M.; et al. Fungal taxonomy and sequence-based nomenclature. *Nat. Microbiol.* **2021**, *6*, 540–548. [CrossRef]
237. Crous, P.W.; Lombard, L.; Sandoval-Denis, M.; Seifert, K.A.; Schroers, H.J.; Chaverri, P.; Gené, J.; Guarro, J.; Hirooka, Y.; Bensch, K.; et al. *Fusarium*: More than a node or a foot-shaped basal cell. *Stud. Mycol.* **2021**, *98*, 1–184. [CrossRef] [PubMed]

238. Della Terra, P.P.; Gonsales, F.F.; de Carvalho, J.A.; Hagen, F.; Kano, R.; Bonifaz, A.; Camargo, Z.P.; Rodrigues, A.M. Development and evaluation of a multiplex qPCR assay for rapid diagnostics of emerging sporotrichosis. *Transbound. Emerg. Dis.* **2021**, *in press*. [CrossRef] [PubMed]
239. Pinheiro, B.G.; Hahn, R.C.; Camargo, Z.P.; Rodrigues, A.M. Molecular tools for detection and identification of *Paracoccidioides* species: Current status and future perspectives. *J. Fungi* **2020**, *6*, 293. [CrossRef] [PubMed]
240. de Carvalho, J.A.; Monteiro, R.C.; Hagen, F.; Camargo, Z.P.; Rodrigues, A.M. Trends in molecular diagnostics and genotyping tools applied for emerging *Sporothrix* species. *J. Fungi* **2022**.
241. Hu, S.; Chung, W.-H.; Hung, S.-I.; Ho, H.-C.; Wang, Z.-W.; Chen, C.-H.; Lu, S.-C.; Kuo, T.-T.; Hong, H.-S. Detection of *Sporothrix schenckii* in clinical samples by a nested PCR assay. *J. Clin. Microbiol.* **2003**, *41*, 1414–1418. [CrossRef] [PubMed]
242. Kano, R.; Nakamura, Y.; Watanabe, S.; Tsujimoto, H.; Hasegawa, A. Identification of *Sporothrix schenckii* based on sequences of the chitin synthase 1 gene. *Mycoses* **2001**, *44*, 261–265. [CrossRef] [PubMed]
243. Kano, R.; Matsuoka, A.; Kashima, M.; Nakamura, Y.; Watanabe, S.; Mizoguchi, M.; Hasegawa, A. Detection of *Sporothrix schenckii* chitin synthase 1 (CHS1) gene in biopsy specimens from human patients with sporotrichosis. *J. Derm. Sci.* **2003**, *33*, 73–74. [CrossRef]
244. Mendoza, M.; Brito, A.; Schaper, D.A.; Spooner, V.A.; Alvarado, P.; Castro, A.; Fernandez, A. Technical evaluation of nested PCR for the diagnosis of experimental sporotrichosis. *Rev. Iberoam. De Micol.* **2012**, *29*, 120–125. [CrossRef]
245. Zhang, M.; Li, F.; Gong, J.; Yang, X.; Zhang, J.; Zhao, F. Development and evaluation of a real-time polymerase chain reaction for fast diagnosis of sporotrichosis caused by *Sporothrix globosa*. *Med. Mycol.* **2019**, *58*, 61–65. [CrossRef]
246. Rodrigues, A.M.; de Hoog, G.S.; Camargo, Z.P. Genotyping species of the *Sporothrix schenckii* complex by PCR-RFLP of calmodulin. *Diagn Microbiol. Infect. Dis.* **2014**, *78*, 383–387. [CrossRef]
247. Rodrigues, A.M.; Najafzadeh, M.J.; de Hoog, G.S.; de Camargo, Z.P. Rapid identification of emerging human-pathogenic *Sporothrix* species with rolling circle amplification. *Front. Microbiol.* **2015**, *6*, 1385. [CrossRef]
248. Ramírez-Soto, M.C.; Aguilar-Ancori, E.G.; Quispe-Ricalde, M.A.; Muñiz-Duran, J.G.; Quispe-Florez, M.M.; Chinen, A. Molecular identification of *Sporothrix* species in a hyperendemic area in Peru. *J. Infect. Public. Health.* **2021**, *14*, 670–673. [CrossRef]
249. Gonsales, F.F.; Fernandes, N.C.C.A.; Mansho, W.; Montenegro, H.; Benites, N.R. Direct PCR of lesions suggestive of sporotrichosis in felines. *Arq. Bras. Med. Vet. Zootec.* **2020**, *72*, 2002–2006. [CrossRef]
250. Zhang, M.; Li, F.; Li, R.; Gong, J.; Zhao, F. Fast diagnosis of sporotrichosis caused by *Sporothrix globosa*, *Sporothrix schenckii*, and *Sporothrix brasiliensis* based on multiplex real-time PCR. *PLoS Negl. Trop. Dis.* **2019**, *13*, e0007219. [CrossRef]
251. Madrid, H.; Gené, J.; Cano, J.; Silvera, C.; Guarro, J. *Sporothrix brunneoviolacea* and *Sporothrix dimorphospora*, two new members of the *Ophiostoma stenoceras-Sporothrix schenckii* complex. *Mycologia* **2010**, *102*, 1193–1203. [CrossRef] [PubMed]
252. Rangel-Gamboa, L.; Martinez-Hernandez, F.; Maravilla, P.; Arenas-Guzman, R.; Flisser, A. Update of phylogenetic and genetic diversity of *Sporothrix schenckii sensu lato*. *Med. Mycol.* **2016**, *54*, 248–255. [CrossRef] [PubMed]
253. de Carvalho, J.A.; Hagen, F.; Fisher, M.C.; de Camargo, Z.P.; Rodrigues, A.M. Genome-wide mapping using new AFLP markers to explore intraspecific variation among pathogenic *Sporothrix* species. *PLoS Negl. Trop. Dis.* **2020**, *14*, e0008330. [CrossRef]
254. Restrepo, C.M.; Llanes, A.; Lleonart, R. Use of AFLP for the study of eukaryotic pathogens affecting humans. *Infect. Genet. Evol.* **2018**, *63*, 360–369. [CrossRef] [PubMed]
255. Roberto, T.N.; De Carvalho, J.A.; Beale, M.A.; Hagen, F.; Fisher, M.C.; Hahn, R.C.; de Camargo, Z.P.; Rodrigues, A.M. Exploring genetic diversity, population structure, and phylogeography in *Paracoccidioides* species using AFLP markers. *Stud. Mycol.* **2021**, *100*, 100131. [CrossRef]
256. Rodrigues, A.M.; Beale, M.A.; Hagen, F.; Fisher, M.C.; Terra, P.P.D.; de Hoog, S.; Brilhante, R.S.N.; de Aguiar Cordeiro, R.; de Souza Collares Maia Castelo-Branco, D.; Rocha, M.F.G.; et al. The global epidemiology of emerging *Histoplasma* species in recent years. *Stud. Mycol.* **2020**, *97*, 100095. [CrossRef]
257. Singhal, N.; Kumar, M.; Kanaujia, P.K.; Virdi, J.S. MALDI-TOF mass spectrometry: An emerging technology for microbial identification and diagnosis. *Front. Microbiol.* **2015**, *6*, 791. [CrossRef]
258. Murray, P.R. What is new in clinical microbiology-microbial identification by MALDI-TOF mass spectrometry: A paper from the 2011 William Beaumont Hospital Symposium on molecular pathology. *J. Mol. Diagn. JMD* **2012**, *14*, 419–423. [CrossRef]
259. Oliveira, M.M.; Santos, C.; Sampaio, P.; Romeo, O.; Almeida-Paes, R.; Pais, C.; Lima, N.; Zancope-Oliveira, R.M. Development and optimization of a new MALDI-TOF protocol for identification of the *Sporothrix* species complex. *Res. Microbiol.* **2015**, *166*, 102–110. [CrossRef] [PubMed]
260. Etchecopaz, A.; Toscanini, M.A.; Gisbert, A.; Mas, J.; Scarpa, M.; Iovannitti, C.A.; Bendezú, K.; Nusblat, A.D.; Iachini, R.; Cuestas, M.L. Sporothrix brasiliensis: A review of an emerging South American fungal pathogen, its related disease, presentation and spread in Argentina. *J. Fungi* **2021**, *7*, 170. [CrossRef] [PubMed]
261. Matos, A.M.F.; Moreira, L.M.; Barczewski, B.F.; de Matos, L.X.; de Oliveira, J.B.V.; Pimentel, M.I.F.; Almeida-Paes, R.; Oliveira, M.G.; Pinto, T.C.A.; Lima, N.; et al. Identification by MALDI-TOF MS of *Sporothrix brasiliensis* isolated from a subconjunctival infiltrative lesion in an immunocompetent patient. *Microorganisms* **2019**, *8*, 22. [CrossRef] [PubMed]
262. Espinel-Ingroff, A.; Abreu, D.P.B.; Almeida-Paes, R.; Brilhante, R.S.N.; Chakrabarti, A.; Chowdhary, A.; Hagen, F.; Cordoba, S.; Gonzalez, G.M.; Govender, N.P.; et al. Multicenter and international study of MIC/MEC distributions for definition of epidemiological cutoff values (ECVs) for species of *Sporothrix* identified by molecular methods. *Antimicrob. Agents Chemother.* **2017**, *61*, e01057-17. [CrossRef] [PubMed]

263. Costa, R.O.; Macedo, P.M.; Carvalhal, A.; Bernardes-Engemann, A.R. Use of potassium iodide in dermatology: Updates on an old drug. *An. Bras. De Dermatol.* **2013**, *88*, 396–402. [CrossRef]
264. Brilhante, R.S.N.; Silva, M.; Pereira, V.S.; de Oliveira, J.S.; Maciel, J.M.; Silva, I.; Garcia, L.G.S.; Guedes, G.M.M.; Cordeiro, R.A.; Pereira-Neto, W.A.; et al. Potassium iodide and miltefosine inhibit biofilms of *Sporothrix schenckii* species complex in yeast and filamentous forms. *Med. Mycol.* **2018**, *57*, 764–772. [CrossRef]
265. Sharma, B.; Sharma, A.K.; Sharma, U. Sporotrichosis: A comprehensive review on recent drug-based therapeutics and management. *Curr. Derm. Rep.* **2022**, *11*, 110–119. [CrossRef]
266. Gremião, I.D.F.; Miranda, L.H.M.d.; Pereira-Oliveira, G.R.; Menezes, R.C.; Machado, A.C.d.S.; Rodrigues, A.M.; Pereira, S.A. Advances and challenges in the management of feline sporotrichosis. *Rev. Iberoam. Micol.* **2022**, S1130-1406(22)00021-3. [CrossRef]
267. Borba-Santos, L.P.; Rodrigues, A.M.; Gagini, T.B.; Fernandes, G.F.; Castro, R.; de Camargo, Z.P.; Nucci, M.; Lopes-Bezerra, L.M.; Ishida, K.; Rozental, S. Susceptibility of *Sporothrix brasiliensis* isolates to amphotericin B, azoles, and terbinafine. *Med. Mycol.* **2015**, *53*, 178–188. [CrossRef]
268. CDDI. Cortellis Drug Discovery Intelligence Database. Available online: https://www.cortellis.com/drugdiscovery/ (accessed on 4 June 2022).
269. Francesconi, G.; Francesconi do Valle, A.C.; Passos, S.L.; de Lima Barros, M.B.; de Almeida Paes, R.; Curi, A.L.; Liporage, J.; Porto, C.F.; Galhardo, M.C. Comparative study of 250 mg/day terbinafine and 100 mg/day itraconazole for the treatment of cutaneous sporotrichosis. *Mycopathologia* **2011**, *171*, 349–354. [CrossRef]
270. Carolus, H.; Pierson, S.; Lagrou, K.; Van Dijck, P. Amphotericin B and other polyenes-discovery, clinical use, mode of action and drug resistance. *J. Fungi* **2020**, *6*, 321. [CrossRef]
271. Institute, C.a.L.S. *Reference Method for Broth Dilution Antifungal Susceptibility Testing of Yeast, M27*, 4th ed.; Clinical and Laboratory Standards Institute: Wayne, PA, USA, 2017; p. 46.
272. Institute, C.a.L.S. *Reference Method for Broth Dilution Antifungal Susceptibility Testing of Filamentous Fungi, Approved Standard, M38*, 3rd ed.; Clinical and Laboratory Standards Institute: Wayne, PA, USA, 2017; Volume 37, p. 46.
273. Fernandez-Silva, F.; Capilla, J.; Mayayo, E.; Guarro, J. Modest efficacy of voriconazole against murine infections by *Sporothrix schenckii* and lack of efficacy against *Sporothrix brasiliensis*. *Mycoses* **2014**, *57*, 121–124. [CrossRef] [PubMed]
274. Mario, D.N.; Guarro, J.; Santurio, J.M.; Alves, S.H.; Capilla, J. In vitro and in vivo efficacy of amphotericin B combined with posaconazole against experimental disseminated sporotrichosis. *Antimicrob. Agents Chemother.* **2015**, *59*, 5018–5021. [CrossRef] [PubMed]
275. Fichman, V.; Valle, A.C.F.d.; de Macedo, P.M.; Freitas, D.F.S.; Oliveira, M.M.E.d.; Almeida-Paes, R.; Gutierrez-Galhardo, M.C. Cryosurgery for the treatment of cutaneous sporotrichosis in four pregnant women. *PLoS Negl. Trop. Dis.* **2018**, *12*, e0006434. [CrossRef] [PubMed]
276. Ferreira, C.P.; do Valle, A.C.; Freitas, D.F.; Reis, R.; Galhardo, M.C. Pregnancy during a sporotrichosis epidemic in Rio de Janeiro, Brazil. *Int. J. Gynaecol. Obs.* **2012**, *117*, 294–295. [CrossRef] [PubMed]
277. Legabão, B.C.; Fernandes, J.A.; de Oliveira Barbosa, G.F.; Bonfim-Mendonça, P.S.; Svidzinski, T.I.E. The zoonosis sporotrichosis can be successfully treated by photodynamic therapy: A scoping review. *Acta Trop.* **2022**, *228*, 106341. [CrossRef]
278. Borba-Santos, L.P.; Gagini, T.; Ishida, K.; de Souza, W.; Rozental, S. Miltefosine is active against *Sporothrix brasiliensis* isolates with in vitro low susceptibility to amphotericin B or itraconazole. *J. Med. Microbiol.* **2015**, *64*, 415–422. [CrossRef]
279. Fischman Gompertz, O.; Rodrigues, A.M.; Fernandes, G.F.; Bentubo, H.D.; de Camargo, Z.P.; Petri, V. Atypical clinical presentation of sporotrichosis caused by *Sporothrix globosa* resistant to itraconazole. *Am. J. Trop. Med. Hyg.* **2016**, *94*, 1218–1222. [CrossRef]
280. Nakasu, C.C.T.; Waller, S.B.; Ripoll, M.K.; Ferreira, M.R.A.; Conceição, F.R.; Gomes, A.D.R.; Osório, L.D.G.; de Faria, R.O.; Cleff, M.B. Feline sporotrichosis: A case series of itraconazole-resistant *Sporothrix brasiliensis* infection. *Braz. J. Microbiol.* **2021**, *52*, 163–171. [CrossRef]
281. Waller, S.B.; Dalla Lana, D.F.; Quatrin, P.M.; Ferreira, M.R.A.; Fuentefria, A.M.; Mezzari, A. Antifungal resistance on *Sporothrix* species: An overview. *Braz. J. Microbiol.* **2021**, *52*, 73–80. [CrossRef]
282. Lyra, M.R.; Sokoloski, V.; de Macedo, P.M.; Azevedo, A.C.P. Sporotrichosis refractory to conventional treatment: Therapeutic success with potassium iodide. *An. Bras. De Dermatol.* **2021**, *96*, 231–233. [CrossRef] [PubMed]
283. Poester, V.R.; Basso, R.P.; Stevens, D.A.; Munhoz, L.S.; de Souza Rabello, V.B.; Almeida-Paes, R.; Zancopé-Oliveira, R.M.; Zanchi, M.; Benelli, J.L.; Xavier, M.O. Treatment of human sporotrichosis caused by *Sporothrix brasiliensis*. *J. Fungi* **2022**, *8*, 70. [CrossRef]
284. Bernardes-Engemann, A.R.; Tomki, G.F.; Rabello, V.B.d.S.; Almeida-Silva, F.; Freitas, D.F.S.; Gutierrez-Galhardo, M.C.; Almeida-Paes, R.; Zancopé-Oliveira, R.M. Sporotrichosis caused by non-wild type *Sporothrix brasiliensis* strains. *Front. Cell. Infect. Microbiol.* **2022**, *12*, 893501. [CrossRef] [PubMed]
285. Lee, Y.; Puumala, E.; Robbins, N.; Cowen, L.E. Antifungal drug resistance: Molecular mechanisms in *Candida albicans* and beyond. *Chem. Rev.* **2021**, *121*, 3390–3411. [CrossRef]
286. Koehler, A.; Pagani, D.M.; da Silva Hellwig, A.H.; Scroferneker, M.L. In-vitro antifungal susceptibility of the genus *Sporothrix* and correlation with treatment options for sporotrichosis: A systematic review. *Rev. Med. Microbiol.* **2021**, *32*, 219–227. [CrossRef]
287. Gagini, T.; Borba-Santos, L.P.; Messias Rodrigues, A.; Pires de Camargo, Z.; Rozental, S. Clotrimazole is highly effective in vitro against feline *Sporothrix brasiliensis* isolates. *J. Med. Microbiol.* **2017**, *66*, 1573–1580. [CrossRef] [PubMed]

288. Fichman, V.; Marques de Macedo, P.; Francis Saraiva Freitas, D.; Carlos Francesconi do Valle, A.; Almeida-Silva, F.; Reis Bernardes-Engemann, A.; Zancopé-Oliveira, R.M.; Almeida-Paes, R.; Clara Gutierrez-Galhardo, M. Zoonotic sporotrichosis in renal transplant recipients from Rio de Janeiro, Brazil. *Transpl. Infect. Dis.* **2021**, *23*, e13485. [CrossRef] [PubMed]
289. Georgopoulos, A.; Petranyi, G.; Mieth, H.; Drews, J. *In vitro* activity of naftifine, a new antifungal agent. *Antimicrob. Agents Chemother.* **1981**, *19*, 386–389. [CrossRef] [PubMed]
290. Artunduaga Bonilla, J.J.; Honorato, L.; Haranahalli, K.; Gremião, I.D.F.; Pereira, S.A.; Guimarães, A.; Baptista, A.R.S.; Patricia, d.M.T.; Rodrigues, M.L.; Miranda, K.; et al. Antifungal activity of Acylhydrazone derivatives against *Sporothrix* spp. *Antimicrob. Agents Chemother.* **2021**, *65*, e02593-20. [CrossRef] [PubMed]
291. Borba-Santos, L.P.; Visbal, G.; Braga, T.G.; Rodrigues, A.M.; De Camargo, Z.P.; Bezerra, L.L.; Ishida, K.; De Souza, W.; Rozental, S. Δ24-sterol methyltransferase plays an important role in the growth and development of *Sporothrix schenckii* and *Sporothrix brasiliensis*. *Front. Microbiol.* **2016**, *7*, 311. [CrossRef]
292. Borba-Santos, L.P.; Nucci, M.; Ferreira-Pereira, A.; Rozental, S. Anti-*Sporothrix* activity of ibuprofen combined with antifungal. *Braz. J. Microbiol.* **2021**, *52*, 101–106. [CrossRef] [PubMed]
293. Borba-Santos, L.P.; Ishida, K.; Calogeropoulou, T.; Souza, W.; Rozental, S. Adamantylidene-substituted alkylphosphocholine TCAN26 is more active against Sporothrix schenckii than miltefosine. *Mem. Do Inst. Oswaldo Cruz* **2016**, *111*, 523–527. [CrossRef]
294. Asquith, C.R.M.; Machado, A.C.S.; de Miranda, L.H.M.; Konstantinova, L.S.; Almeida-Paes, R.; Rakitin, O.A.; Pereira, S.A. Synthesis and identification of pentathiepin-based inhibitors of *Sporothrix brasiliensis*. *Antibiotics* **2019**, *8*, 249. [CrossRef]
295. Gopinath, P.; Yadav, R.K.; Shukla, P.K.; Srivastava, K.; Puri, S.K.; Muraleedharan, K.M. Broad spectrum anti-infective properties of benzisothiazolones and the parallels in their anti-bacterial and anti-fungal effects. *Bioorg. Med. Chem. Lett.* **2017**, *27*, 1291–1295. [CrossRef] [PubMed]
296. Azevedo-França, J.A.; Granado, R.; de Macedo Silva, S.T.; Santos-Silva, G.D.; Scapin, S.; Borba-Santos, L.P.; Rozental, S.; de Souza, W.; Martins-Duarte É, S.; Barrias, E.; et al. Synthesis and biological activity of novel zinc-itraconazole complexes in protozoan parasites and *Sporothrix* spp. *Antimicrob. Agents Chemother.* **2020**, *64*, e01980-19. [CrossRef] [PubMed]
297. de Azevedo-França, J.A.; Borba-Santos, L.P.; de Almeida Pimentel, G.; Franco, C.H.J.; Souza, C.; de Almeida Celestino, J.; de Menezes, E.F.; Dos Santos, N.P.; Vieira, E.G.; Ferreira, A.; et al. Antifungal promising agents of zinc(II) and copper(II) derivatives based on azole drug. *J. Inorg. Biochem.* **2021**, *219*, 111401. [CrossRef]
298. Gagini, T.; Colina-Vegas, L.; Villarreal, W.; Borba-Santos, L.P.; de Souza Pereira, C.; Batista, A.A.; Kneip Fleury, M.; de Souza, W.; Rozental, S.; Costa, L.A.S.; et al. Metal–azole fungistatic drug complexes as anti-*Sporothrix* spp. agents. *New J. Chem.* **2018**, *42*, 13641–13650. [CrossRef]
299. Tandon, V.K.; Kumar, S.; Mishra, N.N.; Shukla, P.K. Micelles catalyzed chemo- and regio-selective one pot and one step synthesis of 2,3,5,6-tetrakis(alkyl and arylsulfanyl)-1,4-benzoquinones and 2,5-diaminosubstituted-1,4-benzoquinones "In-Water" and their biological evaluation as antibacterial and antifungal agents. *Eur. J. Med. Chem.* **2012**, *56*, 375–386. [CrossRef]
300. Brilhante, R.S.; Silva, N.F.; Marques, F.J.; Castelo-Branco Dde, S.; de Lima, R.A.; Malaquias, A.D.; Caetano, E.P.; Barbosa, G.R.; de Camargo, Z.P.; Rodrigues, A.M.; et al. In vitro inhibitory activity of terpenic derivatives against clinical and environmental strains of the *Sporothrix schenkii* complex. *Med. Mycol.* **2015**, *53*, 93–98. [CrossRef]
301. Brilhante, R.S.; Malaquias, A.D.; Caetano, E.P.; Castelo-Branco Dde, S.; Lima, R.A.; Marques, F.J.; Silva, N.F.; Alencar, L.P.; Monteiro, A.J.; Camargo, Z.P.; et al. In vitro inhibitory effect of miltefosine against strains of *Histoplasma capsulatum var. capsulatum* and *Sporothrix* spp. *Med. Mycol.* **2014**, *52*, 320–325. [CrossRef]
302. Borba-Santos, L.P.; Vila, T.; Rozental, S. Identification of two potential inhibitors of *Sporothrix brasiliensis* and *Sporothrix schenckii* in the Pathogen Box collection. *PLoS ONE* **2020**, *15*, e0240658. [CrossRef]
303. Borba-Santos, L.P.; Barreto, T.L.; Vila, T.; Chi, K.D.; Dos Santos Monti, F.; de Farias, M.R.; Alviano, D.S.; Alviano, C.S.; Futuro, D.O.; Ferreira, V.; et al. In vitro and in vivo antifungal activity of buparvaquone against *Sporothrix brasiliensis*. *Antimicrob. Agents Chemother.* **2021**, *65*, e0069921. [CrossRef] [PubMed]
304. Brilhante, R.S.; Pereira, V.S.; Oliveira, J.S.; Lopes, R.G.; Rodrigues, A.M.; Camargo, Z.P.; Pereira-Neto, W.A.; Castelo-Branco, D.S.; Cordeiro, R.A.; Sidrim, J.J.; et al. Pentamidine inhibits the growth of *Sporothrix schenckii* complex and exhibits synergism with antifungal agents. *Future Microbiol.* **2018**, *13*, 1129–1140. [CrossRef] [PubMed]

Journal of
Fungi

MDPI

Review

Update on the Epidemiology, Diagnosis, and Treatment of Coccidioidomycosis

Samantha L. Williams * and Tom Chiller

Mycotic Diseases Branch, Centers for Disease Control and Prevention, Atlanta, GA 30329, USA; tnc3@cdc.gov
* Correspondence: pog3@cdc.gov

Abstract: Coccidioidomycosis is a fungal infection caused by *Coccidioides immitis* and *Coccidioides posadasii*. The dimorphic fungi live in the soils of arid and semi-arid regions of the western United States, as well as parts of Mexico, Central America, and South America. Incidence of disease has risen consistently in recent years, and the geographic distribution of *Coccidioides* spp. appears to be expanding beyond previously known areas of endemicity. Climate factors are predicted to further extend the range of environments suitable for the growth and dispersal of *Coccidioides* species. Most infections are asymptomatic, though a small proportion result in severe or life-threatening forms of disease. Primary pulmonary coccidioidomycosis is commonly mistaken for community-acquired pneumonia, often leading to inappropriate antibacterial treatment and unnecessary healthcare costs. Diagnosis of coccidioidomycosis is challenging and often relies on clinician suspicion to pursue laboratory testing. Advancements in diagnostic tools and antifungal therapy developments seek to improve the early detection and effective management of infection. This review will highlight recent updates and summarize the current understanding of the epidemiology, diagnosis, and treatment of coccidioidomycosis.

Keywords: coccidioidomycosis; *Coccidioides*; Valley fever; endemic mycoses; fungal diseases

Citation: Williams, S.L.; Chiller, T. Update on the Epidemiology, Diagnosis, and Treatment of Coccidioidomycosis. *J. Fungi* **2022**, *8*, 666. https://doi.org/10.3390/jof8070666

Academic Editor: Maria José Soares Mendes Giannini

Received: 5 May 2022
Accepted: 20 June 2022
Published: 25 June 2022

1. Introduction

Coccidioidomycosis, also known as Valley fever, is an infection caused by the inhalation of airborne arthroconidia from the soil-dwelling fungi, *Coccidioides* spp. Though often considered a rare disease, the environmental mycosis is a growing public health concern due to rising case counts and evidence of geographic expansion. Symptoms develop in approximately 40% of cases, frequently resembling other respiratory illnesses with signs such as cough, fever, shortness of breath, and fatigue [1]. Clinical findings may be indistinguishable from community-acquired pneumonia (CAP), which can lead to misdiagnosis and delays in appropriate antifungal treatment [2]. Although the infection is usually self-limiting, many patients require antifungal treatment to resolve illness, and a small subset of infections result in life-threatening severe pulmonary or disseminated disease [3,4]. Direct medical costs combined with potential productivity losses constitute a substantial economic burden [5–8].

In the United States, coccidioidomycosis is known to be endemic in the Southwest, with southern Arizona and the San Joaquin Valley in California comprising hyperendemic zones [9]. *Coccidioides immitis* has been found as far north as Washington state [10,11]. Globally, a combination of case reports and skin test studies also established areas of endemicity in Mexico, parts of South America (Argentina, Bolivia, Brazil, Colombia, Paraguay, Venezuela), and parts of Central America (Guatemala, Honduras), though much remains unknown about coccidioidomycosis in these regions because of limited diagnostic capabilities and reporting [12,13]. The global geographic distribution of *Coccidioides* spp. is displayed in Figure 1. Reported cases of coccidioidomycosis increased steadily in recent years and likely underestimate the true burden of disease.

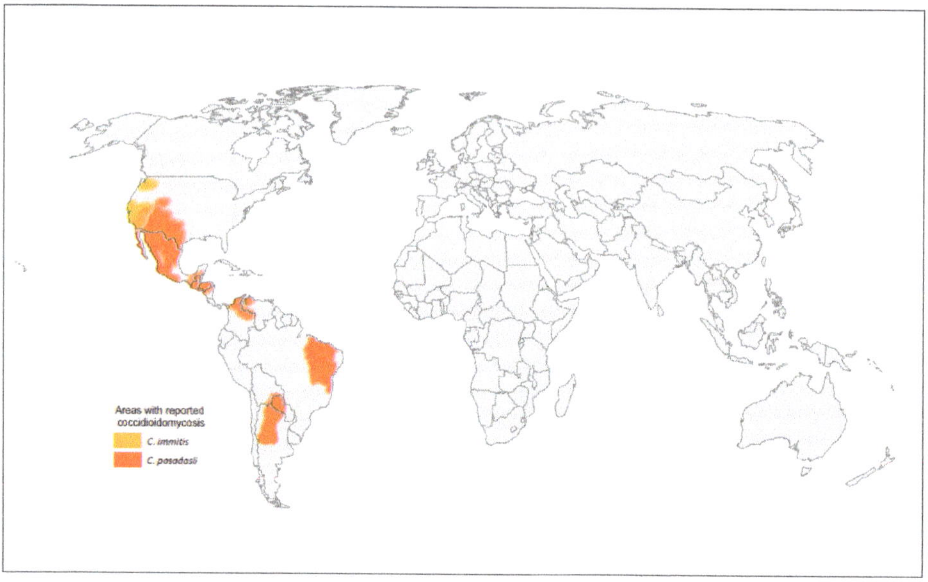

Figure 1. Global geographic distribution of the *Coccidioides* species.

The development of more rapid diagnostics is key to early and accurate diagnosis. Despite improvements in test offerings and performance over time, challenges to coccidioidomycosis diagnosis persist. Results can be difficult to interpret, and sensitivity and specificity may vary based on immunosuppression status and stage of disease. These complexities are compounded by low clinician awareness, particularly outside of known endemic regions [14,15]. Continued developments in antifungal therapy aim to improve patient outcomes and address concerns of toxicities, though questions regarding early treatment and optimal management persist.

This review summarizes the current understanding of the epidemiology, diagnosis, and treatment and management of coccidioidomycosis.

2. Epidemiology

2.1. Increased Number of Reported Cases

Coccidioidomycosis, caused by *Coccidioides immitis* and *Coccidioides posadasii*, is a nationally notifiable disease in the United States, though it is reportable only in 26 states and the District of Columbia [16]. Reportable status is designated by the state or jurisdiction and requires healthcare professionals and laboratories to notify public health departments of cases that meet the Council of State and Territorial Epidemiologists (CSTE) definition. For nationally notifiable diseases or conditions, states voluntarily submit case data to the US Centers for Disease Control and Prevention (CDC) for patients meeting CSTE criteria. The number of cases reported to the CDC rose considerably since 2014, as shown in Figure 2. Following a three-year decline from 2012–2014, case counts more than doubled from 8232 in 2014 to 20,003 in 2019 [17]. Arizona and California, which account for more than 95% of reported cases, showed similar trends in rates of disease. Incidence in Arizona grew from 84.4 cases per 100,000 population to 144.1 per 100,000 from 2014–2019, while California's incidence more than tripled from 6.0 per 100,000 population to 22.5 per 100,000 population in the same timeframe [18,19]. Based on provisional counts, Arizona's incidence increased to 161.1 per 100,000 population in 2020 and stayed relatively even at 159.8 per 100,000 population in 2021, while incidence in California dipped to 16.9 per 100,000 in 2020 before rising to 20.7 per 100,000 in 2021 [20,21].

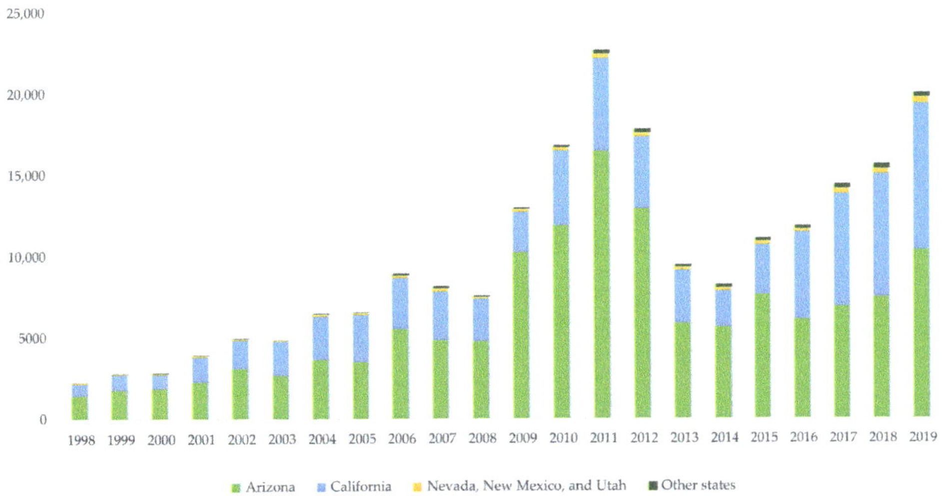

Figure 2. Coccidioidomycosis case counts submitted to the National Notifiable Diseases Surveillance System, 1998–2019. Case counts reported by individual states might differ slightly from those reported by the National Notifiable Diseases Surveillance System because of differences in the timing of reports or surveillance methods.

The cause of the concerning increase is likely multifaceted. Environmental factors favorable to the growth and subsequent dispersal of *Coccidioides* spp. may have contributed to a higher frequency of disease [22,23]. The causative fungus is known to live in arid and semi-arid regions, and statistical models demonstrated the influence of temperature and precipitation patterns on the proliferation of *Coccidioides* spp., though only a weak correlation was found in some highly endemic areas of California [22,24–29]. Periods of precipitation facilitate fungal growth in the environment, while ensuing periods of low precipitation and high temperature create ideal conditions to release fungal spores [28–31]. These effects may be amplified after droughts; incidence in California declined during the 2007–2009 and 2012–2015 droughts but increased markedly in the following two years [23]. Particulate matter of size less than 10 μm (PM10) is also thought to impact coccidioidomycosis incidence, and PM10 concentration rose by 12% in the Southwest and 29% in the West from 2010–2020 [30,32,33].

Population growth may have impacted the rising coccidioidomycosis case counts. Arizona continues to be one of the fastest-growing states, recording a 12% population increase from 2010 to 2020, driven primarily by a 16% increase in the highly-populated Maricopa County; California also experienced a 6% population increase during the same timeframe [34]. Many of the incoming residents are likely immunologically naïve to coccidioidomycosis. Across several endemic regions, desert land was converted to urban and suburban centers to accommodate population influx, resulting in substantial soil disturbance and potential exposure [35,36].

Advances in healthcare practices have expanded the at-risk population for coccidioidomycosis. Prolonged life spans have led to a growing population over 65 years of age [34]. This group has a higher prevalence of chronic disease and is more frequently diagnosed with *Coccidioides* spp. infection [37]. Developments in therapeutics and medical procedures extended survival for patients with weakened immune systems or previously fatal conditions. The number of stem cell transplants increased by 8% from 2015–2019, while solid organ transplantation grew by 45% from 2011–2021 [38,39]. Use of immunosuppressive agents has become more widespread with greater availability [40]. Transplants and immunosuppressants both represent known risk factors for developing severe coccidioidomycosis [41,42].

Additional factors that may have contributed to the increase in coccidioidomycosis case counts include changed or improved reporting practices, increased laboratory testing, or heightened awareness. Patients in Arizona who heard of the disease before seeking healthcare were diagnosed earlier than patients who were unaware of coccidioidomycosis and were also more likely to request testing [43]. However, surveys of representative samples showed low Valley fever awareness, even in regions of known endemicity [44,45]. In California, only 25.0% of respondents living in high-incidence areas knew that *Coccidioides* spp. (termed 'Valley fever fungus' in the survey) existed in their area of residence, and just 3.5% of respondents with a risk factor for severe coccidioidomycosis knew that they were at an elevated risk for severe infection [45]. Tailored messaging to vulnerable populations may increase the knowledge of disease and consequently influence healthcare-seeking behavior and testing practices.

2.2. *Geographic Expansion of Coccidioides Species*

The initial geographic distribution of coccidioidomycosis was established in the 1940s and 1950s through extensive coccidioidin skin tests to assess prevalence [9]. Although many cases reported outside of the traditional areas of endemicity are attributed to travel in endemic regions or reactivation of a latent infection, several outbreaks in California and Utah have indicated that the geographic range is extending northward [46–48]. Whole-genome sequencing confirmed the local acquisition of coccidioidomycosis in 2010 in Washington. A clinical isolate from the patient and soil isolates retrieved from the suspected point of exposure were found to be identical, providing the first evidence of *Coccidioides* endemicity in the state [10,49].

Reasons for the expansion beyond traditionally recognized areas are not certain, though several have been theorized. Climate factors are known to influence environmental dynamics, and it is hypothesized they may create suitable conditions for *Coccidioides* spp. habitation in areas that did not previously support fungal growth. A climate niche model used disease incidence data and climate projections to predict that, by the year 2100, the area of coccidioidomycosis endemicity will more than double and cases will increase by 50%, based on global warming scenarios [24]. Although evidence demonstrated the potential influence of climate change on the future geographic distribution of *Coccidioides* spp., it is unclear whether climate contributed to the current observed expansion. Climate change may additionally trigger an increase in severe weather events, such as wildfires, which have been linked to infection and cause substantial disruption and dispersal of soil [50,51]. Dust storms (haboobs) were previously thought to drive increases in coccidioidomycosis incidence, but recent studies found no significant association [52–54]. Air sampling studies in Arizona even suggest that dust storms may diminish the concentration of arthroconidia in the air [54].

Some research also suggests that rodents may contribute to the geographic distribution by serving as a reservoir for coccidioidomycosis, though conflicting results called this theory into question. The hypothesis, first proposed in the 1940s, was largely discarded following contradictory results from large-scale soil sampling [55,56]. However, several studies since found a correlation between the amount of *Coccidioides* spp. recovered from soil obtained from rodent burrows compared with surrounding topsoil [57–59]. Contemporary genomic analysis also spurred a resurgence of the theory, as the findings indicated a higher proportion of animal versus plant tissue-associated genes [60]. Notably, recent systematic soil sampling in Washington did not yield any association between rodent burrows and the presence of *Coccidioides* [49]. The rodent hypothesis offers a conceivable explanation for the fungus' patchy distribution in endemic regions, but more research is needed to better understand the role of rodents in the *Coccidioides* life cycle. Continued surveillance will be essential to monitor trends in case counts and geographic expansion.

2.3. Risk Factors

Coccidioidomycosis can affect anyone who is exposed to the causative fungus, but certain groups may be at higher risk of infection or severe disease. People with weakened immune systems have demonstrated greater susceptibility to disease. Elevated risk exists for severe coccidioidomycosis among people living with HIV/AIDS, particularly those with low CD4 counts, though incidence among this population decreased with the advent of antiretroviral therapy [61–63]. Transplant recipients and patients receiving immunosuppressive medications such as corticosteroids, chemotherapy, or tumor necrosis factor inhibitors represent additional at-risk groups owing to cellular immunodeficiencies [64,65].

Reactivation of a latent infection is also a concern among transplant recipients. Transplant centers in endemic areas may administer antifungal prophylaxis to prevent recurrence of coccidioidomycosis, often up to one-year post-transplantation [4]. Both universal and targeted programs aimed at patients with positive serologic results or a history of *Coccidioides* spp. infection demonstrated encouraging results [66–69]. However, no consensus exists regarding drug type or duration [69,70]. Further research is needed to define optimal prophylaxis strategies among certain patient groups.

Pregnancy is an established risk factor for coccidioidomycosis. Evidence suggests a correlation between the severity of coccidioidomycosis and the length of gestation, with the most severe disease occurring in the late stages of pregnancy and the immediate postpartum period. Illness in pregnant women is further complicated by treatment challenges because azole antifungals are known teratogens [71,72]. Epidemiologic studies showed that people of African American or Filipino descent are at a heightened risk for infection, particularly severe or disseminated disease, though reasons are unknown [73–75].

Occupational hazards have been documented for workers exposed to dust from soil disturbance. Jobs involving digging, excavation, or soil disruption in endemic areas are considered to pose the greatest risk of coccidioidomycosis; common examples include construction, archaeology, agriculture, firefighting, and mining, gas, or oil extraction [76]. Although the aforementioned activities are more commonly associated with *Coccidioides* spp. exposure, cases related to minimal soil disturbance have been reported, such as an outbreak among cast and crew on an outdoor television set, as well as extremely high rates among employees and inmates at state prisons in California [77,78].

Researchers have long postulated that males are at an elevated risk of *Coccidioides* spp. infection based on the observed differences in disease incidence by sex [79–81]. Some claim that the disparity is likely because of the disproportionate representation of males in occupations or recreational activities associated with dust or soil exposure. Recent data furthered the assertion of sex as an independent risk factor through a survey of human patients, veterinary patients, and nonhuman primates [82]. Results showed significantly higher rates of infection, severe disease, and greater average maximum serum complement fixation (CF) titers among human males compared with females. Significant differences in incidence emerged at age 19 and remained until age 80. Additionally, nonhuman primate and unaltered canine males exhibited a significantly increased risk for coccidioidomycosis compared with their female counterparts, though no significant sex differences were observed between castrated male dogs and female dogs [82]. Findings from this study suggest that biological mechanisms may contribute to the reported differences in *Coccidioides* spp. infection between males and females.

2.4. Coccidioidomycosis and COVID-19

Several reports described a co-infection of *Coccidioides* spp. and severe acute respiratory syndrome novel coronavirus (SARS-CoV-2), the causative agent of coronavirus disease 2019 (COVID-19) [83–90]. Common risk factors include older age, diabetes, immunosuppression, African American or Latino heritage, and smoking [91]. The interaction between the two infections remains largely unclear. Underlying respiratory illness and chronic lung disease are associated with severe COVID-19, suggesting that patients with chronic pulmonary coccidioidomycosis may be at an elevated risk of severe COVID-19 [92–94]. SARS-CoV-2 infection may also increase the risk of coccidioidomycosis reactivation because of immune dysregulation within the host, though at present only one case report is indicative of reactivation [90,95,96].

The COVID-19 pandemic may have lengthened delays in diagnosis as a result of similar symptoms between SARS-CoV-2 and *Coccidioides* spp. infection. A survey of infectious disease physicians indicated that testing practices did not change because of COVID-19, though primary care doctors were not among the respondents to this survey [97]. It is also unknown whether health-seeking behavior or access to healthcare during the pandemic affected coccidioidomycosis case counts. Further research is needed to understand the relationship between coccidioidomycosis and COVID-19.

3. Diagnosis

3.1. Diagnostic Challenges

Healthcare providers are faced with a variety of challenges in relation to coccidioidomycosis diagnosis, though they generally fall into two categories: (1) technical efficiency of laboratory testing and (2) provider knowledge and behavior. Nonspecific symptoms resemble other respiratory illnesses, and laboratory test results may be difficult to interpret or logistically challenging. A delayed or missed diagnosis can consequently lead to adverse patient outcomes in the absence of proper antifungal treatment. Presenting symptoms are often mistaken for bacterial or viral pneumonia, yet current Infectious Disease Society of America (IDSA) diagnostic guidelines for CAP do not recommend testing for *Coccidioides* spp. infection, despite evidence that the fungal disease accounts for up to one third of CAP etiologies in some endemic areas [98,99].

According to recent enhanced surveillance, 70% of patients had another condition diagnosed before being tested for coccidioidomycosis, and the median duration from seeking healthcare to diagnosis was 38 days [14]. Another study found that 70% of patients received antibiotics in the three months before their first positive coccidioidal test, and those patients were prescribed a median of three antibiotic courses [100]. Implications of inappropriate antibacterial treatment can include drug resistance and unnecessary costs.

A variety of laboratory diagnostic tests are available to detect *Coccidioides* spp. infection, but testing rates remain low. Performance measures and considerations for the various laboratory tests are described in Table 1. A survey of healthcare providers revealed that only 3.7% reported "frequently" testing CAP patients for coccidioidomycosis, and 15.0% tested "sometimes". Even in Arizona and California, just 32.4% and 7.4% of providers, respectively, reported frequent testing [15]. Just over 60% of healthcare providers surveyed in Arizona were confident in their ability to diagnose *Coccidioides* infection, and just under 60% claimed to be knowledgeable about laboratory tests used for detecting coccidioidal antibodies [101].

A recent assessment of practice patterns in an Arizona healthcare system showed that 73% of coccidioidomycosis diagnoses were made during hospital admission. Nearly half of those patients had at least one prior healthcare encounter related to their symptoms, but coccidioidal serology was only obtained for 29% of them during those visits [102]. Their subsequent hospitalizations suggest that retesting may have been beneficial to counteract any false negatives in the early stages of the disease.

3.2. Serology

Serologic antibody testing is the most frequently used diagnostic tool for coccidioidomycosis. Antibody development may trail illness onset by several weeks; serial testing is therefore recommended, should symptoms persist following an initial negative test result. Consideration of other laboratory tests may be warranted for immunosuppressed patients, as sensitivities for coccidioidal antibody tests are generally lower among this population.

Enzyme immunoassay (EIA) tests are widely available and offer rapid results, detecting coccidioidal antibodies within hours. EIA testing of both immunoglobulin (Ig) M and IgG levels can be highly sensitive (59–88%) and specific (68–90%), though results are variable [103–105]. Results for IgG EIA alone are generally preferred to those for IgM alone, as IgM EIA tests are known to show false positives and should be interpreted with caution; IgG EIA test sensitivities range from 47–87% with specificities of 90–96%, while IgM EIA test sensitivities range from 22–61% with specificities of 70–99% [104–106]. A lateral flow assay (LFA) has been developed to offer a fast and simple alternative for antibody detection, but initial data have shown markedly lower sensitivity compared to EIA [107].

Alternative serologic tests for antibody detection include immunodiffusion (ID) and complement fixation (CF), which are less sensitive than EIA but more specific and commonly serve as confirmatory tests [103]. ID tests measure IgG or IgM but can take several days to return results, limiting their utility for quick diagnosis. Quantitative CF results are valuable measures to assess disease severity and progression; higher CF titers may indicate dissemination, and increasing titers are associated with clinical deterioration. However, CF tests measure only IgG antibodies, and cross-reactivity with other dimorphic fungi may influence results [108–110]. All coccidioidomycosis antibody tests involve specialty equipment, and CF testing also requires a high level of specialized training. Methods for ID and CF vary across laboratories, leading to a considerable variability of results. Standardization is needed to increase the reliability and comparability of these diagnostic tools.

3.3. Antigen Detection

Antigen testing may inform the detection of *Coccidioides* in early stages of disease, particularly among immunocompromised patients [111,112]. A commercially-available test for serum, urine, or cerebrospinal fluid (CSF) samples has demonstrated high specificity. Sensitivity is moderate among immunocompromised populations but drops considerably among immunocompetent patients, and CSF antigen testing is sensitive only in patients with coccidioidal meningitis. Cross-reactivity with blastomycosis and histoplasmosis may complicate diagnosis in the absence of additional testing [109,111,113].

3.4. Microscopy and Culture

Identification of *Coccidioides* in clinical specimens by culture remains the gold standard for coccidioidomycosis diagnosis, though biosafety concerns and potential challenges visualizing the spherules may limit the use of this method. Additionally, culture growth may take up to a week, and sensitivity is dependent on specimen quality.

Histopathology and cytopathology represent other traditional methods used to identify the organism in clinical specimens. Sensitivities of histopathology (23–84%) and cytopathology (15–75%) also vary greatly [114].

3.5. Additional Laboratory Diagnostic Methods

Encouraging results from polymerase chain reaction (PCR) testing demonstrate the potential value of molecular methods as a diagnostic tool for coccidioidomycosis. At present, PCR tests are not commonly used directly on clinical specimens, and the site of specimen collection may impact test performance [115,116]. A study to evaluate performance of the serum (1→3) β-d-glucan (BG) assay showed sensitivity (43.9%) and specificity (91.1%) comparable to BG testing for other invasive mycoses, such as aspergillosis and candidiasis. However, sensitivity was lower among patients with acute pulmonary coccidioidomycosis (19.1%), specificity was determined against healthy controls, and BG values correlated poorly with serum coccidioidal CF titers [117]. The utility of BG testing as a clinical diagnostic tool is limited by its inability to detect specific pathogens.

Table 1. Performance and considerations for coccidioidomycosis laboratory diagnostic tests.

Test	Sensitivity	Specificity ‡	Considerations
Serology			
Antibody			Antibody production may lag behind symptom onset. Sensitivity is often lower in immunosuppressed patients.
EIA IgG or IgM [103–105]	59–88%	68–96%	Rapid performance time within hours. Often used as a screening test, later confirmed by ID or CF. IgM only may lead to more false positives than IgG only.
EIA IgG [103–105]	47–87%	89–97%	
EIA IgM [103–105]	22–61%	70–99%	
ID § [103,118]	60–91%	99–100%	Results may take several days to receive. Some specialized training is required. Methods are not standardized across laboratories.
CF § [103,108,109,118]	65–98%	80–98%	Titers may offer prognostic value of disease progression. Measurement of IgG only. Highly specialized training is required. Methods are not standardized across laboratories.
LFA § [117,118]	31–99%	92–98%	Rapid 1-h performance time.
Antigen			
Urine and serum [113]	57%	99%	May detect *Coccidioides* in the early stages of the disease [112]. May be preferred to antibody tests for immunocompromised patients. Substantial cross-reactivity with other dimorphic fungi.
Urine [111,113]	37–71%	99%	
Serum [119]	73%	100%	
Microscopy and culture			
Culture [114]	23–93%	High	Considered the gold standard of coccidioidomycosis diagnosis. Biosafety level 3 lab needed for safe isolation of *Coccidioides*. Culture growth may take up to a week. Sensitivity is heavily dependent on specimen quality.
Histopathology [114]	23–84%	High	
Cytology [114]	15–75%	High	
Additional laboratory methods			
PCR [115,116]	56–75%	99–100%	Rapid 4-h performance time. Site of specimen collection may influence results.
(1→3) β-d-glucan [117]	44%	91%	Lower sensitivity among patients with acute pulmonary coccidioidomycosis. Values correlate poorly with CF titers. Test cannot detect specific pathogens.

Abbreviations: CF, complement fixation; EIA, enzyme immunoassay; ID, immunodiffusion; IgG, immunoglobulin G; IgM, immunoglobulin M; PCR, polymerase chain reaction. ‡ Specificity is based on published results; estimates may not be directly comparable, as different control populations were used in some cases. § Sensitivity and specificity ranges include testing from outbreak investigations.

4. Treatment

Management of coccidioidomycosis often depends on the severity of disease and clinical history of the patient. Acute pulmonary coccidioidomycosis in immunocompetent hosts typically resolves without antifungal intervention and results in life-long immunity [3]. Regular assessment to monitor symptoms and radiological results may prove sufficient for these patients. Some physicians advise empiric antifungal therapy to shorten symptom duration and prevent dissemination, though there are no data from prospective randomized clinical trials to support the efficacy of early treatment on these outcomes [120].

Antifungal treatment for primary pulmonary coccidioidomycosis is recommended for immunosuppressed hosts or patients with particularly devastating forms of disease. IDSA guidelines outline symptoms that may warrant the initiation of therapy; these symptoms include substantial weight loss, persistent intense night sweats, infiltrates involving more than half of 1 lung or portions of both lungs, prominent or persistent hilar adenopathy, CF titers exceeding 1:16, inability to work, or symptoms that persist for more than 2 months. Additional considerations for early treatment include a history of diabetes, frailty because of old age, comorbidities, or African American or Filipino ancestry, though the IDSA guidelines note that ethnicity and diabetes status should only modestly influence management decisions [4].

4.1. Azoles

The advent of azole therapy significantly influenced the antifungal treatment of coccidioidomycosis. Ketoconazole became the first azole used against *Coccidioides* in 1981, though it is no longer recommended because of concerns of adverse effects and apparently superior efficacy of other drugs [121,122]. Fluconazole is the most commonly prescribed antifungal agent for coccidioidomycosis, available in oral and intravenous formulations. Its low cost, tolerability, and penetration into most body sites make it an appealing option for drug administration [4,123]. Side effects may include alopecia, dry skin, chapped lips, and arthropathy, and effects may become more pronounced with higher dosage [124]. Fluconazole in vitro minimum inhibitory concentrations (MICs) were found to be significantly higher than those of other triazoles in limited reports, though this has not been correlated with patient outcomes [125].

Itraconazole is also commonly used to treat coccidioidomycosis and is available as an oral solution or a capsule. Findings from a randomized double-blind trial showed that itraconazole had a higher efficacy (70% response rate) compared to fluconazole (37% response rate) for patients with skeletal forms of disease, though absorption can be challenging. Relapse rates were also lower for patients treated with itraconazole (18%) as opposed to fluconazole (28%) [126]. Hypertension, hypokalemia, sodium retention, and decreased myocardial contractility constitute reported adverse effects from itraconazole therapy, and consequently the drug is not recommended for patients at risk of heart failure [127–129].

Voriconazole may be prescribed for patients who are intolerant or unresponsive to fluconazole or itraconazole. It can be administered intravenously or through oral formulation and has extensive distribution throughout the body, including CSF penetration. Although voriconazole has exhibited efficacy in cases of coccidioidal meningitis, concerns of drug-drug interactions and toxicities may limit its use [130–132]. Voriconazole has been associated with visual impairments, altered mental status, and harmful cutaneous effects, including photodermatitis, melanoma, and squamous cell carcinoma [133–135].

Posaconazole is an alternate antifungal option for refractory cases of coccidioidomycosis. Originally available only as an oral suspension, the development of an intravenous formulation and delayed response oral tablet markedly improved absorption [136]. Posaconazole penetrates most body sites, with the exception of CSF, and is generally considered to be effective, even demonstrating superior performance to other triazoles in murine models [130,137–141]. Side effects are commonly gastrointestinal in nature, including nausea, vomiting, and diarrhea. Hypokalemia, hypertension, and peripheral edema are also reported to be associated with Posaconazole treatment [98].

Most recently, isavuconazole has been made available in oral and intravenous formulations. It is widely distributed throughout the body, yet although isavuconazole has proven to be effective against other mycoses, data for coccidioidomycosis patients in a clinical setting are limited [142–144]. Gastrointestinal side effects similar to those of Posaconazole have been observed.

4.2. Polyenes

Prior to the introduction of triazole antifungal therapy, amphotericin B served as the primary agent for coccidioidomycosis treatment. Multiple formulations are available intravenously: amphotericin B deoxycholate (AmBd), liposomal amphotericin B (L-AMB), amphotericin B colloidal dispersion (ABCD), and amphotericin B lipid complex (ABLC). All forms are associated with nephrotoxicity, though the frequency of adverse events is generally lower in the lipid variations [145,146]. Use of amphotericin B is now primarily limited to cases that are intolerant or resistant to triazoles. However, intrathecal AmBd may still be administered in patients with coccidioidal meningitis or in their first trimester of pregnancy to avoid potential teratogenic effects of triazoles [4,147].

4.3. Treatment Duration and Follow-Up

Duration of coccidioidomycosis treatment varies based on disease type and progression. Therapy for uncomplicated acute pulmonary infection is commonly discontinued after 3–6 months, whereas patients with severe or chronic forms of disease may require life-long treatment. Antifungal regimens should be routinely assessed for possible adverse effects, drug-drug interactions, and therapeutic drug monitoring if needed. Cessation of treatment is generally prompted by diminished symptoms, improvements of imaging results, and declining CF titers.

Regardless of treatment status, clinical follow-up is essential to evaluate the resolution of signs and symptoms and to identify possible relapse or dissemination. Assessments often incorporate a combination of serologic testing, radiological examination, and patient interview to monitor the course of infection. Follow-up is recommended for at least one year once the patient shows signs of improvement.

5. Conclusions

Our understanding of coccidioidomycosis has no doubt deepened in recent years, but much remains to be learned. Incidence of disease is rising, and the geographic range of *Coccidioides* spp. is growing. Continued and expanded surveillance is needed in the United States and across Central and South America to monitor trends and identify potential new areas of endemicity to inform public health efforts. Increased clinician awareness and knowledge of suitable diagnostic methods are essential to improve early detection and avoid inappropriate treatment and unnecessary medical costs. Furthermore, the development of new and more rapid diagnostic tools, as well as antifungal therapies that target *Coccidioides* spp., is necessary to advance the diagnosis and subsequent resolution of disease. Insights into the epidemiology, diagnosis, and treatment of coccidioidomycosis can be used to guide future prevention and management strategies to minimize morbidity and mortality from this important disease.

Author Contributions: S.L.W. and T.C. have made substantial contributions to this review. All authors have read and agreed to the published version of the manuscript.

Funding: This research received no external funding.

Institutional Review Board Statement: Not applicable.

Informed Consent Statement: Not applicable.

Acknowledgments: The authors wish to acknowledge Brendan Jackson, Mitsuru Toda, Kaitlin Benedict, Dallas Smith, Ana Litvintseva, and Mark Lindsley for their support of this work.

Conflicts of Interest: The authors declare no conflict of interest.

Disclaimer: The findings and conclusions of this paper are those of the authors and do not necessarily represent the official position of the US Centers for Disease Control and Prevention (CDC).

References

1. Smith, C.E.; Whiting, E.G.; Baker, E.E.; Rosenberger, H.G.; Beard, R.R.; Saito, M.T. The Use of Coccidioidin1, 2. *Am. Rev. Tuberc.* **1948**, *57*, 330–360. [CrossRef] [PubMed]
2. Donovan, F.M.; Wightman, P.; Zong, Y.; Gabe, L.; Majeed, A.; Ynosencio, T.; Bedrick, E.J.; Galgiani, J.N. Delays in Coccidioidomycosis Diagnosis and Associated Healthcare Utilization, Tucson, Arizona, USA. *Emerg. Infect. Dis.* **2019**, *25*, 1745–1747. [CrossRef] [PubMed]
3. Smith, C.E.; Beard, R.R. Varieties of Coccidioidal Infection in Relation to the Epidemiology and Control of the Diseases. *Am. J. Public Health Nations Health* **1946**, *36*, 1394–1402. [CrossRef] [PubMed]

4. Galgiani, J.N.; Ampel, N.M.; Blair, J.E.; Catanzaro, A.; Geertsma, F.; Hoover, S.E.; Johnson, R.H.; Kusne, S.; Lisse, J.; MacDonald, J.D.; et al. 2016 Infectious Diseases Society of America (IDSA) Clinical Practice Guideline for the Treatment of Coccidioidomycosis. *Clin. Infect. Dis.* **2016**, *63*, e112–e146. [CrossRef]
5. Benedict, K.; Jackson, B.R.; Chiller, T.; Beer, K.D. Estimation of direct healthcare costs of fungal diseases in the United States. *Clin. Infect. Dis. Off. Publ. Infect. Dis. Soc. Am.* **2019**, *68*, 1791–1797. [CrossRef]
6. Benedict, K.; Whitham, H.K.; Jackson, B.R. Economic Burden of Fungal Diseases in the United States. *Open Forum Infect. Dis.* **2022**, *9*, ofac097. [CrossRef]
7. Wilson, L.; Ting, J.; Lin, H.; MacLean, M.; Peterson, M.W.; Stockamp, N.; Libke, R.; Brown, P. The Rise of Valley Fever: Prevalence and Cost Burden of Coccidioidomycosis Infection in California. *Int. J. Environ. Res. Public Health* **2019**, *16*, 1113. [CrossRef]
8. Grizzle, A.J.; Wilson, L.; Nix, D.E.; Galgiani, J.N. Clinical and Economic Burden of Valley Fever in Arizona: An Incidence-Based Cost-of-Illness Analysis. *Open Forum Infect. Dis.* **2020**, *8*, ofaa623. [CrossRef]
9. Edwards, P.Q.; Palmer, C.E. Prevalence of Sensitivity to Coccidioidin, With Special Reference to Specific and Nonspecific Reactions to Coccidioidin and to Histoplasmin. *Dis. Chest.* **1957**, *31*, 35–60. [CrossRef]
10. Marsden-Haug, N.; Goldoft, M.; Ralston, C.; Limaye, A.P.; Chua, J.; Hill, H.; Jecha, L.; Thompson, G.R.; Chiller, T. Coccidioidomycosis Acquired in Washington State. *Clin. Infect. Dis.* **2013**, *56*, 847–850. [CrossRef]
11. Benedict, K.; Thompson, G.R.; Deresinski, S.; Chiller, T. Mycotic Infections Acquired outside Areas of Known Endemicity, United States. *Emerg. Infect. Dis.* **2015**, *21*, 1935–1941. [CrossRef]
12. Laniado-Laborín, R.; Arathoon, E.G.; Canteros, C.; Muñiz-Salazar, R.; Rendon, A. Coccidioidomycosis in Latin America. *Med. Mycol.* **2019**, *57* (Suppl. 1), S46–S55. [CrossRef]
13. Ashraf, N.; Kubat, R.C.; Poplin, V.; Adenis, A.A.; Denning, D.W.; Wright, L.; McCotter, O.; Schwartz, I.S.; Jackson, B.R.; Chiller, T.; et al. Re-drawing the Maps for Endemic Mycoses. *Mycopathologia* **2020**, *185*, 843–865. [CrossRef]
14. Benedict, K.; Ireland, M.; Weinberg, M.P.; Gruninger, R.J.; Weigand, J.; Chen, L.; Perez-Lockett, K.; Bledsoe, C.; Denny, L.; Cibulskas, K.; et al. Enhanced Surveillance for Coccidioidomycosis, 14 US States, 2016. *Emerg. Infect. Dis.* **2018**, *24*, 1444–1452. [CrossRef]
15. Benedict, K.; Li, Y.; Molinari, N.A.M.; Jackson, B.R. Health Care Providers' Testing Practices for Coccidioidomycosis and Histoplasmosis in Patients With Community-Acquired Pneumonia—United States, 2020. *Open Forum Infect. Dis.* **2021**, *8*, ofab020. [CrossRef]
16. Reportable Fungal Diseases by State Fungal Diseases CDC. Published 19 November 2021. Available online: https://www.cdc.gov/fungal/fungal-disease-reporting-table.html (accessed on 1 April 2022).
17. Valley Fever Statistics Coccidioidomycosis Types of Fungal Diseases Fungal CDC. Published 15 October 2021. Available online: https://www.cdc.gov/fungal/diseases/coccidioidomycosis/statistics.html (accessed on 26 March 2022).
18. *Epidemiologic Summary of Valley Fever (Coccidioidomycosis) in California, 2019*; California Department of Public Health: Sacramento, CA, USA, 2019. Available online: https://www.cdph.ca.gov/Programs/CID/DCDC/CDPH%20Document%20Library/CocciEpiSummary2019.pdf (accessed on 12 April 2022).
19. Valley Fever 2019 Annual Report; Arizona Department of Health Services. Available online: https://www.azdhs.gov/documents/preparedness/epidemiology-disease-control/valley-fever/reports/valley-fever-2019.pdf (accessed on 12 April 2022).
20. Coccidioidomycosis in California Provisional Monthly Report, January–February 2022; California Department of Public Health. Available online: https://www.cdph.ca.gov/Programs/CID/DCDC/CDPH%20Document%20Library/CocciinCAProvisionalMonthlyReport.pdf (accessed on 12 April 2022).
21. Year-to-Date Communicable Disease Summary; Arizona Department of Health Services: 2022. Available online: https://www.azdhs.gov/documents/preparedness/epidemiology-disease-control/disease-data-statistics-reports/data-statistics-archive/2021/2021-ytd-communicable-disease-summary.pdf (accessed on 12 April 2022).
22. Gorris, M.E.; Cat, L.A.; Zender, C.S.; Treseder, K.K.; Randerson, J.T. Coccidioidomycosis Dynamics in Relation to Climate in the Southwestern United States. *GeoHealth* **2018**, *2*, 6–24. [CrossRef]
23. Head, J.R.; Sondermeyer-Cooksey, G.; Heaney, A.K.; Yu, A.T.; Jones, I.; Bhattachan, A.; Campo, S.; Wagner, R.; Mgbara, W.; Phillips, S.; et al. Influence of Meteorological Factors and Drought on Coccidioidomycosis Incidence in California, 2000–2020. *Epidemiology* **2022**. [CrossRef]
24. Gorris, M.E.; Treseder, K.K.; Zender, C.S.; Randerson, J.T. Expansion of Coccidioidomycosis Endemic Regions in the United States in Response to Climate Change. *GeoHealth* **2019**, *3*, 308–327. [CrossRef]
25. Baptista-Rosas, R.C.; Hinojosa, A.; Riquelme, M. Ecological Niche Modeling of Coccidioides spp. in Western North American Deserts. *Ann. N. Y. Acad. Sci.* **2007**, *1111*, 35–46. [CrossRef]
26. Kolivras, K.N.; Comrie, A.C. Modeling valley fever (coccidioidomycosis) incidence on the basis of climate conditions. *Int. J. Biometeorol.* **2003**, *47*, 87–101. [CrossRef]
27. Tamerius, J.D.; Comrie, A.C. Coccidioidomycosis Incidence in Arizona Predicted by Seasonal Precipitation. *PLoS ONE* **2011**, *6*, e21009. [CrossRef]
28. Talamantes, J.; Behseta, S.; Zender, C.S. Fluctuations in Climate and Incidence of Coccidioidomycosis in Kern County, California. *Ann. N. Y. Acad. Sci.* **2007**, *1111*, 73–82. [CrossRef]
29. Zender, C.S.; Talamantes, J. Climate controls on valley fever incidence in Kern County, California. *Int. J. Biometeorol.* **2006**, *50*, 174–182. [CrossRef]

30. Comrie, A.C. Climate Factors Influencing Coccidioidomycosis Seasonality and Outbreaks. *Environ. Health Perspect.* **2005**, *113*, 688–692. [CrossRef]
31. Park, B.J.; Sigel, K.; Vaz, V.; Komatsu, K.; McRill, C.; Phelan, M.; Colman, T.; Comrie, A.C.; Warnock, D.W.; Galgiani, J.N.; et al. An Epidemic of Coccidioidomycosis in Arizona Associated with Climatic Changes, 1998–2001. *J. Infect. Dis.* **2005**, *191*, 1981–1987. [CrossRef]
32. US EPA, O. Particulate Matter (PM10) Trends. Published 19 July 2016. Available online: https://www.epa.gov/air-trends/particulate-matter-pm10-trends (accessed on 1 April 2022).
33. Kollath, D.R.; Mihaljevic, J.R.; Barker, B.M. PM10 and Other Climatic Variables Are Important Predictors of Seasonal Variability of Coccidioidomycosis in Arizona. *Microbiol. Spectr.* **2022**, *10*, e0148321. [CrossRef]
34. U.S. Census Bureau QuickFacts. Available online: https://www.census.gov/quickfacts/fact/table/maricopacountyarizona,CA,AZ,US/AGE775220 (accessed on 26 March 2022).
35. Colson, A.J.; Vredenburgh, L.; Guevara, R.E.; Rangel, N.P.; Kloock, C.T.; Lauer, A. Large-Scale Land Development, Fugitive Dust, and Increased Coccidioidomycosis Incidence in the Antelope Valley of California, 1999–2014. *Mycopathologia* **2017**, *182*, 439–458. [CrossRef]
36. Guevara, R.E.; Motala, T.; Terashita, D. The Changing Epidemiology of Coccidioidomycosis in Los Angeles (LA) County, California, 1973–2011. *PLoS ONE* **2015**, *10*, e0136753. [CrossRef]
37. Leake, J.A.D.; Mosley, D.G.; England, B.; Graham, J.V.; Plikaytis, B.D.; Ampel, N.M.; Perkins, B.A.; Hajjeh, R.A. Risk Factors for Acute Symptomatic Coccidioidomycosis among Elderly Persons in Arizona, 1996–1997. *J. Infect. Dis.* **2000**, *181*, 1435–1440. [CrossRef]
38. Center for International Blood and Marrow Transplant Research Transplant Activity Report Covering 2009–2013. Available online: https://bloodstemcell.hrsa.gov/data/donation-and-transplantation-statistics/transplant-activity-report (accessed on 26 March 2022).
39. Organ Procurement and Transplantation Network: View Data Reports. Available online: https://optn.transplant.hrsa.gov/data/view-data-reports/ (accessed on 26 March 2022).
40. Casadevall, A. Fungal Diseases in the 21st Century: The Near and Far Horizons. *Pathog. Immun.* **2018**, *3*, 183–196. [CrossRef]
41. Brown, J.; Benedict, K.; Park, B.J.; Thompson, G.R. Coccidioidomycosis: Epidemiology. *Clin. Epidemiol.* **2013**, *5*, 185–197. [CrossRef] [PubMed]
42. Odio, C.D.; Marciano, B.E.; Galgiani, J.N.; Holland, S.M. Risk Factors for Disseminated Coccidioidomycosis, United States. *Emerg. Infect. Dis.* **2017**, *23*, 308–311. [CrossRef] [PubMed]
43. Tsang, C.A.; Anderson, S.M.; Imholte, S.B.; Erhart, L.M.; Chen, S.; Park, B.J.; Christ, C.; Komatsu, K.K.; Chiller, T.; Sunenshine, R.H. Enhanced Surveillance of Coccidioidomycosis, Arizona, USA, 2007–2008. *Emerg. Infect. Dis.* **2010**, *16*, 1738–1744. [CrossRef] [PubMed]
44. Benedict, K.; Molinari, N.A.M.; Jackson, B.R. Public Awareness of Invasive Fungal Diseases—United States, 2019. *Morb. Mortal. Wkly. Rep.* **2020**, *69*, 1343–1346. [CrossRef]
45. Hurd-Kundeti, G.; Sondermeyer Cooksey, G.L.; Jain, S.; Vugia, D.J. Valley Fever (Coccidioidomycosis) Awareness—California, 2016–2017. *Morb. Mortal. Wkly. Rep.* **2020**, *69*, 1512–1516. [CrossRef]
46. Werner, S.B.; Pappagianis, D. Coccidioidomycosis in Northern California—An Outbreak among Archeology Students near Red Bluff. *Calif. Med.* **1973**, *119*, 16–20.
47. Petersen, L.R.; Marshall, S.L.; Barton, C.; Hajjeh, R.A.; Lindsley, M.D.; Warnock, D.W.; Panackal, A.A.; Shaffer, J.B.; Haddad, M.B.; Fisher, F.S.; et al. Coccidioidomycosis among Workers at an Archeological Site, Northeastern Utah. *Emerg. Infect. Dis.* **2004**, *10*, 637–642. [CrossRef]
48. Werner, S.B.; Pappagianis, D.; Heindl, I.; Mickel, A. An Epidemic of Coccidioidomycosis among Archeology Students in Northern California. *N. Engl. J. Med.* **1972**, *286*, 507–512. [CrossRef]
49. Factors Influencing Distribution of Coccidioides Immitis in Soil, Washington State, 2016-PubMed. Available online: https://pubmed.ncbi.nlm.nih.gov/34730378/ (accessed on 23 March 2022).
50. Kobziar, L.N.; Thompson, G.R. Wildfire smoke, a potential infectious agent. *Science* **2020**, *370*, 1408–1410. [CrossRef]
51. Laws, R.L.; Jain, S.; Cooksey, G.S.; Mohle-Boetani, J.; McNary, J.; Wilken, J.; Harrison, R.; Leistikow, B.; Vugia, D.J.; Windham, G.C.; et al. Coccidioidomycosis outbreak among inmate wildland firefighters: California, 2017. *Am. J. Ind. Med.* **2021**, *64*, 266–273. [CrossRef]
52. Tong, D.Q.; Wang, J.X.L.; Gill, T.E.; Lei, H.; Wang, B. Intensified dust storm activity and Valley fever infection in the southwestern United States. *Geophys. Res. Lett.* **2017**, *44*, 4304–4312. [CrossRef]
53. Comrie, A.C. No Consistent Link Between Dust Storms and Valley Fever (Coccidioidomycosis). *GeoHealth* **2021**, *5*, e2021GH000504. [CrossRef]
54. Gade, L.; McCotter, O.Z.; Bowers, J.R.; Waddell, V.; Brady, S.; Carvajal, J.A.; Sunenshine, R.; Komatsu, K.K.; Engelthaler, D.M.; Chiller, T.; et al. The detection of Coccidioides from ambient air in Phoenix, Arizona: Evidence of uneven distribution and seasonality. *Med. Mycol.* **2020**, *58*, 552–559. [CrossRef]
55. Emmons, C.W. Coccidioidomycosis in Wild Rodents. A Method of Determining the Extent of Endemic Areas. *Public Health Rep. (1896–1970)* **1943**, *58*, 1–5. [CrossRef]

56. *Health ASD of Proceedings of Symposium on Coccidioidomycosis: Held at Phoenix, Ariz.-Feb. 11–13, 1957*; U.S. Department of Health, Education, and Welfare Public Health Service, Bureau of State Services, Communicable Disease Center: Washington, DC, USA, 1957.
57. Kollath, D.R.; Miller, K.J.; Barker, B.M. The mysterious desert dwellers: *Coccidioides immitis and Coccidioides posadasii*, causative fungal agents of coccidioidomycosis. *Virulence* **2019**, *10*, 222–233. [CrossRef]
58. Soil Ecology of Coccidioides Immitis at Amerindian Middens in California Applied Microbiology. Available online: https://journals.asm.org/doi/abs/10.1128/am.27.2.379-388.1974 (accessed on 23 March 2022).
59. Cordeiro, R.A.; Brilhante, R.S.N.; Rocha, M.F.G.; Fechine, M.A.B.; Camara, L.M.C.; Camargo, Z.P.; Sidrim, J.J.C. Phenotypic characterization and ecological features of Coccidioides spp. from Northeast Brazil. *Med. Mycol.* **2006**, *44*, 631–639. [CrossRef]
60. Sharpton, T.J.; Stajich, J.E.; Rounsley, S.D.; Gardner, M.J.; Wortman, J.R.; Jordar, V.S.; Maiti, R.; Kodira, C.D.; Neafsey, D.E.; Zeng, Q.; et al. Comparative genomic analyses of the human fungal pathogens *Coccidioides* and their relatives. *Genome Res.* **2009**, *19*, 1722–1731. [CrossRef]
61. Ampel, N.M.; Dols, C.L.; Galgiani, J.N. Coccidioidomycosis during human immunodeficiency virus infection: Results of a prospective study in a coccidioidal endemic area. *Am. J. Med.* **1993**, *94*, 235–240. [CrossRef]
62. Woods, C.W.; McRill, C.; Plikaytis, B.D.; Rosenstein, N.E.; Mosley, D.; Boyd, D.; England, B.; Perkins, B.A.; Ampel, N.M.; Hajjeh, R.A. Coccidioidomycosis in Human Immunodeficiency Virus–Infected Persons in Arizona, 1994–1997: Incidence, Risk Factors, and Prevention. *J. Infect. Dis.* **2000**, *181*, 1428–1434. [CrossRef]
63. Masannat, F.Y.; Ampel, N.M. Coccidioidomycosis in Patients with HIV-1 Infection in the Era of Potent Antiretroviral Therapy. *Clin. Infect. Dis.* **2010**, *50*, 1–7. [CrossRef]
64. Blair, J.E.; Logan, J.L. Coccidioidomycosis in Solid Organ Transplantation. *Clin. Infect. Dis.* **2001**, *33*, 1536–1544. [CrossRef]
65. Bergstrom, L.; Yocum, D.E.; Ampel, N.M.; Villanueva, I.; Lisse, J.; Gluck, O.; Tesser, J.; Posever, J.P.; Miller, M.; Araujo, J.; et al. Increased risk of coccidioidomycosis in patients treated with tumor necrosis factor α antagonists. *Arthritis Rheum.* **2004**, *50*, 1959–1966. [CrossRef]
66. Blair, J.E.; Douglas, D.D.; Mulligan, D.C. Early results of targeted prophylaxis for coccidioidomycosis in patients undergoing orthotopic liver transplantation within an endemic area. *Transpl. Infect. Dis.* **2003**, *5*, 3–8. [CrossRef]
67. Kahn, A.; Carey, E.J.; Blair, J.E. Universal fungal prophylaxis and risk of coccidioidomycosis in liver transplant recipients living in an endemic area. *Liver Transpl.* **2015**, *21*, 353–361. [CrossRef]
68. Keckich, D.W.; Blair, J.E.; Vikram, H.R.; Seville, M.T.; Kusne, S. Reactivation of Coccidioidomycosis Despite Antifungal Prophylaxis in Solid Organ Transplant Recipients. *Transplantation* **2011**, *92*, 88–93. [CrossRef]
69. Truong, C.N.; Nailor, M.D.; Walia, R.; Cherrier, L.; Nasar, A.; Goodlet, K.J. Universal lifelong fungal prophylaxis and risk of coccidioidomycosis in lung transplant recipients living in an endemic area. *Clin. Infect. Dis. Off. Publ. Infect. Dis. Soc. Am.* **2021**, *74*, ciab752. [CrossRef]
70. Al-Obaidi, M.M.; Nematollahi, S.; Nix, D.E.; Zangeneh, T.T. Remarks on the universal lifelong coccidioidomycosis prophylaxis in lung transplant recipients. *Clin. Infect. Dis.* **2021**, *74*, ciab878. [CrossRef]
71. Bercovitch, R.S.; Catanzaro, A.; Schwartz, B.S.; Pappagianis, D.; Watts, D.H.; Ampel, N.M. Coccidioidomycosis During Pregnancy: A Review and Recommendations for Management. *Clin. Infect. Dis.* **2011**, *53*, 363–368. [CrossRef]
72. Crum, N.F.; Ballon-Landa, G. Coccidioidomycosis in Pregnancy: Case Report and Review of the Literature. *Am. J. Med.* **2006**, *119*, e11–e993. [CrossRef]
73. Crum, N.F.; Lederman, E.R.; Stafford, C.M.; Parrish, J.S.; Wallace, M.R. Coccidioidomycosis: A Descriptive Survey of a Reemerging Disease. Clinical Characteristics and Current Controversies. *Medicine* **2004**, *83*, 149–175. [CrossRef]
74. Sondermeyer Cooksey, G.L.; Nguyen, A.; Vugia, D.; Jain, S. Regional Analysis of Coccidioidomycosis Incidence—California, 2000–2018. *Morb. Mortal. Wkly. Rep.* **2020**, *69*, 1817–1821. [CrossRef] [PubMed]
75. Chang, D.C.; Anderson, S.; Wannemuehler, K.; Engelthaler, D.M.; Erhart, L.; Sunenshine, R.H.; Burwell, L.A.; Park, B.J. Testing for Coccidioidomycosis among Patients with Community-Acquired Pneumonia. *Emerg. Infect. Dis.* **2008**, *14*, 1053–1059. [CrossRef] [PubMed]
76. Valley Fever (Coccidioidomycosis), Jobs at Risk NIOSH CDC. Published 11 March 2022. Available online: https://www.cdc.gov/niosh/topics/valleyfever/risk.html (accessed on 26 March 2022).
77. de Perio, M.A.; Niemeier, R.T.; Burr, G.A. Coccidioides Exposure and Coccidioidomycosis among Prison Employees, California, United States. *Emerg. Infect. Dis.* **2015**, *21*, 1031–1033. [CrossRef] [PubMed]
78. Wilken, J.A.; Marquez, P.; Terashita, D.; McNary, J.; Windham, G.; Materna, B. Coccidioidomycosis Among Cast and Crew Members at an Outdoor Television Filming Event—California, 2012. *MMWR Morb. Mortal Wkly Rep.* **2014**, *63*, 20.
79. Durry, E.; Pappagianis, D.; Werner, S.B.; Hutwagner, L.; Sun, R.K.; Maurer, M.; McNeil, M.M.; Pinner, R.W. Coccidioidomycosis in Tulare County, California, 1991: Reemergence of an endemic disease. *J. Med. Vet. Mycol.* **1997**, *35*, 321–326. [CrossRef]
80. Sondermeyer, G.; Lee, L.; Gilliss, D.; Tabnak, F.; Vugia, D. Coccidioidomycosis-associated Hospitalizations, California, USA, 2000–2011. *Emerg. Infect. Dis.* **2013**, *19*, 1590–1597. [CrossRef]
81. Einstein, H.E.; Johnson, R.H. Coccidioidomycosis: New Aspects of Epidemiology and Therapy. *Clin. Infect. Dis.* **1993**, *16*, 349–356. [CrossRef]
82. McHardy, I.; Reagan, K.L.; Sebastian, J.F.; Barker, B.; Bays, D.J.; Dandekar, S.; Cohen, S.H.; Jennings, K.E.; Sykes, J.; Thompson, G.R. Sex Differences in Susceptibility to Coccidioidomycosis. *Open Forum Infect. Dis.* **2022**, *9*, ofab543. [CrossRef]

83. Shah, A.S.; Heidari, A.; Civelli, V.F.; Sharma, R.; Clark, C.S.; Munoz, A.D.; Ragland, A.S.; Johnson, R.H. The Coincidence of 2 Epidemics, Coccidioidomycosis and SARS-CoV-2: A Case Report. *J. Investig. Med. High Impact Case Rep.* **2020**, *8*, 2324709620930540. [CrossRef]
84. Chang, C.C.; Senining, R.; Kim, J.; Goyal, R. An Acute Pulmonary Coccidioidomycosis Coinfection in a Patient Presenting with Multifocal Pneumonia With COVID-19. *J. Investig. Med. High Impact Case Rep.* **2020**, *8*, 2324709620972244. [CrossRef]
85. Chen, J.C.; Wong, D.; Rabi, S.; Worswick, S.; DeClerck, B.; Gibb, J. All That Coughs Is Not COVID-19: A Delayed Diagnosis of Disseminated Coccidioidomycosis Following Severe Acute Respiratory Syndrome Coronavirus 2 Infection. *Open Forum Infect. Dis.* **2021**, *8*, ofab246. [CrossRef]
86. Nielsen, M.C.; Reynoso, D.; Ren, P. The Brief Case: A Fatal Case of SARS-CoV-2 Coinfection with *Coccidioides* in Texas—Another Challenge We Face. Burnham, C.A.D., Ed. *J. Clin. Microbiol.* **2021**, *59*, e00163-21. [CrossRef]
87. Ko, J.; Lee, M.M. A Case of Disseminated Coccidioidomycosis in a Patient with a Prolonged Course of COVID-19 Pneumonia. In *TP98. TP098 FUNGUS AMONG-US-RARE FUNGAL CASE REPORTS*; American Thoracic Society International Conference Abstracts; American Thoracic Society: New York, NY, USA, 2021; p. A3997. [CrossRef]
88. Mathew, J.; Cherukuri, S.V.; Dihowm, F. SARS-CoV-2 with concurrent coccidioidomycosis complicated by refractory pneumothorax in a Hispanic male: A case report and literature review. *World J. Respirol.* **2021**, *11*, 1–11. [CrossRef]
89. Patel, B.; Jarrett, B.; Bixby, B. Diagnostic Error and Cognitive Bias in the Era of COVID-19: Don't Forget About Endemic Diseases. *Chest* **2020**, *158*, A541–A542. [CrossRef]
90. Krauth, D.S.; Jamros, C.M.; Rivard, S.C.; Olson, N.H.; Maves, R.C. Accelerated Progression of Disseminated Coccidioidomycosis Following SARS-CoV-2 Infection: A Case Report. *Mil. Med.* **2021**, *186*, usab132. [CrossRef]
91. Heaney, A.K.; Head, J.R.; Broen, K.; Click, K.; Taylor, J.; Balmes, J.R.; Zelner, J.; Remais, J.V. Coccidioidomycosis and COVID-19 Co-Infection, United States, 2020. *Emerg. Infect. Dis.* **2021**, *27*, 1266–1273. [CrossRef]
92. Shi, Y.; Yu, X.; Zhao, H.; Wang, H.; Zhao, R.; Sheng, J. Host susceptibility to severe COVID-19 and establishment of a host risk score: Findings of 487 cases outside Wuhan. *Crit. Care.* **2020**, *24*, 108. [CrossRef]
93. CDC COVID-19 Response Team. Preliminary Estimates of the Prevalence of Selected Underlying Health Conditions among Patients with Coronavirus Disease 2019—United States, 12 February–28 March 2020. *Morb. Mortal. Wkly. Rep.* **2020**, *69*. [CrossRef]
94. Yang, J.; Zheng, Y.; Gou, X.; Pu, K.; Chen, Z.; Guo, Q.; Ji, R.; Wang, H.; Wang, Y.; Zhou, Y. Prevalence of comorbidities and its effects in patients infected with SARS-CoV-2: A systematic review and meta-analysis. *Int. J. Infect. Dis.* **2020**, *94*, 91–95. [CrossRef]
95. Qin, C.; Zhou, L.; Hu, Z.; Zhang, S.; Yang, S.; Tao, Y.; Xie, C.; Ma, K.; Shang, K.; Wang, W.; et al. Dysregulation of immune response in patients with COVID-19 in Wuhan, China. *Clin. Infect. Dis. Off. Publ. Infect. Dis. Soc. Am.* **2020**, *71*, ciaa248. [CrossRef]
96. Beaman, L.; Benjamini, E.; Pappagianis, D. Activation of macrophages by lymphokines: Enhancement of phagosome-lysosome fusion and killing of Coccidioides immitis. *Infect. Immun.* **1983**, *39*, 1201–1207. [CrossRef]
97. Benedict, K.; Williams, S.; Beekmann, S.E.; Polgreen, P.M.; Jackson, B.R.; Toda, M. Testing Practices for Fungal Respiratory Infections and SARS-CoV-2 among Infectious Disease Specialists, United States. *J. Fungi.* **2021**, *7*, 605. [CrossRef]
98. Olson, G.; Davis, A.M. Diagnosis and Treatment of Adults With Community-Acquired Pneumonia. *JAMA* **2020**, *323*, 885. [CrossRef]
99. Valdivia, L.; Nix, D.; Wright, M.; Lindberg, E.; Fagan, T.; Lieberman, D.; Stoffer, T.; Ampel, N.M.; Galgiani, J.N. Coccidioidomycosis as a Common Cause of Community-acquired Pneumonia. *Emerg. Infect. Dis.* **2006**, *12*, 958–962. [CrossRef]
100. Chi, G.C.; Benedict, K.; Beer, K.D.; Jackson, B.R.; McCotter, O.; Xie, F.; Lawrence, J.M.; Tartof, S.Y. Antibiotic and antifungal treatment among persons with confirmed coccidioidomycosis—Southern California, 2011. *Med. Mycol.* **2020**, *58*, 411–413. [CrossRef]
101. Chen, S.; Erhart, L.M.; Anderson, S.; Komatsu, K.; Park, B.; Chiller, T.; Sunenshine, R. Coccidioidomycosis: Knowledge, attitudes, and practices among healthcare providers—Arizona, 2007. *Med. Mycol.* **2011**, *49*, 649–656. [CrossRef]
102. Pu, J.; Donovan, F.M.; Ellingson, K.; Leroy, G.; Stone, J.; Bedrick, E.; Galgiani, J.N. Clinician Practice Patterns That Result in the Diagnosis of Coccidioidomycosis Before or During Hospitalization. *Clin. Infect. Dis.* **2021**, *73*, e1587–e1593. [CrossRef]
103. Malo, J.; Holbrook, E.; Zangeneh, T.; Strawter, C.; Oren, E.; Robey, I.; Erickson, H.; Chahal, R.; Durkin, M.; Thompson, C.; et al. Enhanced Antibody Detection and Diagnosis of Coccidioidomycosis with the MiraVista IgG and IgM Detection Enzyme Immunoassay. *J. Clin. Microbiol.* **2017**, *55*, 893–901. [CrossRef]
104. Malo, J.; Holbrook, E.; Zangeneh, T.; Strawter, C.; Oren, E.; Robey, I.; Erickson, H.; Carranza-Chahal, R.; Durkin, M.; Thompson, C.; et al. Comparison of three anti-coccidioides antibody enzyme immunoassay kits for the diagnosis of coccidioidomycosis. *Med. Mycol.* **2020**, *58*, 774–778. [CrossRef]
105. Grys, T.E.; Brighton, A.; Chang, Y.H.; Liesman, R.; Bolster LaSalle, C.; Blair, J.E. Comparison of two FDA-cleared EIA assays for the detection of Coccidioides antibodies against a composite clinical standard. *Med. Mycol.* **2019**, *57*, 595–600. [CrossRef] [PubMed]
106. Kuberski, T.; Herrig, J.; Pappagianis, D. False-Positive IgM Serology in Coccidioidomycosis. *J. Clin. Microbiol.* **2010**, *48*, 2047–2049. [CrossRef] [PubMed]
107. Donovan, F.M.; Ramadan, F.A.; Khan, S.A.; Bhaskara, A.; Lainhart, W.D.; Narang, A.T.; Mosier, J.M.; Ellingson, K.D.; Bedrick, E.J.; Saubolle, M.A.; et al. Comparison of a Novel Rapid Lateral Flow Assay to Enzyme Immunoassay Results for Early Diagnosis of Coccidioidomycosis. *Clin. Infect. Dis.* **2021**, *73*, e2746–e2753. [CrossRef] [PubMed]
108. Huppert, M. Serology of coccidioidomycosis. *Mycopathol. Mycol. Appl.* **1970**, *41*, 107–113. [CrossRef] [PubMed]

109. Blair, J.E.; Coakley, B.; Santelli, A.C.; Hentz, J.G.; Wengenack, N.L. Serologic testing for symptomatic coccidioidomycosis in immunocompetent and immunosuppressed hosts. *Mycopathologia* **2006**, *162*, 317–324. [CrossRef]
110. Mendoza, N.; Blair, J.E. The Utility of Diagnostic Testing for Active Coccidioidomycosis in Solid Organ Transplant Recipients. *Am. J. Transplant.* **2013**, *13*, 1034–1039. [CrossRef]
111. Durkin, M.; Connolly, P.; Kuberski, T.; Myers, R.; Kubak, B.M.; Bruckner, D.; Pegues, D.; Wheat, L.J. Diagnosis of Coccidioidomycosis with Use of the *Coccidioides* Antigen Enzyme Immunoassay. *Clin. Infect. Dis.* **2008**, *47*, e69–e73. [CrossRef]
112. Galgiani, J.N.; Grace, G.M.; Lundergan, L.L. New Serologic Tests for Early Detection of Coccidioidomycosis. *J. Infect. Dis.* **1991**, *163*, 671–674. [CrossRef]
113. Kassis, C.; Durkin, M.; Holbrook, E.; Myers, R.; Wheat, L. Advances in Diagnosis of Progressive Pulmonary and Disseminated Coccidioidomycosis. *Clin. Infect. Dis.* **2021**, *72*, 968–975. [CrossRef]
114. Wheat, L.J.; Knox, K.S.; Hage, C.A. Approach to the Diagnosis of Histoplasmosis, Blastomycosis and Coccidioidomycosis. *Curr Curr. Treat. Options Infect. Dis.* **2014**, *6*, 337–351. [CrossRef]
115. Dizon, D.; Mitchell, M.; Dizon, B.; Libke, R.; Peterson, M.W. The utility of real-time polymerase chain reaction in detecting Coccidioides immitis among clinical specimens in the Central California San Joaquin Valley. *Med. Mycol.* **2019**, *57*, 688–693. [CrossRef]
116. Vucicevic, D.; Blair, J.E.; Binnicker, M.J.; McCullough, A.E.; Kusne, S.; Vikram, H.R.; Parish, J.M.; Wengenack, N.L. The utility of Coccidioides polymerase chain reaction testing in the clinical setting. *Mycopathologia* **2010**, *170*, 345–351. [CrossRef]
117. Thompson, G.R.; Bays, D.J.; Johnson, S.M.; Cohen, S.H.; Pappagianis, D.; Finkelman, M.A. Serum (1→3)-β-d-Glucan Measurement in Coccidioidomycosis. *J. Clin. Microbiol.* **2012**, *50*, 3060–3062. [CrossRef]
118. Caceres, D.H.; Chiller, T.; Lindsley, M.D. Immunodiagnostic Assays for the Investigation of Fungal Outbreaks. *Mycopathologia* **2020**, *185*, 867–880. [CrossRef]
119. Durkin, M.; Estok, L.; Hospenthal, D.; Crum-Cianflone, N.; Swartzentruber, S.; Hackett, E.; Wheat, L.J. Detection of *Coccidioides* Antigenemia following Dissociation of Immune Complexes. *Clin. Vaccine Immunol.* **2009**, *16*, 1453–1456. [CrossRef]
120. Messina, J.A.; Maziarz, E.K.; Galgiani, J.; Truong, J.T.; Htoo, A.K.; Heidari, A.; Johnson, R.H.; Narang, A.T.; Donovan, F.M.; Ewell, M.; et al. A randomized, double-blind, placebo-controlled clinical trial of fluconazole as early empiric treatment of coccidioidomycosis pneumonia (Valley Fever) in adults presenting with community-acquired pneumonia in endemic areas (FLEET-Valley Fever). *Contemp. Clin. Trials. Commun.* **2021**, *24*, 100851. [CrossRef]
121. Sugar, A.M.; Alsip, S.G.; Galgiani, J.N.; Graybill, J.R.; Dismukes, W.E.; Cloud, G.A.; Craven, P.C.; Stevens, D.A. Pharmacology and toxicity of high-dose ketoconazole. *Antimicrob. Agents Chemother.* **1987**, *31*, 1874–1878. [CrossRef]
122. Pont, A.; Graybill, J.R.; Craven, P.C.; Galgiani, J.N.; Dismukes, W.E.; Reitz, R.E.; Stevens, D.A. High-Dose Ketoconazole Therapy and Adrenal and Testicular Function in Humans. *Arch. Intern. Med.* **1984**, *144*, 2150–2153. [CrossRef]
123. Arndt, C.A.S.; Walsh, T.J.; McCully, C.L.; Balis, F.M.; Pizzo, P.A.; Poplack, D.G. Fluconazole Penetration into Cerebrospinal Fluid: Implications for Treating Fungal Infections of the Central Nervous System. *J. Infect. Dis.* **1988**, *157*, 178–180. [CrossRef]
124. Brewer, A.C.; Huber, J.T.; Girardo, M.E.; Kosiorek, H.E.; Burns, M.W.; Stewart, T.D.; Blair, J.E. Cutaneous effects associated with fluconazole in patients treated for coccidioidomycosis. *Int. J. Dermatol.* **2019**, *58*, 250–253. [CrossRef]
125. Thompson, G.R.; Barker, B.M.; Wiederhold, N.P. Large-Scale Evaluation of In Vitro Amphotericin B, Triazole, and Echinocandin Activity against Coccidioides Species from U.S. Institutions. *Antimicrob. Agents Chemother.* **2017**, *61*, e02634-16. [CrossRef]
126. Galgiani, J.N. Comparison of Oral Fluconazole and Itraconazole for Progressive, Nonmeningeal Coccidioidomycosis: A Randomized, Double-Blind Trial. *Ann. Intern. Med.* **2000**, *133*, 676. [CrossRef]
127. Hoffmann, W.J.; McHardy, I.; Thompson, G.R., III. Itraconazole induced hypertension and hypokalemia: Mechanistic evaluation. *Mycoses* **2018**, *61*, 337–339. [CrossRef]
128. Lestner, J.M.; Roberts, S.A.; Moore, C.B.; Howard, S.J.; Denning, D.W.; Hope, W.W. Toxicodynamics of Itraconazole: Implications for Therapeutic Drug Monitoring. *Clin. Infect. Dis.* **2009**, *49*, 928–930. [CrossRef]
129. Sharkey, P.K.; Rinaldi, M.G.; Dunn, J.F.; Hardin, T.C.; Fetchick, R.J.; Graybill, J.R. High-dose itraconazole in the treatment of severe mycoses. *Antimicrob. Agents Chemother.* **1991**, *35*, 707–713. [CrossRef] [PubMed]
130. Kim, M.M.; Vikram, H.R.; Kusne, S.; Seville, M.T.; Blair, J.E. Treatment of Refractory Coccidioidomycosis With Voriconazole or Posaconazole. *Clin. Infect. Dis.* **2011**, *53*, 1060–1066. [CrossRef] [PubMed]
131. Cortez, K.J.; Walsh, T.J.; Bennett, J.E. Successful Treatment of Coccidioidal Meningitis with Voriconazole. *Clin. Infect. Dis.* **2003**, *36*, 1619–1622. [CrossRef] [PubMed]
132. Freifeld, A.; Proia, L.; Andes, D.; Baddour, L.M.; Blair, J.; Spellberg, B.; Arnold, S.; Lentnek, A.; Wheat, L.J. Voriconazole Use for Endemic Fungal Infections. *Antimicrob. Agents Chemother.* **2009**, *53*, 1648–1651. [CrossRef] [PubMed]
133. Epaulard, O.; Villier, C.; Ravaud, P.; Chosidow, O.; Blanche, S.; Mamzer-Bruneel, M.; Thiebaut, A.; Leccia, M.; Lortholary, O. A Multistep Voriconazole-Related Phototoxic Pathway May Lead to Skin Carcinoma: Results From a French Nationwide Study. *Clin. Infect. Dis.* **2013**, *57*, e182–e188. [CrossRef]
134. Haylett, A.K.; Felton, S.; Denning, D.W.; Rhodes, L.E. Voriconazole-induced photosensitivity: Photobiological assessment of a case series of 12 patients. *Br. J. Dermatol.* **2013**, *168*, 179–185. [CrossRef]
135. Williams, K.; Mansh, M.; Chin-Hong, P.; Singer, J.; Arron, S.T. Voriconazole-Associated Cutaneous Malignancy: A Literature Review on Photocarcinogenesis in Organ Transplant Recipients. *Clin. Infect. Dis.* **2014**, *58*, 997–1002. [CrossRef]

136. Pham, A.N.; Bubalo, J.S.; Lewis, J.S., II. Comparison of posaconazole serum concentrations from haematological cancer patients on posaconazole tablet and oral suspension for treatment and prevention of invasive fungal infections. *Mycoses* **2016**, *59*, 226–233. [CrossRef]
137. Anstead, G.M.; Corcoran, G.; Lewis, J.; Berg, D.; Graybill, J.R. Refractory Coccidioidomycosis Treated with Posaconazole. *Clin. Infect. Dis.* **2005**, *40*, 1770–1776. [CrossRef]
138. Catanzaro, A.; Cloud, G.A.; Stevens, D.A.; Levine, B.E.; Williams, B.L.; Johnson, R.H.; Rendon, A.; Mirels, L.F.; Lutz, J.E.; Holloway, M.; et al. Safety, Tolerance, and Efficacy of Posaconazole Therapy in Patients with Nonmeningeal Disseminated or Chronic Pulmonary Coccidioidomycosis. *Clin. Infect. Dis.* **2007**, *45*, 562–568. [CrossRef]
139. Ruping, M.J.G.T.; Albermann, N.; Ebinger, F.; Burckhardt, I.; Beisel, C.; Muller, C.; Vehreschild, J.J.; Kochanek, M.; Fatkenheuer, G.; Bangard, C.; et al. Posaconazole concentrations in the central nervous system. *J. Antimicrob. Chemother.* **2008**, *62*, 1468–1470. [CrossRef]
140. Schein, R.; Homans, J.; Larsen, R.A.; Neely, M. Posaconazole for Chronic Refractory Coccidioidal Meningitis. *Clin. Infect. Dis.* **2011**, *53*, 1252–1254. [CrossRef]
141. González, G.M.; Tijerina, R.; Najvar, L.K.; Bocanegra, R.; Rinaldi, M.; Loebenberg, D.; Graybill, J.R. In Vitro and In Vivo Activities of Posaconazole against *Coccidioides immitis*. *Antimicrob. Agents Chemother.* **2002**, *46*, 1352–1356. [CrossRef]
142. Donnelley, M.A.; Zhu, E.S.; Thompson, G.R. Isavuconazole in the treatment of invasive aspergillosis and mucormycosis infections. *Infect. Drug Resist.* **2016**, *9*, 79–86. [CrossRef]
143. Thompson, G.R.; Rendon, A.; Ribeiro dos Santos, R.; Queiroz-Telles, F.; Ostrosky-Zeichner, L.; Azie, N.; Maher, R.; Lee, M.; Kovanda, L.; Engelhardt, M.; et al. Isavuconazole Treatment of Cryptococcosis and Dimorphic Mycoses. *Clin. Infect. Dis. Off. Publ. Infect. Dis. Soc. Am.* **2016**, *63*, 356–362. [CrossRef]
144. Isavuconazole versus voriconazole for primary treatment of invasive mould disease caused by Aspergillus and other filamentous fungi (SECURE): A phase 3, randomised-controlled, non-inferiority trial Elsevier Enhanced Reader. *Lancet* **2015**, *387*, 760–769. [CrossRef]
145. Hamill, R.J. Amphotericin B Formulations: A Comparative Review of Efficacy and Toxicity. *Drugs* **2013**, *73*, 919–934. [CrossRef]
146. Saravolatz, L.D.; Ostrosky-Zeichner, L.; Marr, K.A.; Rex, J.H.; Cohen, S.H. Amphotericin B: Time for a New "Gold Standard". *Clin. Infect. Dis.* **2003**, *37*, 415–425. [CrossRef]
147. Stevens, D.A.; Shatsky, S.A. Intrathecal amphotericin in the management of coccidioidal meningitis. *Semin. Respir. Infect.* **2001**, *16*, 263–269. [CrossRef]

Journal of *Fungi*

MDPI

Review

Lobomycosis Epidemiology and Management: The Quest for a Cure for the Most Neglected of Neglected Tropical Diseases

Franciely G. Gonçalves [1,2,3], Patrícia S. Rosa [4], Andrea de F. F. Belone [4], Léia B. Carneiro [3], Vania L. Q. de Barros [3], Rosineide F. Bispo [3], Yally A. da S. Sbardelott [3], Sebastião A. V. M. Neves [5], Amy Y. Vittor [6], William J. Woods [3] and Gabriel Z. Laporta [1,*]

1 Graduate Research and Innovation Program, Centro Universitario FMABC, Santo André 09060-870, SP, Brazil; francielygg@hotmail.com
2 Research Lab at Centro Universitario UNINORTE, Rio Branco 69915-901, AC, Brazil
3 State Dermatology Program of Acre, Secretariat of Health in the State of Acre (SESACRE), Rio Branco 69917-650, AC, Brazil; leia.borges@bol.com.br (L.B.C.); vannyla_ice@hotmail.com (V.L.Q.d.B.); rosebispo_8@hotmail.com (R.F.B.); yallysbardlott@hotmail.com (Y.A.d.S.S.); william.ac@brturbo.com.br (W.J.W.)
4 Division of Research, Lauro de Souza Lima Institute, Bauru 17034-971, AP, Brazil; prosa@ilsl.br (P.S.R.); abelone@ilsl.br (A.d.F.F.B.)
5 School of Medicine, Federal University of Acre, Rio Branco 69920-900, AC, Brazil; tiaoviana.sebastiao@gmail.com
6 Division of Infectious Diseases and Global Medicine, Department of Medicine, College of Medicine, University of Florida, Gainesville, FL 32610, USA; amy.vittor@medicine.ufl.edu
* Correspondence: gabriel.laporta@fmabc.br

Citation: Gonçalves, F.G.; Rosa, P.S.; Belone, A.d.F.F.; Carneiro, L.B.; de Barros, V.L.Q.; Bispo, R.F.; Sbardelott, Y.A.d.S.; Neves, S.A.V.M.; Vittor, A.Y.; Woods, W.J.; et al. Lobomycosis Epidemiology and Management: The Quest for a Cure for the Most Neglected of Neglected Tropical Diseases. *J. Fungi* **2022**, *8*, 494. https://doi.org/10.3390/jof8050494

Academic Editors: Alessandro C. Pasqualotto and Pierre Couppié

Received: 30 March 2022
Accepted: 7 May 2022
Published: 10 May 2022

Abstract: Lobomycosis is a chronic disease caused by *Lacazia loboi*, which is endemic to the Amazon rainforest, where it affects forest dwellers in Brazil. There is no disease control program and no official therapeutic protocol. This situation contributes to an unknown disease prevalence and unmet needs of people disabled by this disease who seek access to treatment. This review provides an update on the subject with an emphasis on therapeutic advances in humans. All relevant studies that addressed epidemiology, diagnosis, or therapeutics of lobomycosis were considered. Seventy-one articles published between 1931 and 2021 were included for a narrative literature review on the epidemiology and quest for a cure. An effective therapy for lobomycosis has been found following decades of research led by the State Dermatology Program of Acre in the Amazon rainforest, where the largest number of cases occur. This discovery opened new avenues for future studies. The main recommendations here, addressed to the Brazilian Ministry of Health, are for lobomycosis to become a reportable disease to ensure that disease prevalence is measured, and that it be prioritized such that affected individuals may access treatment free-of-charge.

Keywords: keloidal blastomycosis; Jorge Lobo's Disease; *Lacazia*; Lacaziosis; lobomycosis

1. Introduction

Lobomycosis is caused by the as-of-yet unculturable fungus *Lacazia loboi*, which penetrates the skin following traumatic lesions and reaches down to the subcutaneous tissue, thus causing keloid-like nodular lesions in exposed body areas such as the ears, legs, and arms [1]. It affects humans as well as dolphins. In humans, accidental trauma in association with plants, thorns or insect bites is considered a precursor to skin lesions due to *L. loboi*, though not in all cases. As animal-to-human transmission of *Lacazia* has not been confirmed; its main mode of transmission is hypothesized to be traumatic contact with certain tree trunks of tropical rainforests [2]. While the disease occurs throughout Central and South America, it is mainly found in the Amazon rainforest of Brazil. It was first described in 1931 by the dermatologist Jorge Oliveira Lobo in Recife, Brazil, in a case report of a man who had worked several years in the Amazon and developed keloidal skin lesions in the lumbar and gluteal regions. The first dolphin case was reported from the Atlantic coast of Florida in 1971. Lobomycosis in dolphins has also been reported in Europe in 1983 and a decade later in Brazil [3–7].

Reporting of lobomycosis is not a priority for the Brazilian Ministry of Health, therefore disease prevalence is unknown. However, an increase in new cases has been observed over the years at the State Dermatology Program of Acre (SDPA), Rio Branco County, Acre state, Brazil [1,8]. A large hidden prevalence amongst forest people dwelling in remote areas has been suggested as the probable cause of this rise in cases. There is consensus in the literature that effective drug therapy for the treatment of lobomycosis is lacking, and that surgical resection of keloidal skin lesions is often followed by disease recurrence. In more recent years, however, a scientific breakthrough was achieved by observing the outcomes of a natural experiment. This natural experiment consisted of patients co-infected with leprosy and lobomycosis who were treated with the standard therapy for multibacillary leprosy at the SDPA. Not only did symptoms of leprosy disappear during treatment, but resolution of lesions due to lobomycosis was also observed [9,10]. Here we appraise the literature on the epidemiology, diagnosis and management of lobomycosis in a narrative literature review, with an emphasis on therapies.

2. Materials and Methods

A narrative literature review was undertaken using PubMed as the primary search engine for the selection of studies. Due to the lack of experimental studies, a systematic review was not feasible. The following medical subject headings (MeSH) were applied to "All Fields" in PubMed Central: Lobomycosis OR Lacaziosis OR Jorge Lobo's Disease OR Keloidal Blastomycosis AND Epidemiology; AND Diagnosis; AND Therapeutics; AND France; AND Italy; AND Africa; AND Mexico. The first author (FGG) selected all articles relevant to the three axes of interest in this review: epidemiology, diagnosis, and therapeutics.

To expand the search, a complementary strategy for selection of studies was carried out. Google Scholar was used to search for additional references listed in PhD dissertations by an expert on lobomycosis [11].

3. Results and Discussion

3.1. Case Reports, New Cases, and the Hidden Prevalence of Lobomycosis

The source of lobomycosis infection is unknown. It was long believed that the disease was restricted to Brazil's Amazon region because of its hot and humid climate, and that male forest workers were the main host population. However, the disease is not only present in Brazil, but also in forested areas of other countries in South and Central America [12–18]. Intriguingly, lobomycosis also occurs in dolphins [6]. While it would be reasonable to suppose that sylvatic animals such as the sloth, new world monkeys, or the puma might be susceptible to the disease, this has not been borne out.

One of the major shortcomings in our understanding of lobomycosis epidemiology stems from the lack of systematic collection of disease occurrence statistics (e.g., prevalence, incidence). This is not only due to underreporting, which is common in other neglected tropical diseases, but also because lobomycosis is not a reportable disease in Brazil. This has resulted in a large hidden disease burden, insufficient access to necessary health services, and a rising number of patients presenting with advanced clinical manifestations [13,19].

Our understanding of the epidemiology of lobomycosis thus relies heavily on case reports. The first reported case of lobomycosis was from a resident in the Amazon Basin who worked as a rubber tapper in the 1920s. This case was described by the eponymous dermatologist Jorge Oliveira Lobo in 1931, in Recife County in northeastern Brazil [3]. The second reported case was of a 55-year-old man who was also engaged in forest activities (extraction activities, fishing) in Amazonas state, Brazil [20]. Overall, the majority of cases have been reported from the Amazonian forest regions in Brazil (Acre, Amazonas, and Pará states), Peru, Colombia, Bolivia, Ecuador, Venezuela, Suriname, French Guiana, and

Guyana [1]. While lobomycosis cases have been reported in other countries, including the United States, Canada, Mexico, Spain, France, Panama, Costa Rica and South Africa, the exposure was hypothesized to have occurred in the Amazon Basin [1,6,16,18,21–23].

A longitudinal investigation of lobomycosis was undertaken amongst the Kaiabi indigenous tribe from the 1950s to the 1980s. This tribe is in the northern Mato Grosso state in the Xingu Indigenous Park, which is a well-studied indigenous territory in Brazil. Two adult male brothers were the initial subjects of investigation in 1953. By 1966, 12 cases had been recorded, followed by 15 cases in 1973, 53 cases in 1982, 56 cases in 1986, and 60 cases in 1994 [2,24–30].

Over the years, case reports of lobomycosis were accumulating from several countries, with the Amazon Basin as the likely source of infection. A total 418 cases had been reported by 1996, of which 255 (61%) occurred in Brazilians [31]. A pioneering case series carried out by Opromolla et al. reported on 40 cases of lobomycosis in Acre state, Brazil, raising the total number of cases reported in Brazilians to 295 [32]. By 2000, 47 additional cases were reported, bringing the total to 465 cases. Of the 465 cases, 295 (63%) occurred in Brazilians. Out of these, 60 (20%) occurred amongst the Kaiabi indigenous tribe [33].

From the scattered epidemiological records constructed mainly by case reports, it would seem that the disease is widespread throughout the Amazon Basin, with some hotspots occurring within this biome. While this observation may be an artifact due to underreporting, given that most infectious diseases show some level of spatial clustering, it would stand to reason that the risk of acquiring lobomycosis would not be uniform throughout the Amazon. The increasing number of new cases arising in Acre state, Brazil, further supports this hypothesis. By way of comparison, 249 lobomycosis cases were reported in Acre state between 1998 to 2008 by the SDPA, while 23 new cases were reported in 1996–2005 in Pará state by the Dermatology Service of the Federal University of Pará state [1,10]. Both Amazonian states have large workforces engaging in forest activities, including the extraction of Açai fruit and Brazil nuts. These activities depend on the conserved habitats of tropical rainforests, but workers do not have direct contact with trees. Although Açai crops are adjacent to continuous forests, workers stay in the crops' area only. Brazil nut extraction is based on collecting nuts that have fallen on the ground, so there is no need to climb trees. Rubber extraction, however, requires direct contact with the trees, as the rubber tappers move from one rubber tree to the next to collect the sap. The large number of rubber tappers in Acre state may explain why this state has the largest number of lobomycosis cases.

Of the 490 lobomycosis cases reported worldwide by 2006 [1], the distribution was as follows: 318 cases (65%) in Brazil, 50 (10%) in Colombia, 34 (7%) in Suriname, 23 (5%) in Venezuela, 21 (4%) in Costa Rica, 16 (3%) in French Guiana, 13 (3%) in Panama, 4 (0.5%) in Peru, 3 (0.5%) in Bolivia, 2 (0.5%) in Ecuador and in Guyana, and one (0.25%) in Mexico, Europe, the United States, and Canada [1]. While the Amazon basin was the probable source of infection for nearly all cases, at least two cases, reported from South Africa, may have been acquired outside of the Amazon basin [23]. The first case was of a 65-year-old man with keloidal skin lesions of the feet, arms, and face, with a travel history to Mexico. The second case was of a 20-year-old swimmer and diver who frequently visited Palestine and the United Kingdom [23]. In the former case, it can be interpreted that the actual distribution of *L. loboi* may be greater than expected, expanding beyond the Amazon basin towards North America. In the latter case, it could be speculated that *L. loboi*-contaminated water may confer risk of infection to swimmers or divers.

While the pathogenesis of lobomycosis is poorly understood, the natural history of the disease in humans is characterized by a long incubation period and a slowly progressive, chronic infection. This is likely the result of the subversion of the local immune response [1]. The lengthy incubation period makes it difficult to pinpoint the exact time and location of exposure. Nonetheless, cases of lobomycosis in ecotourists show that transmission can be caused by an acute event (e.g., trauma involving forest debris) dating back years. For example, a 55-year-old Italian man showed infiltrative nodular lesions on the left tibia in

August 2016 [19]. The likely origin of this infection was attributed to an exposure to *Lacazia* fungi several years before during a five-day trek in the forests of the Canaima National Park, Bolivar state, Venezuela [19].

The absolute case numbers of Lobomycosis have risen over the years, adding more information, but also more puzzles with regards to its mode of transmission. For instance, a case reported five years ago had the probable location of transmission noted as unknown [34]. The case was of a 36-year-old male farmer who presented with keloid lesions of the left ear. He lived as a farmer in Brazil's Minas Gerais state, which is located no less than 300 km from the Amazon Basin. He had received visits from Amazonian people at his farm prior to the onset of lobomycosis, raising the question of whether an infected human could have been the source of transmission. Additionally, six new cases of lobomycosis were reported in the Colombian National Army [35]. The disease was probably acquired while in the service in the eastern Colombian Amazon [35]. The duration of illness was between two and 15 years, which suggests that these soldiers were exposed at different times in the same jungle area [35]. Although this jungle area was identified as an infection site for the case series [35], the unknown is where in the environment *Lacazia* fungi reside over the years. Furthermore, a second case of lobomycosis with Mexico as the probable site of infection was reported recently [36], building on the initial case in a South African who had travelled to Mexico [23]. This second individual was a farmer and beekeeper living in southwestern Mexico presenting with multiple nodular lesions in the right ear [36]. In Greece, a histologically confirmed case was reported in a 64-year-old woman without any travel history to Central or South America, marking the first such case from Europe [37]. On the contrary, lobomycosis cases declined amongst the Kaiabi Indians, dropping to only three new cases in the last 20 years [38].

An important outcome from the follow up of lobomycosis cases amongst the Kaiabi Indians between 1965 and 2019 is that a spontaneous cure has not been observed [38]. This means that every new case becomes a prevalent case over time [39]. This, combined with probable high numbers of undiagnosed cases, leads to a mounting burden of disease. The known prevalence today totals 907 cases of lobomycosis in the world [40]. Out of this ($N = 907$), 496 cases (55%) were reported in Acre state, Brazil [40]. In this state, 207 new cases of lobomycosis were reported between 2009–2021. Of these 207 cases, 19 cases occurred in women (9%), the youngest of which involved a 10-year-old child, while the oldest case was of a 106-year-old, and the majority of the individuals affected lived in forested areas [40]. In addition to the known 907 reported cases globally so far [40], it is expected that a larger number remain undiagnosed, contributing to a "hidden prevalence" of lobomycosis.

3.2. *Lobomycosis in Dolphins and Zoonotic Potential*

It had long been believed that lobomycosis was a human disease of Latin American origin until the disease was reported in bottlenose dolphins (*Tursiops truncates*) off the coast of Florida in 1971 [41–43]. Due to human interaction with this geographically widespread dolphin species, lobomycosis cases resulting from dolphin-to-human transmission have been reported [43–45]. At least two cases were confirmed as zoonotic transmission of lobomycosis. Both cases involved work-related contact with a sick dolphin in which the human patients presented with lesions on their hands months after contact [43–45]. Additionally, the possibility of animal-to-human transmission was further implicated when one of the co-authors here (PSR) acquired the disease upon handling experimental mice inoculated with live yeast-like cells from a lobomycosis patient [46]. However, reports of zoonotic transmission are rare and may be more likely in immunocompromised individuals.

Lobomycosis in the bottlenose dolphin is as widespread as the geographical range of the bottlenose dolphin itself, with a prevalence as high as 16% [47–58]. Reports showed lobomycosis in bottlenose dolphins off the coasts of Florida, North Carolina (Atlantic Ocean) and Texas (Gulf of Mexico) in the United States, the coasts of Spain and France (Bay of Biscay), and the Brazilian Atlantic coast of Rio Grande do Sul state (Tramandi

River) [47–58]. The disease has also been reported in the Guiana dolphin (*Sotalia guianensis*) [44]. Interestingly, while the disease occurs readily among dolphins in regions that are not endemic to humans, there is a complete absence of the disease in freshwater dolphins in human-endemic regions, including the Orinoco River in Venezuela and the Amazon River in Brazil [48,49].

Unfortunately, reports of lobomycosis in dolphins are often based on indirect means, such as photographic evidence of lobomycosis-like disease lesions [50]. More than 20 bottlenose dolphins carrying lobomycosis-like disease (LLD) have been photographed in the Indian River Lagoon, Florida [50,51]. Estimates show a LLD prevalence of 3.9% in Guiana dolphins in the Paranagua River Estuary, Paraná state, Brazil [52]. Another dolphin species, *Tursiops aduncus*, was also reported to LLD in the Indian Ocean [53]. The range of LLD prevalence among bottlenose dolphins was estimated as 13.2–16.1% in waters in Central America [54–56]. In southern Belize, the first LLD case in Atlantic spotted dolphin, *Stenella frontalis*, has been recorded recently [57]. Photography showing LLD correlates well with lobomycosis in dolphins, as supported by its 75% sensitivity and 100% specificity in comparison with histologic examination of lesion biopsies [50].

3.3. Clinical Presentation and Diagnosis of Lobomycosis

The diagnosis of lobomycosis is challenging, as the lesions are often mistaken for cutaneous leishmaniasis, nontuberculous mycobacterial infections including leprosy, sporotrichosis, or other dermatological mycoses [39]. Distinguishing clinical features of lobomycosis include slowly progressing keloidal nodules, which may ulcerate or develop a verrucous appearance over time. Other presentations include hypo- or hyperpigmented macules and papules. Lesions may be pruritic or cause a burning sensation, and they can be isolated or disseminated, and are usually localized in the lower limbs, followed by the ears, upper limbs, and head [59,60]. In the disseminated form, body deformities, an intense pruritus, and ulcerations are commonly observed [61].

Biopsy for histological analysis is considered the gold standard [1,10,17,19,32,59]. *Lacazia* fungal cells are identified by staining with haematoxylin-eosin and Gomori-Grocott methenamine silver stains [32]. Analysis of haematoxylin-eosin dyed papillary dermis by light microscopy at 100× magnification reveals hyperkeratosis, collagen fibroplasia, vascular neoformation, and diffuse inflammatory infiltrate with lymphocytes, epithelioid cells, giant cells, and hemosiderin-laden histiocytes [39]. Reticular dermis dyed by Gomori-Grocott methenamine silver and analysed at 200–400× magnification shows round thick-double-walls yeasts occurring singly or in interconnected chains [39]. These histopathological features can be used for the diagnosis of lobomycosis [32,62].

Vinyl adhesive tape (also known as the Scotch test) can also be used for diagnosis after observation of ulcerated lesions [63]. This test is based on the transepidermal elimination of *L. loboi*, in which fungus is eliminated through the horny layer of the epidermis [62]. This technique consists of the application of vinyl adhesive tape to the scale-encrusted aspect of the lesion, followed by the application of this tape to a glass side present with potassium hydroxide (KOH) and dimethyl sulfoxide (DMSO), and subsequent examination by light microscopy. While the diagnostic accuracy has not been well studied, Miranda et al. confirmed lobomycosis via this technique in five of five patients, and were able to distinguish it from other tropical neglected mycoses (chromoblastomycosis and paracoccidiodomycosis) [63]. Different from lobomycosis, paracoccidioidomycosis and chromoblastomycosis can be cultured in the laboratory [63].

Molecular testing has also been successfully employed for the diagnosis of lobomycosis [19,60,61]. Amplification and direct sequencing of fungal ribosomal RNA genes yielded the diagnosis of lobomycosis in a European man who had travelled to the Amazon region of Venezuela [19]. Another approach has been to amplify the gp43-like gene [60]. However, as molecular testing may not always be available in endemic regions, clinical and microscopic diagnosis remain the most used approaches to identify cases [60,61].

3.4. The Quest for a Cure for Lobomycosis

Over the years, numerous antifungal and antibiotic regimens have been attempted with generally unsatisfactory outcomes, falling short of total remission. One studies used sulfadimethoxine 1000 mg/day for an 80-year-old patient in Venezuela in 1961 [64] or sulfamethoxypyridazine 500 mg/day for two cases of 50-year-olds in French Guyana in 1962 [65]. The former study showed partial remission of skin infiltrations and nodules [64], while the latter showed no clear resolution of skin lesions [65] (Table 1). An experimental approach using ketoconazole 400 mg/day showed a decrease in the number of *Lacazia* fungi and mild to moderate remission of skin lesions [66]. In another study (1980), ketoconazole 200 mg/day for six months given to a 45-year-old farmer in Brazil resulted in an unsatisfactory outcome with no cure [67] (Table 1).

After therapeutic studies undertaken by Opromolla et al. in the 1990s [32,63], clofazimine and itraconazole were considered for lobomycosis treatment. Treatment with clofazimine and itraconazole for one year in a 46-year-old Brazilian male was reported to result in the total remission of skin lesions [68] (Table 1). The success of this treatment regimen may be attributed to this patient's localized facial lesion [68], which is less complex than treating the disseminated forms of the disease [10,69]. In another case involving a localized skin lesion of the left ear, a 29-year-old male forest ranger in Peru was treated with posaconazole 400 mg twice a day for 27 months [70]. Although it resulted in the remission of skin lesions, *Lacazia* fungi were still viable after treatment [70]. However, even after four years of follow up of this case, the disease had not recurred [70] (Table 1). In another successful case, a patient was initially treated with itraconazole 200 mg/day and cryotherapy for seven months [34]. As lobomycosis re-appeared, he had further surgery along with clofazimine (100 mg/day), itraconazole (200 mg/day) and cryotherapy with liquid nitrogen for two years [34]. At the time of the study's end, the complete remission of skin lesions and the absence of fungi in a biopsy were seen [34] (Table 1).

Notably, combination therapy with itraconazole clofazimine, rifampin, dapsone, and surgical excision resulted in a clinical cure in both localized and disseminated forms of the disease [46,63,71]. To date, the most promising approach to lobomycosis was found by chance while treating leprosy patients co-infected with lobomycosis in the Leprosy Elimination Program carried out by SDPA in Acre state, Brazil. Ten co-infected patients were treated with the standard protocol for multibacillary (MB) leprosy with multiple drug therapy (MDT; rifampin, clofazimine, and dapsone) as recommend by the World Health Organization (WHO) [10]. Patients reported reduced itching and softening of the skin lesions. Patients were periodically evaluated, and followed up with biopsy and fungal viability assessments [10]. All lesions showing atrophy were then surgically excised, with no further disease recurrence, resulting in 10 of 10 patients cured [10]. Recently, SDPA reported a randomized clinical trial with multibacillary multidrug therapy (MDT/MB/WHO) and surgery resulting in a likelihood ratio of 2.5 (CI 95% 1.4–4.4) of cure compared with untreated or incompletely treated patients (controls) [71] (Table 1). In this trial, out of 80 patients treated with MDT/MB/WHO, 72 (90%) showed improvement, and 20 (25%) were considered cured [71]. MDT/MB/WHO alone is effective in most lobomycosis cases (Figure 1A,B). MDT/MB/WHO associated with surgery can even lead to cure in the disease's disseminated forms (Figure 1C,D).

Lastly, a 57-year-old Brazilian man showing the disseminated form of lobomycosis was treated with posaconazole (400 mg/twice daily) for 30 months [72]. This patient had previously undergone treatment with itraconazole, dapsone, and clofazimine with no remarkable success. A regimen of posaconazole decreased skin lesions in size and healed some of them, with no side effects [72] (Table 1). This further shows that posaconazole is a potential adjuvant drug for lobomycosis therapeutics.

Table 1. Summary of studies on the treatment of lobomycosis, 1960–2021.

Drug	Dosing (Daily) *	Duration @	n Patient	Follow-Up Time %	Outcome #	Surgery Used	Side Effects $	% of cure	Ref
Sulfadimethoxine	0.5–2 g	11 d	1	Not done	2	No	No	-	[64]
Sulfadimethoxine	0.25–0.5 g	18 d	2	Not done	1	No	No	-	[65]
Ketoconazole	0.2–0.4 g	90 d	1	Not done	2B	No	No	-	[66]
Ketoconazole	0.2 g	180 d	1	Not done	1	No	No	-	[67]
Clofazimine Itraconazole	0.1 g 0.1 g	1 y	1	2 years	3C	No	Yes 1	-	[68]
Clofazimine Dapsone Itraconazole	0.05 g 0.1 g 0.2 g	1 y	1	Not available	3C	Yes	No	-	[46]
Posaconazole	0.8 g	27 m	1	4 years	3B	No	Yes 2	-	[70]
Itraconazole Clofazimine Cryotherapy with liquid nitrogen	0.2 g 0.1 g every 3 months	2 y	1	Not available	3C	Yes	No	-	[34]
Clofazimine Dapsone Clofazimine Rifampin Dapsone Itraconazole	0.05 g 0.1 g 0.3 g/m 0.6 g/m 0.3 g/m 0.2 g	4 y	103	2 years	3C	Yes	Yes 1	25%	[71]
Posaconazole	0.8 g	30 m	1	Not available	2	No	No	-	[72]

*: doses are in grams (g) and daily, except when informed in months (m). @: duration of treatment in days (d), months (m), or years (y). %: Follow-up time after the treatment's end. #: outcomes of skin lesions: 1 = no resolution, 2 = partial resolution, and 3 = clinical cure. #: outcomes of fungal viability: A = unchanged; B = decreased; and C = clinical cure. $: side effects: Yes 1 = skin pigmentation due to the use of clofazimine; Yes 2 = headache.

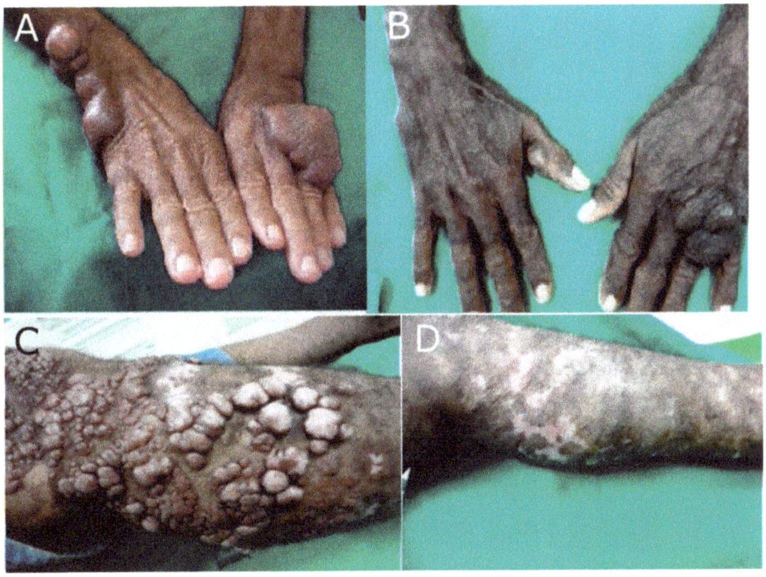

Figure 1. WHO/MDT/MB standard protocol for lobomycosis treatment, SDPA, Acre state, 2020 [71]. (**A**) localized lobomycosis before and (**B**) after WHO/MDT/MB treatment for four years. (**C**) disseminated lobomycosis before and (**D**) after WHO/MDT/MB treatment for four years plus lesion resection twice a week for one year.

The WHO/MDT/MB therapy (clofazimine, rifampin, dapsone) in Gonçalves et al. [71] was associated with less than $100 in medical expenditure per patient. Thus, this triple drug therapy in combination with surgery not only leads to remission of skin lesions, but also does so at reduced cost (Figure 1). We conclude by advocating for the inclusion of lobomycosis in the Brazilian Ministry of Health in the list of reportable diseases and the WHO program for control of neglected tropical diseases.

4. Conclusions

Lobomycosis is the most neglected of the neglected tropical diseases, with a rising number of new cases among residents of the Amazon rainforest. The state of Acre in Brazil has the highest prevalence of lobomycosis in the world, possibly due to the long history of economic development based on rubber tapping. Lobomycosis can cause physical disability, disproportionately affecting low-income heads-of-household who depend on manual labour to provide for their families. This narrative review highlights the role of triple drug therapy with clofazimine, rifampin and dapsone with adjunctive surgical excision as a cost-effective and proven cure for lobomycosis. We recommend that lobomycosis be included in the list of reportable diseases and for the adoption of multibacillary multidrug therapy for the standard treatment of this disease.

Author Contributions: Conceptualization, F.G.G., P.S.R., A.d.F.F.B., S.A.V.M.N., A.Y.V., W.J.W. and G.Z.L.; methodology, F.G.G., L.B.C., V.L.Q.d.B., R.F.B. and Y.A.d.S.S.; writing—original draft preparation, F.G.G. and G.Z.L.; writing—review and editing, A.Y.V., P.S.R., A.d.F.F.B., S.A.V.M.N., W.J.W., and G.Z.L. All authors are in accordance with this article's view. All authors have read and agreed to the published version of the manuscript.

Funding: This work was supported by Brazilian National Council for Scientific and Technological Development [CNPq grant 307432/2019-0 to GZL].

Institutional Review Board Statement: The Ethical Research Committee of the Acre State Health Department approved the randomized clinical trial [71] whose outcomes are highlighted in Figure 1 and Table 1 (No. 448/2010).

Informed Consent Statement: Patients have consented to the use of data associated with them for scientific purposes.

Data Availability Statement: All data used here are contained within the main text.

Conflicts of Interest: The authors declare no conflict of interest.

References

1. de Brito, A.C.; Quaresma, J.A.S. Lacaziose (doença de Jorge Lobo): Revisão e atualização. *An. Bras. Dermatol.* **2007**, *82*, 461–474. [CrossRef]
2. Lacaz, C.S.; Baruzzi, R.G.; Rosa, M.C.B. *Doença de Jorge Lobo*; EDUSP: São Paulo, Brazil, 1986; pp. 1–92.
3. Lobo, J. Um caso de blastomicose, produzido por uma espécie nova, encontrada em Recife. *Rev. Med. Pernamb.* **1931**, *1*, 763–775.
4. Guedes, A.C.M. Pele e anexos. In *Bogliolo Patologia*, 7th ed.; Brasileiro Filho, G., Ed.; Guanabara Koogan: Rio de Janeiro, Brazil, 2006; Volume 1, pp. 1200–1287.
5. Gompertz, O.F.; Gambale, W.; Paula, C.R.; Correa, B. Micoses subcutâneas. In *Microbiologia*, 5th ed.; Trabulsi, L.R., Alterthum, F., Eds.; Atheneu: São Paulo, Brazil, 2008; pp. 511–515.
6. Symeres, W.S. A possible case of Lobo's diseases acquired in Europe from a bottle-nosed dolphin (*Tursiops truncates*). *Bull. Société Pathol. Exot. Fil.* **1983**, *76*, 777–784.
7. Simões-Lopes, P.C.; Paula, G.S. First case of lobomycosis in a bottlenose dolphin from southern Brazil. *Mar. Mamm. Sci.* **1993**, *9*, 329–331. [CrossRef]
8. Silva, A.R.; Pinheiro, G.S.; Matos, W.B.; Couto, M.J.A.; Gonçalves, E.G.R. Doença de Jorge Lobo: Primeiro caso registrado no Estado do Maranhão. *Cad. Pesqui.* **2013**, *20*, 64–67. [CrossRef]
9. Francesconi, V.A.; Klein, A.P.; Santos, A.P.B.G.; Ramasawmy, R.; Francesconi, F. Lobomycosis: Epidemiology, clinical presentation, and management options. *Ther. Clin. Risk Manag.* **2014**, *10*, 851–860. [CrossRef] [PubMed]
10. Woods, W.J.; Belone, A.F.; Carneiro, L.B.; Rosa, P.S. Ten years' experience with Jorge Lobo's disease in the state of Acre, Amazon region, Brazil. *Rev. Inst. Med. Trop. São Paulo* **2010**, *52*, 273–278. [CrossRef]
11. Floriano, M.C. Doença de Jorge Lobo Entre os Índios Caiabu: Epidemiologia, Análise Histopatológica e Imuno-Histoquímica e Imuno-Histoquímica Nas Diferentes Apresentações Clínicas. Ph.D. Thesis, Universidade Federal de São Paulo, São Paulo, Brazil, 2014.

12. Borelli, D. The reservoir area of lobomycosis. Comments on the work of Dr Carlos Peña on 2 Colombian cases. *Mycopathol. Mycol. Appl.* **1969**, *37*, 145–149. [CrossRef]
13. Guimaraes, F.N.; Macedo, D.G. Blastomycosis in the Amazon Valley (keloidian and South American). *Hosp. (Rio J.)* **1950**, *38*, 223–253.
14. Rodriquez-Toro, G. Lobomycosis. *Int. J. Dermatol.* **1993**, *32*, 324–332. [CrossRef]
15. Ramos-E-Silva, M.; Aguiar-Santos-Vilela, F.; Cardoso-de-Brito, A.; Coelho-Carneiro, S. Lobomycosis. Literature review and future perspectives. *Actas Dermo-Sifiliogr.* **2009**, *100* (Suppl. 1), 92–100. [CrossRef]
16. Lupi, O.; Tyring, S.K.; McGinnis, M.R. Tropical dermatology: Fungal tropical diseases. *J. Am. Acad. Dermatol.* **2005**, *53*, 931–951. [CrossRef] [PubMed]
17. Paniz-Mondolfi, A.E.; Jaimes, O.R.; Jones, L.D. Lobomycosis in Venezuela. *Int. J. Dermatol.* **2007**, *46*, 180–185. [CrossRef] [PubMed]
18. Saint-Blancard, P.; Maccari, F.; Le Guyadec, T.; Lanternier, G.; Le Vagueresse, R. Lobomycosis: A mycosis seldom observed in metropolitan France. *Ann. Pathol.* **2000**, *20*, 241–244.
19. Beltrame, A.; Danesi, P.; Farina, C.; Orza, P.; Perandin, F.; Zanardello, C.; Rodari, P.; Staffolani, S.; Bisoffi, Z. Molecular confirmation of lobomycosis in an Italian traveler acquires in the Amazon Region of Venezuela. *Am. J. Trop. Med. Hyg.* **2017**, *97*, 1757–1760. [CrossRef]
20. Fialho, A.S.; Lacaz, C.S. Jorge Lobo blastomycosis. *Folha Med.* **1950**, *31*, 36–40.
21. Burns, R.A.; Roy, J.S.; Woods, C.; Padhye, A.A.; Warnock, D.W. Report of the first human case of Lobomycosis in United States. *J. Clin. Microb.* **2000**, *38*, 1283–1285. [CrossRef]
22. Elsayed, S.; Kuhn, S.M.; Barber, S.; Church, D.L.; Adams, S.; Kasper, R. Human case of lobomycosis. *Emerg. Infect. Dis.* **2004**, *10*, 715–718. [CrossRef]
23. Al-Daraji, W.I.; Husain, E.; Robson, A. Lobomycosis in African patients. *Br. J. Dermatol.* **2008**, *159*, 231–266. [CrossRef]
24. Nery-Guimarães, F. Inoculações em hamsters da blastomicose sul-americana (doença de Lutz), da blastomicose queloidiforme (doença de Lobo) e da blastomicose dos índios do Tapajós-Xingu. *Hosp. (Rio J.)* **1964**, *66*, 581–593.
25. Baruzzi, R.G.; D'Andretta, C., Jr.; Carvalhal, S.; Ramos, O.L.; Pontes, P.L. Ocorrência de blastomicose queloidiana entre índios Caiabi. *Rev. Inst. Med. Trop. São Paulo* **1967**, *9*, 135–142. [PubMed]
26. Machado, P.A.; Silveira, D.F. Piraip, a falsa lepra dos Caiabis. *Rev. Bras. Leprol.* **1966**, *34*, 60.
27. Baruzzi, R.G.; Castro, R.M.; D'Andretta, C., Jr.; Carvalhal, S.; Ramos, O.L.; Pontes, P.L. Occurrence of Lobo's blastomycosis among "Caiabi" brazilian Indians. *Int. J. Dermatol.* **1973**, *12*, 95–98. [CrossRef]
28. Baruzzi, R.G.; Marcopito, L.F.; Vicente, L.S.; Michalany, N.S. Jorge Lobo's disease (keloidal blastomycosis) and tinea imbricata in Indians from the Xingu National Park, Central Brazil. *Trop. Dr.* **1982**, *12*, 13–15. [CrossRef] [PubMed]
29. Baruzzi, R.G.; Rodrigues, D.A.; Michalany, N.S.; Salomão, R. Squamous-cell carcinoma and lobomycosis (Jorge Lobo's disease). *Int. J. Dermatol.* **1989**, *28*, 183–185. [CrossRef]
30. Baruzzi, R.G.; Marcopito, L.F. Doença de Jorge Lobo. In *Tratado de Infectologia*; Veronesi, R., Focaccia, R., Eds.; Atheneu: São Paulo, Brazil, 1996; pp. 1116–1119.
31. Pradinaud, R. Loboa loboi. In *Topley and Wilson's Microbiology and Microbial Infections*; Collier, L., Balows, A., Sussman, M., Eds.; Oxford University Press: New York, NY, USA, 1998; pp. 585–594.
32. Opromolla, D.V.A.; Taborda, P.R.O.; Taborda, V.B.A.; Viana, S.; Furtado, J.F. Lobomicose: Relato de 40 casos novos. *An. Bras. Dermatol.* **1999**, *74*, 135–141.
33. Lacaz, C.S.; Porto, E.; Martins, J.E.C.; Heins-Vaccari, E.M.; Melo, N.T. *Tratado de Micologia Médica*, 9th ed.; Savier: São Paulo, Brazil, 2002; pp. 462–478.
34. Araújo, M.G.; Santos, S.N.M.B.; Guedes, A.C.M.; Cirilo, N.S.; Aguiar, C.R. Lobomycosis: A therapeutic challenge. *An. Bras. Dermatol.* **2018**, *93*, 279–281. [CrossRef]
35. Arenas, C.M.; Rodriguez-Toro, G.; Ortiz-Florez, A.; Serrato, I. Lobomycosis in soldiers, Colombia. *Emerg. Infect. Dis.* **2019**, *25*, 654–660. [CrossRef]
36. Pech-Ortiz, L.; Maya-Aranda, S.; Hernández-Castro, R.; Xicohtencatl-Cortes, J.; Tirado-Sánchez, A.; Bonifaz, A. Lacaziosis (lobomycosis) from Southern Mexico: A case confirmed by molecular biology. *Mycopathologia* **2020**, *185*, 737–739. [CrossRef]
37. Papadavid, E.; Dalamaga, M.; Kapniari, I.; Pantelidaki, E.; Papageorgiou, S.; Pappa, V.; Tsirigotis, P.; Dervenoulas, I.; Stavrianeas, N.; Rigopoulos, D. Lobomycosis: A case from Southeastern Europe and review of the literature. *J. Dermatol. Case Rep.* **2012**, *6*, 65–69. [CrossRef]
38. Florian, M.C.; Rodrigues, A.; de Mendonça, S.B.M.; Colombo, A.L.; Tomimori, J. Epidemiologic and clinical progression of lobomycosis among Kaiabi Indians, Brazil, 1965–2019. *Emerg. Infect. Dis.* **2020**, *26*, 930–936. [CrossRef]
39. Oliveira, F.D.S.; Lopes, N.P.; Talhari, C.; Schettini, A. What is your diagnosis? Keloidal cord-like lesion on the leg. *An. Bras. Dermatol.* **2020**, *95*, 386–389. [CrossRef]
40. Gonçalves, F.G. (Serviço Estadual de Dermatologia do Acre, Programa Estadual de Dermatologia do Acre, Programa Estadual de Controle da Hanseníase, SESACRE/FUNDHACRE, Rio Branco, Acre, Brazil). Personal communication, 2022.
41. Migaki, G.; Valerio, M.G.; Irvine, B.; Garner, F.M. Lobos disease in an Atlantic bottle-nosed dolphin. *J. Am. Vet. Med. Assoc.* **1971**, *159*, 578–582.
42. Caldwell, D.K.; Caldwell, M.C.; Woodard, J.C.; Ajello, L.; Kaplan, W.; McLure, H.M. Lobomycosis as a disease of the Atlantic bottlenosed dolphin (*Tursiops truncatus* Montagu, 1821). *Am. J. Trop. Med. Hyg.* **1975**, *24*, 105–114. [CrossRef]
43. Wiersema, J.P.; Niemel, P.L. Lobo's disease in Surinam patients. *Trop. Geogr. Med.* **1965**, *17*, 89–111.
44. De Vries, G.A.; Laarman, J.J. A case of Lobo's disease in the dolphin *Sotalia guianensis*. *Aquat. Mamm.* **1973**, *1*, 26–33.

45. Norton, S.A. Dolphin-to-human transmission of lobomycosis. *J. Am. Acad. Dermatol.* **2006**, *55*, 723–724. [CrossRef]
46. Rosa, P.S.; Soares, C.T.; Belone, A.; Vilela, R.; Ura, S.; Filho, M.C.; Mendoza, L. Accidental Jorge Lobo's disease in a worker dealing with *Lacazia loboi* infected mice: A case report. *J. Med. Case Rep.* **2009**, *3*, 67. [CrossRef]
47. Moreno, I.B.; Ott, P.H.; Tavares, M.; Oliveira, L.R.; Borba, M.R.; Driemeier, D.; Nakashima, S.B.; Heinzelmann, L.S.; Siciliano, S.; Van Bressen, M.F. Mycotic dermatitis in common bottlenose Dolphins (*Tursiops truncatus*) from Southern Brazil, with confirmed record of lobomycosis disease. In Proceedings of the Annual Meeting of the International Whaling Commission (IWC), IWC Scientific Committee, Santiago, Chile, 23–26 June 2008. Paper SC/60/DW1.
48. Paniz-Mondolfi, A.E.; Sander-Hoffmann, L. Lobomycosis in inshore and estuarine dolphins. *Emerg. Infect. Dis.* **2009**, *15*, 672–673. [CrossRef]
49. Reif, J.S.; Mazzoil, M.S.; McCulloch, S.D.; Varela, R.A.; Goldstein, J.D.; Fair, P.A.; Bossart, G.D. Lobomycosis in Atlantic bottlenose dolphins from the Indian River Lagoon, Florida. *J. Am. Vet. Med. Assoc.* **2006**, *228*, 104–108. [CrossRef]
50. Murdoch, M.E.; Reif, J.S.; Mazzoil, M.; McCulloch, S.D.; Fair, P.A.; Bossart, G.D. Lobomycosis in bottlenose dolphins (*Tursiops truncatus*) from the Indian River Lagoon, Florida: Estimation of prevalence, temporal trends, and spatial distribution. *EcoHealth* **2008**, *5*, 289–297. [CrossRef]
51. Durden, W.N.; St Leger, J.; Stolen, M.; Mazza, T.; Londono, C. Lacaziosis in bottlenose dolphins (*Tursiops truncatus*) in the Indian River Lagoon, Florida, USA. *J. Wildl. Dis.* **2009**, *45*, 849–856. [CrossRef]
52. Bressem, M.F.V.; Santos, M.C.O.; Oshima, J.E.F. Skin diseases in Guiana dolphins (*Sotalia guianensis*) from the Paranaguá estuary, Brazil: A possible indicator of a compromised marine environment. *Mar. Environ. Res.* **2009**, *67*, 63–68. [CrossRef]
53. Kiszka, J.; Van Bressem, M.F.; Pusineri, C. Lobomycosis-like disease and other skin conditions in Indo-Pacific bottlenose dolphins *Tursiops aduncus* from the Indian Ocean. *Dis. Aquat. Org.* **2009**, *84*, 151–157. [CrossRef]
54. Paniz-Mondolfi, A.; Talhari, C.; Hoffman, L.S.; Connor, D.L.; Talhari, S.; Bermudez-Villapol, L.; Hernadez-Perez, M.; Van Bressem, M.F. Lobomycosis: An emerging disease in humans and delphinidae. *Mycoses* **2012**, *55*, 298–309. [CrossRef]
55. Brito, A.C. Lacaziose (Doença de Jorge Lobo). In *Compêndio de Micologia Médica*, 2nd ed.; Zaitz, C., Campbell, I., Marques, A.S., Ruiz, L.R.B., Framil, V.M.S., Eds.; Editora Guanabara Koogan: Rio de Janeiro, Brazil, 2012; pp. 219–231.
56. Bessesen, B.L.; Oviedo, L.; Hart, L.B.; Herra-Miranda, D.; Pacheco-Polanco, J.D.; Baker, L.; Saborío-Rodriguez, G.; Bermúdez-Villapol, L.; Aceved-Gutiérrez, A. Lacaziosis-like disease among bottlenose dolphins *Tursiops truncatus* photographed in Golf Dulce, Costa Rica. *Dis. Aquat. Org.* **2014**, *107*, 173–180. [CrossRef]
57. Ramos, E.A.; Castelblanco-Martínez, D.N.; Garcia, J.; Arias, J.R.; Foley, J.R.; Audley, K.; Van Waerebeek, K.; Van Bressem, M.F. Lobomycosis-like disease in common bottlenose dolphins *Tursiops truncatus* from Belize and Mexico: Bridging the gap between the Americas. *Dis. Aquat. Org.* **2018**, *128*, 1–12. [CrossRef]
58. Rotstein, D.S.; Burdett, L.G.; McLellan, W.; Schwacke, L.; Rowles, T.; Terio, K.A.; Schultz, S.; Pabst, A. Lobomycosis in offshore bottlenose dolphins (*Tursiops truncatus*), North Carolina. *Emerg. Infect. Dis.* **2009**, *15*, 588–590. [CrossRef]
59. Pradinaud, R. Entre le Yucatan, la Floride et la Guyane Française, la lobomycose existe-t-elle aux Antilles. *Bull. Société Pathol. Exot.* **1984**, *77*, 392–400.
60. Vilela, R.; Mendoza, L.; Rosa, P.S.; Belone, A.F.; Madeira, S.; Opromolla, D.V.; Resende, M.A. Molecular model for studying the uncultivated fungal pathogen *Lacazia loboi*. *J. Clin. Microbiol.* **2005**, *43*, 3657–3661. [CrossRef]
61. Nogueira, L.; Mendes, L.; Rodrigues, C.A.; Santos, M.; Talhari, S.; Talhari, C. Lobomycosis and squamous cell carcinoma. *An. Bras. Dermatol.* **2013**, *88*, 293–295. [CrossRef] [PubMed]
62. Miranda, M.F.R.; Silva, A.J.G. Vinyl adhesive tape effective for direct microscopy diagnosis of chromomycosis, lobomycosis, and paracoccidiodomycosis. *Diagn. Microbiol. Infect. Dis.* **2005**, *52*, 39–43. [CrossRef] [PubMed]
63. Opromolla, D.V.A.; Belone, A.F.F.; Taborda, P.R.O.; Taborda, V.B.A. Correlação clínico-patológica em 40 casos novos de lobomicose. *An. Bras. Dermatol.* **2000**, *75*, 425–434.
64. Reyes, O.; Goihman, M.; Goldstein, C. Blastomicosis queloidiana o enfermidad de Jorge Lobo. *Dermatol. Venez.* **1960**, *2*, 245–255.
65. Silverie, R.; Ravisse, P.; Vilar, J.P.; Moulins, C. La blastomycose chéloidienne ou maladie de Jorge Lobo en Guyane Française. *Bull. Société Pathol. Exot.* **1963**, *56*, 29–35.
66. Cucê, L.C.; Wroclawski, E.L.; Sampaio, S.A. Treatment of paracoccidioidomycosis, candidiasis, chromomycosis, lobomycosis and mycetoma with ketoconazole. *Int. J. Dermatol.* **1980**, *19*, 405–408. [CrossRef]
67. Lawrence, D.N.; Ajello, L. Lobomycosis in Western Brazil: Report of a clinical trial with ketoconazole. *Am. J. Trop. Med. Hyg.* **1986**, *35*, 162–166. [CrossRef]
68. Fischer, M.; Chrusciak-Talhari, A.; Reinel, D.; Talhari, S. Successful treatment with clofazimine and itraconazole in a 46-year-old patient after 32 years duration of disease. *Hautarzt* **2002**, *53*, 677–681. [CrossRef]
69. Talhari, S.; Talhari, C. Lobomycosis. *Clin. Dermatol.* **2012**, *30*, 420–424. [CrossRef]
70. Bustamante, B.; Seas, C.; Salomon, M.; Bravo, F. Lobomycosis successfully treated with posaconazole. *Am. J. Trop. Med. Hyg.* **2013**, *88*, 1207–1208. [CrossRef]
71. Gonçalves, F.G.; Rosa, P.S.; Belone, A.F.F.; Carneiro, L.B.; Barros, V.L.Q.; Bispo, R.F.; Sbardellot, Y.A.S.; Neves, S.A.V.M.; Vittor, A.Y.; Woods, W.J.; et al. Multidrug therapy for leprosy can cure patientes with lobomycosis in Acre State, Brazil: A proof of therapy study. *Am. J. Trop. Med. Hyg.* **2021**, *104*, 634–639. [CrossRef] [PubMed]
72. Pasqualotto, A.C.; Jaskulski Filho, S.D.; Sena, M.G.; Santos, A.; Severo, M. Posaconazole for lobomycosis. *Braz. J. Infect. Dis.* **2021**, *25*, 101576. [CrossRef] [PubMed]

MDPI AG
Grosspeteranlage 5
4052 Basel
Switzerland
Tel.: +41 61 683 77 34

Journal of Fungi Editorial Office
E-mail: jof@mdpi.com
www.mdpi.com/journal/jof

www.ingramcontent.com/pod-product-compliance
Lightning Source LLC
Chambersburg PA
CBHW061521210926
51850CB00008B/1996
* 9 7 8 3 7 2 5 8 1 4 4 3 5 *